The Muskellunge Symposium: A Memorial Tribute to E.J. Crossman

Developments in environmental biology of fishes 26

Series Editor
DAVID L.G. NOAKES

The Muskellunge Symposium: A Memorial Tribute to E.J. Crossman

Guest Editors:

James S. Diana, University of Michigan, Ann Arbor, MI, USA
Terry L. Margenau, Wisconsin Department of Natural Resources, Spooner, WI, USA

Reprinted from *Environmental Biology of Fishes*, Volume 79 (1–2) 2007
with addition of extra material

Springer

A C.I.P catalogue record for this book is available from the Library of Congress

ISBN 978-1-4020-6048-9
ISBN 978-1-4020-6049-6

Published by Springer,
P.O. Box 17, 3300 AA Dordrecht, The Netherlands

Sold and distributed in North, Central and South America
By Springer,
101 Philip Drive, Norwell, MA 02061, USA

In all other countries, sold and distributed
By Springer,
P.O. Box 322, 3300 AH Dordrecht, The Netherlands

http://www.springer.com
Printed on acid-free paper

Printed in the Netherlands

The Muskellung Symposium: A Memorial Tribute to E.J. Crossman

Editors

James S. Diana, University of Michigan, Ann Arbor, MI USA
Terry L. Margenau, Wisconsin Department of Natural Resources Spooner, WI, USA

October 2005
MUSKIE
Dr. E. J. Crossman
1929-2003
Dr. Ed Crossman
Muskie Symposium
2005

Preface

This book represents the outcome of a symposium held in Indianapolis, Indiana on 28–30 October 2005. The symposium was an international program on the ecology and management of muskellunge and was held in memory of Dr. Edwin J. Crossman. The symposium structure was largely similar to the original International Musky Symposium, held in LaCrosse, Wisconsin in 1984. It was intended to bring together academic researchers, management biologists, and practical musky anglers into a common venue to review the current status of knowledge on muskellunge and their management.

The symposium was dedicated to the memory of Edwin J. Crossman. Dr. Crossman passed away on 21 December 2003, at the age of 74. While 74 is an advanced age for many people, Ed's passing occurred far too young. He was retired from the University of Toronto and the Royal Ontario Museum, but still an active researcher in fish systematics and muskellunge ecology. Dr. Crossman was influential in developing the knowledge of muskellunge we have today and encouraging many people to follow in his footsteps in the studies of muskellunge and their relatives.

The concept of a second musky symposium began with conversations between members of Muskies, Inc. and the Esocid Technical Committee (ETC) of the American Fisheries Society. As plans progressed, Steve Pallo of the ETC and Scott Law of Muskies, Inc. collaborated to set dates, determine a venue, and procure support for the symposium. As plans developed, we were enlisted to develop the presentations and ultimately edit the manuscripts. The intent of the symposium was to present current information on the ecology and management of muskellunge to the angling public, and at the same time, use the interests and desires of anglers to further educate the research and management community. While we are probably biased in this regard, the symposium was a great success at accomplishing this goal.

We had intentions to both include oral presentations and develop a published proceeding of high quality, as a tribute to the strong outreach skills and excellent scientific integrity of E. J. Crossman. In the hope of rapid publication, we requested manuscripts within one month of the symposium and contracted with *Environmental Biology of Fishes* for co-publication in an issue of the journal, as well as in a book. Not all papers presented at the symposium were published in the journal—some due to choices by the authors; others because of previous publication. In this book, all abstracts from papers presented but not published are also included.

The review process for manuscripts was of highest quality. Jim Diana served as the special editor of the contributions, while Terry Margenau, Dave Wahl, and Jerry Younk served as the editorial team. We appreciate the contributions of the many professionals who reviewed manuscripts for this proceeding. David Noakes, the editor for *Environmental Biology of Fishes,* made final decisions on acceptance of the manuscripts, as well as final editorial suggestions. We thank all of the people involved for their hard work in bringing these proceedings to fruition.

This symposium would not have occurred without the generous support and efforts of many organizations and individuals. Primary sponsorship was provided by Muskies, Inc., Gander Mountain, and the Hoosier Muskie Hunters Chapter of Muskies, Inc. These primary sponsors subsidized both the meeting itself and the publication of this book. Secondary sponsorship came from many musky clubs from around the United States, Muskies Canada and various chapters within Muskies, Inc. The Hoosier Muskie Hunters provided tremendous logistic support, as well as continual interest in the symposium. We have never been at a professional meeting where everyone attended all of the sessions before this symposium, which indicates the strong interests of musky anglers in the science and methods important for the management of this unique fish. The interest in this subject was also demonstrated by the state agencies involved, since they provide the time and travel costs of their employees that attended.

Muskellunge are unique fish in several ways, and to many people they represent a lasting symbol of the wild nature of our northern woods. They have always been rare in inland fish communities, with typical densities less than 0.5 fish per hectare. They grow to large sizes that are unique among fish species in small inland lakes, and as such, are an important predator that is a keystone species of northern lakes, controlling prey types and sizes by their predation on large fish. They are an important game fish in the region, but even more important as a symbol of the magnificence of nature. They have been introduced to numerous reservoirs outside of their native range to provide trophy fishing opportunities, and are also an important component of the fish community and fishery in the Laurentian Great Lakes. The reintroduction of muskellunge throughout their original range is an important part of the current management practices for many states and provinces.

Ed Crossman spent his life studying and promoting his "noble muskellunge." He was instrumental in the lives of virtually all of the authors in this volume, either by his important publications in the field or by his personal interactions. His critical but helpful review of our work, his continual drive for excellence, and his friendly and outgoing demeanor produced an excellent role model for all of us. He is missed by those who knew him well. We hope this volume does justice to his memory, and we believe he would be pleased to see the future of his fish highlighted in this book and emphasized in the work of so many scientists and managers.

James S. Diana and Terry L. Margenau

Preface

The Muskies, Inc. Perspective

The International Musky Symposium began as vision for Muskies, Inc., Chapter 42 (Hoosier Muskie Hunters) to plan, host and financially support an event that would gather scientists, researchers, professional musky anglers and musky enthusiasts at one venue to mutually share their working knowledge and experiences as a way to "Build for the Future." The symposium would follow the fall Board of Directors meeting of Muskies, Inc. (MI) and therefore provide the International Board of Directors and MI chapter and regional vice presidents an opportunity to attend the event. This diverse group of highly educated musky managers, professionals and anglers would review current research projects and poster presentations as well as participate in workshops and panel discussions. This vision became a reality on the weekend of October 28–30, 2005 in Indianapolis, Indiana.

How could an individual chapter of a fishing organization plan for such an event that would gather the fishing public as well as fishery biologists and researchers from around the U.S. and Canada to Indianapolis? We started by contacting our own organization of MI and research directors Steve Budnik and Ron Mazur, who pointed us in the right direction and kept us moving forward. They suggested contacting the American Fisheries Society (AFS) for the technical part of the program.

AFS, through the Esocid Technical Committee (ETC), became the point of contact to help find individuals needed to fill the technical program. The original contact was through Steve Pallo, from the Illinois Department of Natural Resources (DNR), who was chair of the ETC at the time. Steve quickly recruited Terry Margenau from Wisconsin DNR and Jim Diana from University of Michigan to round out the symposium technical committee. These men contributed their time and efforts to ensure the success of the symposium. The symposium committee placed a "call for papers" in *Fisheries, an* AFS publication. The symposium technical program quickly filled as papers were sent to the committee for review and approval. Additionally, the committee asked Jack Wingate to share the success stories from fisheries management in Minnesota concerning the stocking program, fish size limits and genetics. Bill James, Chief of Fisheries – Indiana Department of Fish and Wildlife, enthusiastically committed to present the Indiana DNR strategy of developing its muskellunge fisheries.

Since the goal of the symposium included professional anglers, we contacted Jim Saric at *Musky Hunter* and Pete Mania at *Esox Angler* magazines to help fill the anglers' program. These editors have contacts with musky fishing guides from across the U.S. and Canada who write fishing articles for their publications. The time spent on the water by these professional anglers lends itself to practical insight into daily, weekly and seasonal fishing patterns that interest musky anglers. Rob Kim from *Esox Angler,* as well as Jim Saric and Steve Heiting from *Musky Hunter* provided angling perspectives critical to researchers and managers making prudent decisions concerning fisheries. Larry Ramsell, MI historian, reviewed musky catch records, and Russ Wayre, author and educator, completed the anglers program by reviewing top angling waters in the U.S. and Canada.

Muskies, Inc., Gander Mountain and The Hoosier Muskie Hunters underwrote the symposium as primary sponsors. The individual chapters of MI, various individuals and Muskies Canada also contributed financial support to the symposium. Pete Barber and Jim Beaty of MI helped find sponsors for the event. Gary Dew, president of the Shawnee chapter of MI, developed a raffle for a new Triton boat, embossed with a special MI decal and logo. Ken Karbon from MI auctioned items at the Saturday evening banquet. Proceeds from the boat raffle and auction supported the symposium. Scott Unison, past president of Hoosier Muskie Hunters, served as symposium treasurer.

During the three-day symposium, Kenton Smith, the symposium program chairman, and his talented volunteer staff (Fred Boso, Bob Briney, Chris Craig, Paul Cruse, John Fawcett, Eric Flemming, Ken Irwin, Charlie Johnson, Kathy Lee, Jason Ronnebohm, John Routt, Jeremy Sivis, and John Unison) coordinated 14 general sessions, 18 workshops, poster presentations, a panel discussion, all meals and coffee breaks and a special presentation by Mrs. Margaret Crossman. Thanks also to Steve Worral and his staff at *Muskies First* for documenting the event with photography and video.

A special thanks to Vance Bell and Tony Gray, Hoosier Muskie Hunters presidents, for their commitment to the symposium in the wake of an already busy calendar. Thanks also to Muskies Canada and their President, Rob Howitt, for attending the event. Finally, I appreciate the MI International Presidents, Greg Wells and Dave Cates, who unswervingly supported the vision of the symposium. Their personal commitment to the event guaranteed it would become a reality.

At a planning meeting for the 1984 International Musky Symposium, Gil Hamm, founder of MI, expressed these insightful words: "These are the things Muskies, Inc. needs to be associated with if we have any hope of good sport fishing in the future." Gil's comments are true today. This book is dedicated to those individuals whose vision is to "Build for the Future."

Thank you to my wife, Lisa, and our sons, Josh and Tyler, for allowing me to serve as the chairman of the Dr. Ed Crossman International Musky Symposium. Certainly our boys will be the next generation of anglers to experience all nature has to offer, especially musky fishing.

Scott Law

Sponsors

Building for the Future
Dr. E.J. Crossman Muskie Symposium
October 28–30, 2005

Chapters:
01 – Twin Cities Chapter
02 – Fargo-Moorhead Chapter
03 – Chicagoland Chapter
05 – Pomme de Terre Chapter
06 – First Wisconsin Chapter
07 – South Side Muskie Hawks Chapter
08 – Capital City Chapter
09 – West Virginia Chapter
11 – Mississippi Valley Chapter
12 – Headwaters Chapter
13 – Hayward Lakes Chapter
14 – South of Border Chapter
15 – Star of the North Chapter
16 – Three Rivers Chapter
17 – Quad County Hawg Hunters Chapter
20 – Between the Lakes Chapter
21 – North Metro Chapter
22 – New Jersey Chapter
24 – Brainerd Lakes Chapter
27 – Central Illinois Chapter
28 – Shawnee Muskie Hunters Chapter
29 - Upper Great Plains Chapter
30 – God's Country Chapter
32 – Flatlanders Chapter
35 – Milwaukee Chapter
36 – St. Cloud Chapter
39 – Fox River Valley Chapter
41 – Central Ohio Chapter
42 – Hoosier Muskie Hunters Chapter
44 – Colorado Chapter
45 – Kentucky Chapter
46 – Bemidji – Cass Lake Chapter
47 – Michigan Muskie Alliance Chapter
49 – Webster Lake Musky Club Chapter
52 – Daniel Boone Chapter

Corporations:
1. Gander Mountain, Muskies, Inc.

Others:

1. WI Muskie Alliance, PMTT

Individuals:

1. Greg & Ellen Wells, Danny Kurtilla, Jim & Carla Beaty, Jim Stella, Ken & Betty Karbon

Panel discussion

Panel members

Steve Budnick, *Muskies, Inc.*
John Casselman, *Queen's University*
James Diana, *University of Michigan*
Bill James, *Indiana Department of Natural Resources*
Rob Kim, *Esox Angler* Magazine
Larry Ramsell, *Musky Historian*
Jim Saric, *Musky Hunter* Magazine
Tim Simonson, *Wisconsin Department of Natural Resources*
Jim Smith, *Muskie Magazine*

Moderator: Steve Heiting, Musky Hunter Magazine Panel
Summarizers: Terry Margenau, Wisconsin Department of Natural Resources
Steve Heiting, *Musky Hunter* Magazine

A diverse group of muskellunge "experts" was assembled during the muskellunge symposium to discuss topics central to the management of muskellunge. The panel represented management biologists, university researchers, and musky angler groups. Collectively, the panel of nine brought over two hundred years of muskellunge management, research, and angling experience to the table. The group's objective was to offer their perspectives on topics discussed during the symposium, in addition to touching on several current topics in muskellunge management. Among the items discussed were the importance of partnerships in management of muskellunge, tournaments, genetics and growth potential, aspects of live release, and of course, the world record muskellunge.

Partnerships. A theme that provided the genesis for the symposium in Indianapolis was communication. Communication between management agencies and musky anglers has long been recognized as a critical component for successful muskellunge management. The business of natural resource management incorporates both biology and sociology, and has to be a shared responsibility between management agencies and the public. The success that the state of Indiana has had in development of a successful muskellunge program serves as a good example of what can be achieved when the two facets work together. Meeting, talking, and working together have resulted in stocked fish, education, and financial assistance to research programs. The end result was going from 2 to about 20 lakes with muskellunge, and a program less than 15 years old has already produced fish over 40 pounds.

Musky anglers and their associated clubs have traditionally offered financial support to government agencies for muskellunge research and management efforts. However, as budget constraints continue, program cutbacks increase. The concept of a musky stamp to generate revenue for muskellunge programs has been considered for some time. Anglers wishing to catch and potentially harvest a muskellunge would be required to purchase a musky stamp in addition to their regular fishing license. Earmarking funds has some potential benefits. One obvious benefit would be the creation of a specific account where monies are used for a specific purpose. A good example would be an inland trout stamp, used by many government agencies to generate funds for trout stream improvement that would likely not otherwise be available. A musky stamp could also provide a communication link to anglers not associated with an organized club, and in essence bring in another group of people who want to catch muskellunge. This link would alert all musky anglers that there is something very specific going on between the professionals and the anglers that is making a difference. However, creation of a musky stamp may mean loss of general funds available for muskellunge management and make muskellunge programs dependent on the revenue generated from a stamp, which is an unknown amount. In addition, the incidental catch of muskellunge by general anglers (not possessing a musky stamp) creates a potential problem. Would these anglers be expected to have purchased a musky stamp with the assumption that they may catch a legal length muskellunge while fishing for panfish or walleye? Regardless, the musky stamp concept warrants consideration as a potential source of funding as we move into the future of muskellunge management.

Muskellunge Tournaments Tournament fishing has been a current and somewhat contentious topic among the musky fraternity and other user groups. Are musky tournaments good, bad, or ugly?

Tournaments, by definition, can range from a club outing with 10 anglers on the water with the stakes being a round of drinks, to having 150 boats on the water fishing for $25,000. Tournaments are about competition – some people like to compete, some people don't. Certainly, fishing tournaments focus a lot of attention on a group of anglers. This can provide either a positive or negative image of those involved, depending on how the tournament is conducted, and thus perceived by other user groups and the local community. One consensus among tournament anglers and organizers is that the fish is the number one priority. This includes handling the fish properly (time out of water, time held for measurement) and that tournaments do not exert too much pressure on any given water body. Many tournaments have taken steps to lower stress and improve post-release survival of muskellunge.

Judge boats are often used in tournaments to measure fish. One particular tournament series in Minnesota has been using digital cameras in combination with judge boats to measure fish. However, while tournament organizers are concerned about the fish, they are also concerned about the fishermen, not only making sure their time is safe, but making them believe that the event is fair and on a level playing field. From an organizational perspective, competitors in a tournament have to feel that no cheating has occurred.

Tournaments can also offer considerable financial benefits. For many clubs, a tournament is their major fundraiser. Tournaments have the potential to bring economic benefit to area communities. Some surveys indicate that an estimated $ 1,000 per team is brought into the local community from the tournament. A large tournament with 150 teams can insert an estimated 150,000 into the community.

Tournament angling has become a popular sport for many species (bass and walleye, as well as muskellunge) and is likely here to stay. Participating in tournaments is a personal preference, and tournaments are certainly not for everyone. Tournament anglers, organizers, management agencies, other user groups, and communities all need to work together to ensure the resource is protected and that tournaments become a positive experience for all involved.

Trophy Muskellunge: How to get them and how to keep them Muskellunge are managed as a trophy species. However, the definition of trophy is relative. The definition varies from person to person and likely changes for each person as that individual gains experience in musky angling – wanting to catch more and bigger fish. For someone new to the sport, or a casual angler, a 35-inch muskellunge is a trophy. However, for more experienced/serious musky anglers the magic number seems to be 50 inches.

There has been considerable debate regarding how to produce trophy muskellunge. Is there a superior strain of muskellunge that will grow large wherever it is stocked? Several opinions exist on how to produce trophy muskellunge. Genetics and the receiving water body are the key components for trophy production. Other important considerations include whether native stocks exist, growth and survival of stocked fish, competition with other species such as northern pike, and if rehabilitation (successful reproduction) is the management goal. Studies have shown there is no "silver bullet" strain of muskellunge that can be expected to perform everywhere. For example, Mississippi River (Leech Lake) muskellunge did not perform well in Illinois environments. In Wisconsin, where natural reproduction occurs, protecting and enhancing the existing natural populations has been a top priority. This effort includes using discretion for strains when stocking does occur, along with the protection of critical habitat.

Where no natural reproduction occurs, or outside the native range of muskellunge, the genetic background of stocked fish has less importance. Indiana has used muskellunge from a variety of sources including Pennsylvania, Wisconsin, Iowa, and Ohio. Because no native populations exist in Indiana, the "cosmopolitan" strain chosen performs well in the hatchery, performs well in the wild, grows fast, gets big, and anglers seem to appreciate it. Where rehabilitation is the management objective, strain selection may be based on slow growth, longevity, and growth potential. Once stocked, the fish becomes a product of the body of water it is in. From this perspective, to a great extent, the body of water controls the growth potential.

Live release, or catch-and-release, is practiced by most avid musky anglers and is an integral component in the successful management of musky fisheries. Not only does the release practice help sustain trophy fisheries, the replacement value of large fish removed from a fishery (in terms of number of fish stocked required to replace that fish) puts an economic component also into the picture. Should musky fisheries be total catchand-release? Panel participants agreed that total catch-and-release takes away the flexibility to manage a fishery. Some fisheries will never produce large fish, and likely won't be a trophy fishery. In addition, people want to catch fish – and have the option of harvesting a fish.

Length limits, if set too high, can in practice create a total catch-and-release fishery. For example, a 54-inch minimum length limit could protect a group of fish to be caught, handled, and actually die of old age, while never attaining the length limit. In contrast, using the concept of minimum ultimate length (see the article in this book by Casselman), 99% of the fish in a population would eventually reach a length that would be susceptible to harvest. Naturally concern arises on how to appropriately use the resource. Is a fish not eventually harvested a wasted resource? Muskies, Inc., an organization that was one of the founders of the live release concept, encourages the catch and release of all muskellunge. However, the organization also recognizes that to release or harvest a fish is an individual decision. It is legal to keep a fish on occasion, and fish do die.

The use of barbless hooks for musky angling has been discussed as a means to improve survival of released fish. Barbless hooks offer some potential benefits for anglers and fish, but also raise some concerns. Barbless hooks would reduce handling time as hook removal would become quicker, and conceptually less tissue damage would occur to the fish removing hooks in comparison to barbed hooks. Conversely, some anglers are wary about the effectiveness of barbless hooks. The thought of losing a muskellunge because the fish threw the hook is unacceptable to some anglers. Too many hours are invested in getting a strike to afford having a fish shake a hook while jumping. In addition, there have been some studies with other fish species that suggest barbless hooks may actually cause more tissue damage by penetrating deeper into the fish than barbed hooks. From a regulatory/enforcement standpoint, mandating barbless hooks for muskellunge might be difficult, though some areas (Manitoba) have gone entirely to using barbless hooks. Another perspective offered was that if you want to harvest fish for food you use barbed hooks, but if you want to be concerned about the fish a barbless hook is a good thing to be thinking about. No scientific research currently exists on the effects of barbed vs. barbless hooks on muskellunge. Information will be necessary in the future before management agencies and anglers can act in the best interest of the fishery.

World Record Muskellunge Discussion of the world record muskellunge can get emotional and most people have an opinion to offer. Is the current world record legitimate?[1] Will the world record ever be broken? Where will the next world record come from? Members of the panel were no exception and offered their thoughts on the topic of world record muskellunge. Feelings regarding the legitimacy of Louie Spray's current world record (69 pounds, 11 ounce) were mixed.

However, one message seemed to stand out – musky anglers and their sport have matured and evolved. Musky anglers of today are very capable of collecting data and providing accurate measurements of the fish they catch. Any muskellunge caught in this day and age that pushes the 60-pound barrier will be heavily scrutinized. Certainly the 61 pound 4-ounce muskellunge caught by Martin Williamson in 2000 serves as a modern day example. Nevertheless, the historic trophy fish are legendary things, and are part of what creates the lore and mystique of the muskellunge and musky fishing. Perhaps the time has come to let the historic fish be as they are, and move on.

For the world record to be broken, or even to register muskellunge in the 60pound category, special circumstances need to occur. The biggest and heaviest fish are going to be female fish that are full of eggs. These fish are at their heaviest during the late fall when egg development is occurring. For example, the Martin Williamson fish was caught in November, and likely had a considerable percent of its body weight in eggs. Fall is also a time when large female fish are feeding, and a recently consumed prey item, while a legal part of the fish, will affect the fish's weight. Another important component is longevity. For a muskellunge to reach large size, 25-30 years are required for that fish to grow. That means the fish needs to live in a body of water for a long time, and likely be handled many times – handled carefully many times. Finally, the next world record will likely come from a body of water that doesn't have a lot of pressure (spatial refuge), because even though we have developed the best methods of handling muskellunge, there is still mortality. Two percent mortality may make the difference in getting fish to 25–30 years of age.

Where will the next world record come from? This question keeps all musky anglers wondering with every cast, and planning during those long winter hours. Perhaps it will be a large water body that receives low angling pressure such as Georgian Bay. Projections on growth and longevity for the St. Lawrence River have estimated that it is possible to have muskellunge weighing over 70 pounds, probably 72–73 pounds. Maybe the skull and jaws recently found and displayed at the Symposium were that of a world class muskellunge. One thing for certain – it was a big fish, and there are likely more out there.

Conclusion: Remembering the Past and Looking to the Future

The biology and science of the muskellunge, along with the knowledge of musky anglers about the fish they pursue has come a long way since the 1984 symposium held in LaCrosse, Wisconsin. The formation of partnerships between anglers and management agencies has been, and will continue to be critical to the management of this important fish. The passion, not only anglers have for muskellunge, but also that the scientists have for the fish they work with has resulted in significant progress in protecting the muskellunge resource. Important work that has been done collectively in these times of shrinking budgets and staff is an accomplishment to be proud of. Yet, there are some areas that can be improved on, and more work to be done. Certain geographic areas (e.g., Michigan) have not developed musky fisheries in their areas and have failed to recognize the potential economic benefits that come with such a program. Patience is an important attribute as we head into the future of musky angling and muskellunge management. Anglers develop patience because it can take forever to catch a musky, but we all want change – for musky fisheries to be better. Scientists develop patience, given it takes these fish time to grow, and studies don't happen overnight.

In conclusion, this symposium in Indianapolis served as a reaffirmation of the value of partnerships that musky anglers and resource management agencies enjoy, and what can happen when you dream, when you have a vision, and when you've got good people in leadership positions who step out and say, "We're going to put this together, not because it's easy, but because we need to do it, because it would be valuable." Ed Crossman would be immensely proud of what has been done at this symposium.

[1]At the time of the muskellunge symposium, the World Record Muskie Alliance (WRMA) had filed a report to the National Fresh Water Fishing Hall of Fame (NFWFHF) recommending that Louie Spray's record muskellunge (69 pounds, 11 ounces) be disqualified based largely on photogrammetrical evidence. Upon review of that report, NFWFHF in January, 2006 upheld Spray's record fish having photo analysis of the fish length done with a different method (single plane/direct scaling geometry). WRMA issued a rebuttal to the decision in March 2006, but this rebuttal has not been reviewed by NFWFHF because protest protocol was not followed. Since that time (April 2006) a new record keeping group has been formed (Modern Day Muskellunge World Record Keeping Program) to maintain modern day muskellunge world records.

Abstract of posters and presentations

Muskellunge management and distribution analysis in a south east Wisconsin urban lake

Susan M. Beyler & Robert C. Anderson[1]

Wisconsin Department of Natural Resources, Waukesha; [1]Wisconsin Lutheran College, Milwaukee (email bob.anderson@wlc.edu)

Synopsis
Pewaukee Lake is a shallow 2,493-acre drainage lake in heavily populated Southeast Wisconsin. Musky management began in 1967 with the introduction of 1,200 musky fingerlings. Through 1981, average annual stocking was 1,625 fingerlings. Average annual stocking from 1982 through 1997 was 3,014 fingerlings. The Pewaukee Lake musky population was evaluated by use of fyke nets and electrofishing in spring 1998. Adult musky density was estimated at 0.52 per acre, both sexes combined. Muskellunge movement and home range size were evaluated based on 40 radio-tagged individuals over a four year period (1998-2002). Both external and internal radio transmitters were attached to muskellunge collected in fyke nets, by electroshocking, and by anglers. Average home range sizes were 43 acres for male and 85 acres for female muskellunge. Home range size and musky movement were greatest during spring and fall. Mean length of the sampled population was 33 inches for males and 37 inches for females. Random, stratified creel surveys were conducted in 1982 and 1998 to assess changes in fishing pressure and angler exploitation. Musky catch rate (per hour) dropped 43 percent and harvest rate dropped 70 percent, while directed angling effort increased 135 percent. Mean length of muskies harvested increased 46 percent, from 29.7 inches in 1982, under a 30-inch minimum length limit, to 43.5 inches in 1998, under a 34-inch size limit.

Missouri's musky program, egg to angler

Dale Cornelius

Missouri Department of Conservation, Lebanon, Missouri (e-mail: dale.cornelius@mdc.mo.gov)

Synopsis
The Missouri Department of Conservation introduced muskellunge into Missouri waters in 1966 when 51,000 small fingerlings (1.5–2.0 inches) and 1,500 large fingerlings (7.0–15.0 inches) were stocked into 7,820-acre Pomme de Terre Reservoir. The original objective of the Missouri Muskie Program was to provide anglers with a trophy fishery utilizing a predator that could prey upon large non-game fishes such as gizzard shad, carp, and various sucker species and continues to be an objective today. In 1967 and 1968, 249 large fingerlings were stocked into Lake of the Ozarks. A musky from this stocking was caught by a bass angler in 1981. This fish measured 49.5 inches, weighed 41 pounds 2 ounces and remains the current Missouri state record. From 1967 through 1975, the average stocking rate at Pomme de Terre Reservoir was one fish per 20 acres. The stocking rate increased to one fish per 5 acres from 1976 through 1982, then to one fish per three acres from 1983 through 1990. Beginning in 1995, Missouri's musky program was expanded to small lakes that met the minimum criteria for stocking musky. Currently, Lost Valley Hatchery at Warsaw, MO conducts spawning and rearing of musky for stocking in lakes Pomme de Terre, Hazel Creek, Fellows, Henry Sever, and Busch Conservation Area in Missouri.

Muskellunge population and fishery dynamics in Green River Lake, Kentucky

Eric Cummins

Kentucky Department of Fish & Wildlife Resources, Bowling Green, (email: ericcummins@bellsouth.net)

Synopsis

Damming of the Green River in 1969 degraded a historic muskellunge fishery necessitating annual stockings muskellunge to reestablish and maintain a muskellunge fishery. Muskellunge population dynamics and angler use were examined to evaluate success of annual stockings of large fingerlings (>305 mm) in establishing a quality fishery. Muskellunge recruited to the 762-mm size limit in 2.6 years. Angler catch rates for all size muskellunge averaged 16.9 hours per fish; catch rate for "keeper" sized muskellunge averaged 28.6 hours per fish. Total catch averaged 0.25 fish per hectare; mean harvest was 0.06 fish per hectare. Average size caught and released (voluntary) was 856-mm; average size harvested was 892-mm. Muskellunge angler catch and release rates averaged 43.8%. The percentage of nonresident anglers targeting muskellunge has doubled since a quality fishery was established with the large fingerling stockings. However, fishing pressure for muskellunge was nearly half to a quarter of that experienced by other muskellunge fisheries with similar characteristics.

Human influences on natural spawning and reproductive success of muskellunge

James S. Diana, Ashley Rust, Sarah Zorn & Terry Margenau[1]

School of Natural Resources and Environment, University of Michigan, Ann Arbor (email: jimd@umich.edu);
[1]*Wisconsin Department of Natural Resources, Spooner*

Synopsis

Natural reproduction of muskellunge *Esox masquinongy* in northern Wisconsin is often insufficient to sustain populations in many lakes where the species is native. Presence of the congeneric northern pike *E. lucius,* poor conditions in spawning sediments, altered hydrologic regimes, and human development are all suspected as contributing to the decline in natural reproduction of muskellunge. The purpose of this study was to further describe muskellunge spawning habitat and determine physical, chemical, biological, and land use characteristics that can be used to distinguish self-sustaining muskellunge lakes from lakes where stocking is required to maintain populations.

Survival of eggs from experimental enclosures was low (< 3%) for eggs reared on sediments or woody substrates, and even for eggs suspended in the water. Poor survival was common in good and poor reproductive lakes. Young fish could be collected throughout the summer and in fall electrofishing surveys, and were far more common in good than poor reproductive lakes. Spawning habitats in good reproductive lakes had softer sediments, with higher dissolved oxygen and more organic nitrogen content, and occurred in lakes with few shoreline alterations. A multiple regression model for this relationship relating development to organic nitrogen content of spawning sediments included deadfall trees per kilometer and percent shore totally developed (adjusted $R^2 = 0.60$, $p < 0.01$). The type of human development on the shoreline, was also correlated with fitness of spawning habitats. Direction of water level change during the spawning period, percent of woody debris on the spawning habitat sediment, number of deadfall trees per kilometer of shore, and percent of shore that was totally developed were the most important variables used in classifying the level of muskellunge reproduction a lake could support. The model developed in this study correctly classified 75% of good reproductive and 89% of poor reproductive lakes. If lake managers wish to use muskellunge stocking programs to re-establish self-sustaining populations, they should critically review each candidate lake by considering the Dombeck et al. (1986) model and the model derived in this study.

A conceptual model of muskellunge spawning habitat in Georgian Bay, Ontario, Canada

B. Farmer, P. Chow-Fraser & T. Seilheimer

Biology Department, McMaster University, Hamilton, Ontario (email: chowfras@mcmaster.ca)

Synopsis

Georgian Bay in Lake Huron, Canada, has a potential to produce a trophy-status muskellunge fishery; however, some evidence suggests that the population is in decline, especially in areas impacted by recreational development. To assist research and management, we developed a conceptual model to predict spawning habitat of muskellunge, using meta-analysis of available peer-reviewed and technical literature.

The model incorporates three primary variables: water temperature (7.5–15°C), dissolved oxygen (> 5 mg/L) at the sediment-water interface, and adequate separation of individual eggs after deposition. The model also assumes that muskellunge spawning occurs in wetlands because of their known association with aquatic vegetation. Secondary variables influencing primary conditions include (1) depth, current and substrate colour (assumed to have an effect on temperature); (2) current, sediment oxygen demand, sediment compactness and plant density (assumed to have an effect on dissolved oxygen concentrations); and (3) particle size and plant density (assumed to have an effect on egg separation). Field data will be used to validate the model and to help clarify the relative importance of each variable, and thus allow for refinement of the model.

Spawning and post-spawning movements of the St. Lawrence River muskellunge *(Esox masquinongy)*

Steven R. LaPan, Albert Schiavone[1], Rodger Klindt[1], Robert G. Werner[2] & John M. Farrell[2]

New York State Department of Environmental Conservation, Cape Vincent Fisheries Station, Cape Vincent (email: srlapan@gw.dec.state.ny.us); [1]New York State Department of Environmental Conservation, Watertown; [2]College of Environmental Science and Forest Biology, State University of New York, Syracuse

Synopsis

Radiotelemetry tracking and long-term tagging studies revealed strong reproductive homing of muskellunge *(Esox masquinongy)* in the Thousand Islands region of the St. Lawrence River. Radiolocation also was useful in the identification of critical spawning and nursery habitats. From 1984 to 1989, 47 adult muskellunge were radio-tagged during the spawning season at 22 sites. Radio-tagged fish were located at least once daily during the spawning period. Capture sites of adult muskellunge and areas frequented by tagged muskellunge during the spawning period were subsequently seined to document the presence or absence of young-of-the-year (YOY) muskellunge. YOY muskellunge were collected at 13 of the 22 adult capture sites and also at an additional 14 sites frequented by radio-tagged adults. Post-spawning muskellunge generally migrated upstream to Lake Ontario, remained in the vicinity of their respective spawning site, or moved into deep water where they could not be located. Radio transmitters that functioned into the following spawning season revealed a high degree of reproductive homing. Strong muskellunge spawning site fidelity was also observed in tagging and recaptures during spring trapnetting surveys. Of 184 muskellunge tagged from 1990 to 2003 at twelve sites separated by over 35 km, 33/34(97%) were recaptured at the original tagging location. The one exception was a male muskellunge caught in an adjacent bay five years after the original tagging. Reproductive homing has important implications for management of muskellunge and their habitats.

Biological characteristics of record class muskellunge populations in Georgian Bay and the North Channel, Lake Huron

A. P. Liskauskas

Ontario Ministry of Natural Resources, Upper Great Lakes Management Unit; (email:1arunas.liskauskas@mnr.gov.on.ca)

Synopsis
The nearshore waters of eastern Georgian Bay and the North Channel of Lake Huron comprise a vast area of diverse aquatic habitat supporting numerous muskellunge populations. Since 1996 muskellunge spawning surveys using live-capture trapnets have been conducted at eight locations distributed throughout the area. Captured fish were biologically sampled and tagged to assist in establishing home ranges and recapture rates in the recreational fishery. More than 500 adult muskellunge were captured during these surveys and several dozen potential spawning locations were identified. Ripe females were captured over a wide range of temperatures (8-22°C) and over a prolonged time frame (4–6 weeks). The average size of muskellunge did not vary significantly between sites with females averaging 1174 mm (46.2 in) and males averaging 998 mm (39.3 in) in total length. The relative abundance of muskellunge varied across sampling sites, ranging from 0.11 to 1.53 fish per trapnet night. The recovery of tagged fish in subsequent years provided evidence for homing to specific spawning areas and of pair bonding. The results from these surveys have already proved beneficial in providing justification for the implementation of a record class designation for these waters (minimum size limit of 137 cm (54 inches)). In addition, the identification of sensitive spawning areas and acquisition of genetic material for establishment of population diversity, should contribute to the sustainable management of this important aquatic resource.

Biological characteristics of a riverine muskellunge population

Scott Morrison, Christopher O'Bara & Joel Harrison[1]

West Virginia Division of Natural Resources, Parkersburg (email: scottmorrison@wvdnr.gov); [1]West Virginia Division of Natural Resources, Elkins

Synopsis
Muskellunge *Esox masquinongy ohioensis* examined to determine movement patterns and habitat preferences in Middle Island Creek, a stream in West Central West Virginia. Since 2002, over 120 muskellunge were collected with pDC- boatmounted electrofishing equipment. Individual fish were measured and inserted with PIT tags prior to release. Release locations were noted with GPS and later included into a GIS based system. Approximately 40 fish have been recaptured during subsequent electrofishing-based surveys and via anglers. Recaptured locations were noted with GPS, included in a GIS based system, and compared with initial capture locations. Movement distances varied both between sexed and size of fish. Growth patterns were also dependent on sex and to some degrees location and year of capture.

Considerations in muskellunge *(Esox masquinongy)* introductions: A management dilemma?

John E. Nelson, Timothy D. Simonson[1] & Terry L. Margenau[2]

Wisconsin Department of Natural Resources, Plymouth (email: John.Nelson@dnr.state.wi.us); [1]Wisconsin Department of Natural Resources, Madison, [2]Wisconsin Department of Natural Resources, Spooner

Synopsis
The increasing popularity of muskellunge fishing has led to a desire by anglers and fisheries managers to expand the range of muskellunge. Often the expansions are proposed close to population centers well outside the original range of the species. Several biological and social considerations are appropriate in deciding if and where the introductions are to take place. In Wisconsin, an Environmental Assessment (EA) must be prepared for initial introductions. The EA then receives public review, with approval dependent on the results of that review. Factors considered in the review include the impact of the introduction on the

existing fish community, the availability of forage, the cost of long term management (stocking), the likelihood of successful development of a fishable population of muskellunge, the social acceptance of a muskellunge fishery and the ethical consideration of range expansion. A review of the original and current range of muskellunge will be presented.

Factors affecting historical year-class strengths of trophy muskellunge *(Esox masquinongy)* populations

Chris J. Robinson & John M. Casselman[1]

Ontario Ministry of Natural Resources, Kawartha Lakes Fisheries Assessment Unit, Lindsay (email: chrisrobinson@nexicom.net); [1]Ontario Ministry of Natural Resource; Research, Science and Technology Branch, Glenora Fisheries Station, Picton

Synopsis
The longevity, relative rarity, and consequent perceived value of muskellunge *(Esox masquinongy)* have made the assessment of population characteristics through traditional fisheries methods difficult for this species. Using a subset of 1351 angler-contributed samples (mean age 10.6 years) collected by the Cleithrum Project since 1978, we are now able to quantitatively examine historical year-class strengths for 15 muskellunge populations, with emphasis on four populations in Ontario. Cleithra were interpreted for age, and proportional year-class strengths calculated for each population to account for multiple, sporadic, years of capture. Temperature, water level, size-limit changes, and the presence of northern pike were examined for correlations with year-class strength. Year-class strength histories had varying correlations with summer (Lake St. Clair: $p = 0.009$, $r^2 = 0.36$; all Ontario populations: $p = 0.044$, $r^2 = 0.13$) or spring (Kawartha lakes populations: $p = 0.044$, $r^2 = 0.15$) temperatures in the first year of life, but no significant relationships were observed with the other factors. In one instance, populations from water bodies of similar characteristics had strongly similar year-class histories (Eagle Lake and Lake of the Woods, Ontario; $p = 0.0016$, $r^2 = 0.546$). The 11 Wisconsin muskellunge populations, which only had summer air temperature examined, had *p*values ranging from 0.014 (Buckatabon Lake) to 0.814 (Lac Vieux Desert). We discuss the implications of these results for future muskellunge management.

Historical trends in body growth of five Ontario muskellunge *(Esox masquinongy)* populations

Chris J. Robinson & John M. Casselman[1]

Ontario Ministry of Natural Resources, Kawartha Lakes Fisheries Assessment Unit, Lindsay (email: chrisrobinson@nexicom.net); [1]Ontario Ministry of Natural Resource; Research, Science and Technology Branch, Glenora Fisheries Station, Picton

Synopsis
Muskellunge *(Esox masquinongy)* management in Ontario has primarily focused on harvest control through minimum size limits, requiring information on the growth potential, responses, and variability of populations. We examined historical muskellunge body growth trends, as recorded in the cleithrum bone, from five muskellunge populations, including two populations with extended data sets (overall $n = 456$). Two measures were digitally collected – an index of annual growth and cleithrum size at age eight - and linearly regressed against log-transformed mean daily summer temperatures. There were no significant correlations between temperature and annual growth ($p = 0.120$ to 0.762). However, cleithrum size at age eight increased significantly over time (Lake St. Clair $p = 0.026$; St. Lawrence River $p < 0.0001$); mean temperature and growth over eight-year periods had a significant but weak relationship in the St. Lawrence River population ($p = 0.020$; $r^2 = 0.055$). Analysis of a significant difference ($p = 0.004$) in growth amongst five-year year-class groupings of the St. Lawrence River population showed muskellunge produced from 1975 to 1989 were 9.3% larger than those produced from 1940 to 1959. Temperature has had a significant but relatively minor direct effect on growth of muskellunge, and factors such as prey composition, behavioral thermoregulation, and handling may have overridden or compensated for temperature effects.

Development of muskellunge fisheries in Lakes Monona and Wingra, Wisconsin

R. Scot Stewart, Kurt Welke & Mike Vogelsang

Wisconsin Department of Natural Resources, Fitchburg, WI (email: Richard.Stewart@dnr.state.wi.us)

Synopsis

When dreaming of muskellunge and the lakes they inhabit, most anglers do not picture urban settings. However, in densely populated Madison, WI, Lakes Monona and Wingra have progressed to regionally renowned fisheries characterized by excellent densities of robust fish. These populations developed due to a unique combination of aggressive stocking involving the state and the public, restrictive size limits, and a rotation of sampling to periodically quantify the populations. Annual stocking of both hybrid and true muskellunge began as early as the mid-70s, with a shift to predominantly true musky in the early 90s. A stocking rate of 2 fish per acre annually led to a population of approximately 4 fish per acre by 1997, in Lake Wingra. The stocking rate was reduced in subsequent years and an updated population estimate for musky in the lake will be complete in the spring of 2005. In Lake Monona, an average stocking rate of 0.6 fish per acre annually resulted in a population of 0.49 fish per acre. The Capital City Chapter of Muskie Inc. and the Oregon Muskie Busters have made substantial financial contributions to supplement state stocking. In 2004, the minimum size limit on both lakes was raised from the statewide 34" limit to a 45" limit. We plan to use our future sampling rotation to evaluate not only musky density and size structure, but also to continue to fine tune stocking rates and watch for possible impacts on other fish populations.

Analysis of 40 years of Lake St. Clair musky fishing records

Michael V. Thomas & Robert Haas

Michigan Department of Natural Resources, Harrison Township (email: thomasmv@michigan.gov)

Synopsis

The Michigan-Ontario Muskie Club (MOMC) has been the major organization representing musky anglers in the Lake St. Clair area since the 1950s. Among other activities, the club sponsored derbies or tournaments each year. Through the years, catch records were diligently maintained by club officers. These records present a historical documentation of trends in the musky population and musky fishery of Lake St. Clair over more than four decades. In this paper, we analyze MOMC catch records for trends in the context of changes in the ecology of the lake and fishing regulations. With permission from the club, we conducted a time series analysis of: 1. the single heaviest fish entered in all club events within each year 2. the ten heaviest fish entered in all club events within each year and 3. the top 250 heaviest fish entered in all club events across the time series. Our analysis revealed a consistent pattern of heavier fish in the catch since 1990. This coincides with significant ecological and regulatory changes during the 1980s. Additionally, voluntary catch and release of musky became widely practiced by Lake St. Clair musky anglers during the 1980s. We submit that the MOMC catch records provide valuable insight into the response of the Lake St. Clair musky population to ecological and regulatory changes. We encourage other fishing clubs and organizations to carefully record their catches. Similarly, we suggest that resource management agencies can glean useful data from well-maintained catch records of fishing groups when they are available.

Muskellunge movement in the Manitowish Chain, Vilas County, Wisconsin

Jordan G. Weeks, Michael J. Hansen & Patrick J. Schmalz

University of Wisconsin-Stevens Point, College of Natural Resources (email: Jordan.G.Weeks@uwsp.edu)

Synopsis

We quantified movement and spawning site fidelity of muskellunge in the Manitowish Chain of 10 interconnected lakes in Vilas County Wisconsin. We marked 314 muskellunge (16–60 fish per lake) with T-bar anchor tags and tabulated tag returns from catches in electrofishing, trap-netting, angler returns and creel surveys. During the first summer after spawning in spring 2004, 3.47% of tagged muskellunge were recovered (0.0–10.5% per lake). Over all lakes, 51.7% of muskellunge (0.0–1.0% per lake) were recovered in the same lake in which they were tagged. We will quantify spawning site fidelity through spring of 2005 and seasonal movement and home range through summer of 2005. Our findings will be used to determine if spearing and angling fisheries should be managed for individual lakes or for the entire set of interconnected lakes within the chain.

Environ Biol Fish (2007) 79:1–3
DOI 10.1007/s10641-007-9184-y

SPECIAL ISSUE CROSSMAN

Dr. E.J. Crossman, a personal perspective: introductory remarks to the Dr. Ed Crossman Muskellunge Symposium

Margaret Crossman

Received: 26 October 2006 / Accepted: 30 November 2006 / Published online: 9 February 2007

Abstract Margaret Crossman gave the introductory address to the Muskellunge Symposium honouring her late husband, Dr. E.J. (Ed) Crossman. Her introductory notes, a personal perspective, are provided.

Dr. Crossman—E. J.—Edwin—"Edwin," a man of several names but a friend to anyone who asked for a helping hand. Life is a marvellous journey through time, and Ed's was a full measure. He started life during the Depression. His father, who worked in Niagara Falls, New York, USA, was fortunate in having 3 days work a week. Because of this, he had time to pursue his love of hunting and fishing and not so favourite, gardening. All this helped fill the family's larder. Eventually Ed joined his father, who taught him sound rules. If you want to find game the next time you go hunting or fishing, do not take more than your share. Neither were you to disturb their habitat if you wanted them to stay about.

Ed was a keen hunter and fisherman, but in later years, he was more interested in being out in the outdoors, listening to the beagles sounding off, doing their job. It was more of a challenge to take the barbs off the hooks and give the fish the chance to throw the hook if it was snagged. Usually fishing was done in silence. I was the navigator and much preferred to paddle or row into favourable fishy spots. One summer, because Ed had to write an English supplementary exam, fishing was combined with quotes from English poems for him to identify. As we worked our way down the stream, even the fish seemed to be receptive too.

At Queen's University, he came under the influence of Wes Curran. Dr. Curran gave him an opportunity to work a summer at the Queen's University Biological Station, north of Kingston, Ontario, Canada. Ed got his first taste of serious research. The station was a hotbed of enthusiasm and cooperation between the disciplines of the research going on in the water, forests, swamps, and fields. It was through Wes Curran that Ed was introduced to the wonderful muskellunge, Nogies Creek, and the Toronto Anglers and Hunters Association (who helped finance his project). The first consumed most of his research life, the latter two encouraged his association with the angling community, not the least among them was to become Muskies Inc. and Muskies Canada.

It was around then that I got into the mix. We were married in June 1952. Our honeymoon started at Nogies Creek, lifting nets the next morning. We had a glorious honeymoon throughout our life, working out the winters so we were ready to get back to the adventures of the fieldwork. Not only graduate students but Natural

M. Crossman (✉)
Scarborough, Ontario, Canada

Resources people shared in Ed's zeal to find out the how, when, and why and put together the what-for. As his family, we were also allowed to participate. We learned the vagaries of minnow traps, the pull of the seine, the care needed with the bottom line, the curse of blood suckers when you are after fish in a vegetative area. For heaven's sake, double-bag specimens and make sure there is air in the second bag. Thank goodness for that air. Once collecting in Quetico Park, we overturned in white water and the day's collection went bobbing downstream. What fun! It was a good thing our son, John, was a good swimmer,

To broaden his knowledge of Canadian fishes, Ed was able to go to the University of British Columbia. There he was fortunate to come under the influence of Peter Larkin and Cas Lindsey. This pair were very serious about their work but mischievous as could be when you least expected it. Here was engrained the ethos—what you do was serious and must be well done but with levity. I well remember tales of feral cats loosed in the fisheries hut to greet the late-working graduate students. But the parry to the forestry gang was dead skate hidden away on their heating pipes. Much labour went into carving and blackening a replica of an Indian arrowhead to be sequestered into a midden for a fellow student who was having no success finding relics. What an uproar when he found it was a fake. Serious work did go on, and we did come home with a Ph.D.

We left behind the beautiful mountains and productive trout streams of B.C. to "thankfully" return to central North America with its four seasons and a job at the Royal Ontario Museum with Bev Scott. At University of Toronto, ichthyology was a must. Enter W.B. Scott into Ed's life. Here the whole world of fish opened up. Bev Scott became Ed's ichthyological beacon. Their personal interests and understanding of the broad spectrum and interest in fishes led to their writing *Freshwater Fishes of Canada*. The people fanned out across the country. Between them, Bev and Ed had a fellowship network of ichthyologists, Natural Resources personnel, anglers, and commercial fishermen. It always amazed me the give and take with such good-heartedness. A mass of information was made available through publication of the gleanings.

Ed travelled a good part of the world in search of information about the muskellunge and its fellow family members. Each trip would bring back tales of the people encountered or of circumstances different from ours. The Arctic tales of the supply plane not coming to pick them up to move on to the next station and all their gear being packed. The fishing pole and good fishing saved them until the plane appeared. In East Europe, a meal with friends went well but no serious talking happened until the restaurant closed and lights were extinguished and all the other guests had departed. Then to get back into the ancient place he was staying, the keeper of the door had given him a gigantic key to unlock the door. It was not one to slip into your pocket or forget to hand in. Wherever Ed went, a supply of peanut butter went too. On a trip to eastern Russia, he felt his supply was safe. Everyone said "*Nyet*" to giving it a taste. However, coming in off the lake one afternoon, he found the cook had been into his peanut butter and "she liked it."

Over the years, I have been Ed's gofer, but he graced me with the title of research assistant. I have been privileged to listen in or read about what was happening with his work. It was fun to watch Ed in his office, phone glued to his ear. For ages, he would listen, answer, puzzle in sequence. "Who was that?"—Wisconsin, Minnesota, Manitoba, Maine, British Columbia. Each query either created a puzzle or added pieces to put the puzzle together.

Ed always had several projects on the go, but it was important to bring them to a conclusion so that you had fulfilled your obligation to your goal and your sponsors. The "noble muskie" was an ongoing study and concern. I felt his goal was to get to know the fish, its history, its lineage, and its environmental requirements. I am pleased to see from the program here at this symposium that you share this and are carrying on the good work. Thank you for letting me share in this gathering and for the honour you have given his family to dedicating this symposium in his name. He felt honoured to count you among his friends and acquaintances.

Finally, I would like to read you a poem, "Oh, Great Spirit," which was used in Native Indian ceremonies (Illustration 1).

Oh, Great Spirit!

Great Spirit, whose voice I hear in the winds
And whose breath gives life to everyone
Hear me
I come to you as one of Your many children
I am weak and small
And need your wisdom and your strength
Let me walk in beauty, and make my eyes ever be-
hold
The red and purple sunsets.
Make my hands respect the things you have made,
And make my ears sharp so I may hear Your voice.
Make me wise, so I may understand what You have
Taught to my people and
The lesson You have hidden in each leaf and each
rock
I ask for wisdom and strength
Not to be superior to my brothers, but be able
To fight my greatest enemy, myself.
Make me ever ready to come before You with
Clean hands and a straight eye.
So as life fades away like the fading sunset,
My spirit may come to You without shame.

Illustration 1. North American Native ceremonial poem "Oh, Great Spirit!"

Illustration 2. Celebrating the 30th anniversary of the publishing of *Freshwater Fishes of Canada*, Canadian Conference for Fisheries Research, Ottawa, January 2003. Ed and friends. Ed Crossman, Bev Scott, Don McPhail, Cas Lindsey, Joe Nelson, Ken Stewart, Dick Beamish (from right to left). (Courtesy D. Noakes)

Illustration 3. EJ in a "listening moment"

Illustration 4. EJ angling for supper, Northwest Territories, Canada

Environ Biol Fish (2007) 79:5–10
DOI 10.1007/s10641-006-9166-5

SPECIAL ISSUE - CROSSMAN

Dr. E.J. (Ed) Crossman's scientific contributions on muskellunge: celebrating a lasting legacy

John M. Casselman

Received: 8 June 2006 / Accepted: 16 October 2006 / Published online: 6 December 2006

Dr. E.J. (Ed) Crossman's first association with fish, and muskellunge, was as he was growing up in Niagara Falls, Ontario, a few blocks from the lower gorge of the Niagara River, an area that he frequented often. We are especially fortunate that from this early association with fish and the river, Ed decided to make ichthyology his career and make muskellunge a particular favourite. His first in-depth involvement with muskellunge began when he conducted his master's at the University of Toronto on "The Maskinonge Population of Nogies Creek, Ontario". (Note the early spelling of the common name of the species—nomenclature became a subject of special interest). Under the guidance of Dr. F.E.J. Fry, this early research had a strong element of field biology, which was initiated by practical hands-on experience with muskellunge, trap-netting and tagging fish in the Nogies Creek Sanctuary near Bobcaygeon, Ontario. Handling the living organism in its natural environment was inspiring and whetted a curiosity about the species. This association continued for almost five decades and involved a diverse range of subjects, with the fish always the focus of special attention. Ed's very first published scientific contribution came from his master's research (Crossman 1956), and it was the first of many on the species.

J. M. Casselman (✉)
Department of Biology, Queen's University,
Kingston, Ontario, Canada K7L 3N6
e-mail: casselmj@biology.queensu.ca

Ed continued his graduate studies with a Ph.D. program at the University of British Columbia but immediately afterwards was recruited by Dr. W.B. (Bev) Scott, who became a life-long friend and colleague, to return to Ontario and join the Department of Ichthyology and Herpetology at the Royal Ontario Museum. He immediately began to develop a research program on muskellunge, assembling a comprehensive well-indexed reference library that formed the basis of future bibliographies and reference material that became well known and were frequently visited by students of the subject. He published widely on esocids and soon became a world authority, but muskellunge was, above all, his species of special interest. Throughout his entire career, until his untimely death in December 2003, he conducted in-depth studies on muskellunge populations across the province of Ontario, ranging from the Niagara River and the Kawartha Lakes in the south to the muskellunge lakes in the northwest (Wabigoon and Eagle). In spite of a broadening interest, he continued studies and based graduate-student programs on his favourite Nogies Creek Sanctuary population and took a personal interest in rejuvenating the Nogies Creek field station and participating in studies there.

In graduate-student programs through the Department of Zoology, University of Toronto,

A collage to celebrate Dr. E.J. (Ed) Crossman's scientific contributions (pictures from EJC's and JMC's personal archives). (**A**) Ed and colleague (Barry Muir) sampling a trap-net catch of fish from Nogies Creek, Ontario (ca. 1955). (**B**) Ed with a Royal Ontario Museum Department of Ichthyology and Herpetology conservation projects vehicle donated by the Toronto Anglers', Hunters' Association (ca. 1960). (**C**) Dr. Crossman's lecturing at University of Toronto (ca. 1975). (**D**) Ed observing and recording spawning behaviour of fish at Lake Simcoe (ca. 1980). (**E**) Ed and his wife, Margaret, in many ways his professional companion, preparing egg-sampling boxes in northwestern Ontario (ca. 1984). (**F**) Ed with a northern pike caught electrofishing in northwestern Ontario (ca. 1984)

he encouraged the study of muskellunge and other esocids, covering a broad range of subjects. In addition, he was active in graduate-student committees at various other universities.

Ed's broad experience developed into a remarkable understanding of the species, and many sought his advice and insights. This added to his wealth of knowledge and built an increasing appreciation for his special fish, which he often referred to as "the noble beast" (Crossman 1978). He delved into many aspects of the science of muskellunge, particularly enjoying the scientific interaction with others, focusing on its size and mystique. He readily admitted that he benefited as much from others as they did from him. But he particularly enjoyed interacting with muskellunge anglers, who sought him out frequently for his practical understanding and unique insights into all aspects of their special fish.

Dr. Crossman's knowledge was well documented by the breadth and depth of his publications. His professional accomplishments were many and diverse, with more than 200 scientific publications, approximately a quarter of which dealt specifically with muskellunge.

In celebrating Ed and his noble fish, it seems fitting to list his 74 published contributions that directly pertained to the species (cited below). Many are major scientific contributions, peer-reviewed, and published in the primary literature. Even though Ed was strongly academic and in many ways a true philosopher, he had a special interest in the practical. He interacted widely with those who used the resource and handled the fish and produced numerous articles of a practical nature, written with a popular slant for anglers. These publications encompass a broad range of biology and transfer a wealth of science. What a diverse assemblage! The only common thread appears to be the species. His publications can be grouped as follows:

Subject	Number	Subject	Number
Age and growth	14	Morphology	4
Bibliography	1	Movement and behaviour	4
Culture and stocking	1	Nomenclature	3
Distribution	6	Physiology	1
Evolution	1	Predator–prey interaction	1
Fisheries and angling	12	Reproduction	2
General reviews	6	Research reviews	6
Genetics and hybridization	4	Taxonomy and systematics	4
Identification	4		

Of the 17 subjects that were the primary focus of Ed's muskellunge writings, age and growth tops the list at 19%, followed by fishing and angling at 16%, with distribution and general and research reviews each at 8%. Five categories compete for the next level of importance, ranging from taxonomy and systematics, morphology, identification, and genetics to studies of the fish in the natural environment, particularly dealing with movement and activity. The broad diversity of subjects emphasizes a wide range of interests from the theoretical to the practical, from the natural environment to laboratory studies.

One of the most relevant and useful publications is the annotated bibliography that he co-authored on the species (Crossman and Goodchild 1978). This summarizes all the muskellunge literature up to that time. But remarkably enough, only approximately one-quarter of his own published contributions precede that date.

We are fortunate that Ed was a prolific writer and encouraged others to be the same, often by assisting at all levels to ensure that the job was done. To see this list of scientific contributions is to appreciate one of the many reasons why this, the Second International Muskie Symposium, co-sponsored by the American Fisheries Society Esocid Technical Committee and Muskies, Inc., was dedicated in his honour. His collaboration with the muskie fishing fraternity was mutually beneficial. He especially enjoyed the association with organized and enlightened anglers such as Muskies, Inc. and Muskies Canada. Indeed, their magazines provided a forum for many of his popular articles on this special fish, such as encouraging catch and release, reducing angling stress, understanding disease problems, and dealing with those pesky Y bones.

For many young professionals interested in the esocids, their first interaction in the scientific community with Ed was when they requested reviews of their fledgling writings and scientific contributions. Many of us benefited much from that. The memories and the numerous red editorial notations remain particularly vivid. Ed always spoke directly yet kindly, a decisive individual with a strong presence. Always willing to help, he assisted many with careful and detailed instructive reviews, constructive comments, and yes, challenging criticism if need be. Demanding a complete understanding to explain the science being elaborated, Ed always queried deeply into the facts no matter the subject and, quite importantly, had a knack for drawing out the truth in the frankest and simplest ways.

My strongest scientific association with Ed involved The Cleithrum Project, a collaboration that I proposed and he encouraged. It deals with the detailed and precise science extracted from an archival museum-type structure, the cleithral bone. It involves anglers in a cooperative and mutually beneficial association. The anglers provided the structure and data on harvested fish, and we were able to extract critically important detailed age-growth science for managing and sustaining quality muskellunge populations and fisheries. In many different ways, The Cleithrum Project involvement was typical of Ed and emphasized his diverse interests and a commitment to helping sustain muskellunge populations and fisheries. Indeed, his last publication reiterated the importance of this ongoing project.

With such an extensive contribution to muskellunge research and management, one should be concerned about the loss of this eminent individual who contributed so much to the understanding and well-being of the species and its fisheries. But his contribution and legacy will live on in many ways, particularly through his scientific publications. They are significant contributions that will endure and will be frequently used and referenced in the future. The Literature Cited in the publications of the proceedings of this symposium underline the significance of this statement. Ed's publications are frequently cited (50 papers with over 675 citations listed in the *Science Citation Index*). Fortunately, we have Ed's prolific writings. These will provide a lasting legacy.

Dr. E.J. (Ed) Crossman's Published Contributions on Muskellunge

Crossman EJ (1956) Growth, mortality, and movements in a sanctuary population of maskinonge (*Esox masquinongy* Mitchill). J Fish Res Board Canada 13(5):599–612

Crossman EJ (1960) Variation in number and asymmetry in branchiostegal rays in the family Esocidae. Can J Zool 38(2):363–375

Crossman EJ (1962) Predator-prey relationships in pikes (Esocidae). J Fish Res Board Canada 19(5):979–980

Crossman EJ (1963) Review: Makee: the life and death of a muskellunge. Can Field Natural 77(3):172

Crossman EJ, Buss K (1965) Hybridization in the family Esocidae. J Fish Res Board Canada 22(5):1261–1292

Crossman EJ, Casselman JM (1969) Identification of northern pike and muskellunge from axial skeletons, scales, and epipleurals. J Fish Res Board Canada 26(1):175–178

Crossman EJ (1969) The Canadian freshwater fish fauna. Ac Rer Natur Mus Slov Bratislava XV(2):137–152

Crossman EJ, Harington CR (1970) Pleistocene pike, *Esox lucius*, and *Esox* sp. from the Yukon Territory and Ontario. Can J Earth Sci 7(4):1130–1138

Crossman EJ (1971) Pike. Related forms and history. App. 2:292–307. In: Pike, by Fred Buller, McDonald, Co. Ltd., London, 320 pp

Beamish RJ, Merrilees MJ, Crossman EJ (1971) Karotypes and DNA values for members of the suborder Esocoidei (Osteichthyes: Salmoniformes). Chromosoma (Berl) 34:436–447

Merrilees MJ, Crossman EJ (1973) Surface pits in the family Esocidae. 1. Struct Types J Morphol 141(3):307–320

Merrilees MJ, Crossman EJ (1973) Surface pits in the family Esocidae. 2. Epidermal dermal interaction and evidence for aplasia of the lateral line sensory system. J Morphol 141(3):321–344

Scott WB, Crossman EJ (1973) Freshwater fishes of Canada. Fish Res Board Canada Bull 184:966 pp

McAllister DE, Crossman EJ (1973) A guide to the freshwater sport fishes of Canada. Nat Mus Canada Nat Mus Nat Sci Nat Hist Ser 1:91 pp

Merrilees MJ, Crossman EJ (1974) Species and geographic variability of surface pits in the Esocidae. Copeia 1974(4):893–909

Crossman EJ (1975) Musky! ROM research on an important sport fish. Rotunda 1975 (Winter) 8(1):36–40

Crossman EJ (1976) Quetico fishes. Royal Ontario Museum Life Sciences Miscellaneous Publication, 86 pp

Crossman EJ (1977) Displacement and home range movements of muskellunge determined by ultrasonic tracking. Environ Biol Fish 1(2):145–158

Crossman EJ, Meade JW (1977) Artificial hybrids between Amur pike *Esox reicherti* and North American esocids. J Fish Res Board Canada 34(12):2338–2343

Crossman EJ, Goodchild CD (1978) An annotated bibliography of the muskellunge *Esox masquinongy* (Osteichthyes: Salmoniformes). Royal Ontario Museum Life Sciences Miscellaneous Publication, 131 pp

Crossman EJ (1978) How can you tell a musky from a pike? Muskies, Inc. Annual, pp 10–11

Crossman EJ, Minor JD (1977) Tracking studies of the muskellunge in shallow heavily vegetated water. Underwater Telemetry Newslett 7(2):4–5

Minor JD, Crossman EJ (1978) Home range and seasonal movements of muskellunge as determined by radiotelemetry. American Fisheries Society Special Publication 11, Proc. Symp. Selected Coolwater Species, pp 146–153

Crossman EJ (1978) Taxonomy and distribution of North American esocids. American Fisheries Society Special Publication 11, Proc. Symp. Selected Coolwater Species, pp 13–26

Crossman EJ (1979) Piking in North America. Pike Angler's Club Mag 5:7–9

Crossman EJ (1979) The record of the age of your pike. Pike Angler's Club Mag 5:9–11

Crossman EJ (1979) Authentication or debunking old fish stories. Muskies Inc 13(4):6–8

Crossman EJ, Casselman JM (1979) The Cleithrum Aging Project. Muskies Canada 1(6):3–8

Crossman EJ, Casselman JM (1979) The Cleithrum Aging Project. Muskies Inc Newslett 13(7):12–17

McNeil FI, Crossman EJ (1979) Fin clips in the evaluation of stocking programs for muskellunge *Esox masquinongy*. Trans Am Fish Soc 108(4):335–343

Crossman EJ (1979) Trip tip, central Ontario muskies. In-Fisherman, 26 (Aug.–Sept.), pp 96–97

Crossman EJ, Van Meter HD (1979) Annotated list of the fishes of the Lake Ontario watershed. Great Lakes Fishery Commission, Technical Report 36, 25 pp

Crossman EJ (1980) Account on N.A. species of family Esocidae. In: Lee DS, Gilbert CR, Hocutt CH, Jenkins RE, McAllister DE, Stauffer JR (eds) Atlas of North American freshwater fishes. N.C. State Museum of Natural History, pp 131–138 (850 pp)

Beggs GL, Holeton GF, Crossman EJ (1980) Some physiological consequences of angling stress in muskellunge, *Esox masquinongy* Mitchill. J Fish Biol 17(6):649–659

Crossman EJ (1981) The fish. In: Guide to becoming a muskie master. Magnus Publishing Ltd., Toronto, Chapter 1, pp 7–21 (128 pp)

Crossman EJ (1984) Research on muskellunge at Wabigoon Lake, Dryden, Ontario. Muskie (Newslett Muskies Inc) 18(7):7–8

Crossman EJ (1984) Report on Wabigoon muskellunge and pike research. Muskie 18(11):8

Lebeau B, Pageau G, Crossman EJ (1984) The muskellunge as a multiple spawner. International Muskellunge Symposium. La Crosse, Wisconsin. Abstract published 1986, American Fisheries Society Special Publication 15

Crossman EJ, Campbell S, Munroe L (1984) What's in a name? The history of the nomenclature of the muskellunge. International Muskellunge Symposium. La Crosse, Wisconsin. Abstract published 1986, American Fisheries Society Special Publication 15

Crossman EJ (1985) Muskellunge research help needed. Muskie 19(4):5

Crossman EJ (1986) Ontario muskellunge research progress report. Muskie 20(1):17–19

Crossman EJ (1986) The name for the muskellunge-northern pike hybrid. Muskie 20(6):16–17. (same article in: Newslett Muskies Canada 8(6) July 1986:4–8. See follow-up report in Muskie 20(11):4.)

Lebeau B, Jolicoeur P, Pageau G, Crossman EJ (1986) Asymptotic growth, egg production and trivariate allometry in *Esox masquinongy* Mitchill. Growth 50:185–200

Crossman EJ (1986) The noble muskellunge: a review. In: Hall GE (ed) Managing Muskies. American Fisheries Society Special Publication 15, pp 1–13

Casselman JM, Crossman EJ, Ihssen PE, Reist JD, Booke H (1986) Identification of muskellunge, northern pike, and their hybrids. In: Hall GE (ed) Managing Muskies. American Fisheries Society Special Publication 15, pp 14–46

Casselman, JM, Crossman EJ (1986) Size, age, and growth of trophy muskellunge and muskellunge-northern pike hybrids – the Cleithrum Project, 1979–1983. In: Hall GE (ed) Managing Muskies. American Fisheries Society Special Publication 15, pp 93–110

Crossman EJ (1987) Muskellunge research in northwestern Ontario 1986. A special tribute to Larry Ramsell. Muskie 21(5):18–19

Crossman EJ (1988) Piking in North America. In: Rickards B (ed) The best of Pikelines. Boydell Press, Woodbridge, England, pp 55–57 (120 pp) (reprinted from various past issues of Pikelines)

Crossman EJ (1989) What big teeth you have. Can Sportfish Mag 6:51–56

Crossman EJ (1989) The tiger musky or norlunge. Can Sportfish Mag 7:36–37

Crossman EJ, Casselman JM (1989) Cleithrum Project Report. Muskies Canada 11(10)

Crossman EJ, Casselman JM (1990) Cleithrum Project. Muskie 24(1):7

Crossman EJ (1990) Reproductive homing in muskellunge, *Esox masquinongy*. Can J Fish Aquat Sci 47:1803–1812

Mandrak NE, Crossman EJ (1992) A checklist of Ontario freshwater fishes annotated with distribution maps. Royal Ontario Museum, 176 pp

Crossman EJ, Casselman JM, Ruthven MC (1992) Cleithrum Project. Muskie 26(10):4 & 7 and Muskies Canada Newslett 15(10):12

Mandrak NE, Crossman EJ (1992) Postglacial dispersal of freshwater fishes into Ontario. Can J Zool 70(11):2247–2259

Crossman EJ (1993) The Amur pike, cousin of the muskellunge. Can Sportfish Mag 28, Winter:24

Crossman EJ (1994) Live release of muskellunge. Muskie 26(5):10

Ráb P, Crossman EJ (1994) Chromosomal NOR phenotypes in North American pikes and pickerels, genus Esox, with notes on the Umbridae (Euteleostei: Esocae). Can J Zool 72(11):1951–1956

Crossman EJ (1996) Taxonomy and distribution. In: Craig JF (ed) Pike biology and exploitation. Chapman and Hall, London, England, pp 1–11

Crossman EJ (1998) The Cleithrum Project: new facts from old bones. In: 1996 Ramsell L, Hamlin W (eds) Musky Almanac. Musky Hunters Publications, Minoqua, Wisconsin, pp 122–128

Crossman EJ (1995) Evolution and natural history. In: Musky Country, The book of North America's pre-

mier big game fish. Willow Creek Press, Minoqua, Wisconsin, pp 28–43
Crossman EJ, Casselman JM (1996) The Cleithrum Project: an update to 1955. Muskies Canada Release J 19(1):2 pp
Crossman EJ, Casselman JM (1996) The Cleithrum Project: an update to 1955. Muskie April 1966:7–9 and 21
Casselman JM, Crossman EJ, Robinson CJ (1996) Assessing sustainability of trophy muskellunge fisheries. In: Kerr SJ, Olver CH (eds) Managing muskies in the '90s. Workshop Proceedings, Ontario Ministry of Natural Resources. Southern Region, Science and Technology Transfer Workshop Proceedings WP-007, pp 29–39 (169 pp)
Crossman EJ, Casselman JM (1996) The Cleithrum Project: an update to 1995. In: Kerr SJ, Olver CH (eds) Managing muskies in the '90s. Workshop Proceedings, Ontario Ministry of Natural Resources. Southern Region Science and Technology Transfer Unit Workshop Proceedings WP-007, pp 147–151 (169 pp)
Casselman JM, Robinson CJ, Crossman EJ (1999) Growth and ultimate length of muskellunge from Ontario water bodies. N Am J Fish Manage 19:271–290
Crossman EJ (1999) The CP and other comments. Muskies Canada Inc Release J 22(5):16–17
Crossman EJ (1999) More recent information on angling stress. Pikelines No. 86, November:17–18
Crossman EJ (2000) Angling stress and catch and release. Muskie, June:21–22
Crossman EJ (2000) Interesting new opportunities for northeastern muskellunge. Muskie, September:30–31
Crossman EJ (2000) Muskellunge in Maine and New Brunswick. Am Fish Soc Introduced Fish Sec Newslett 18(2):7
Crossman EJ (2001) Catch and release. Pike Muskie Mag Winter 2000–2001:9–10
Crossman EJ, Casselman JM (2003) The Cleithrum Project is alive and well! Muskie, October:7

Environ Biol Fish (2007) 79:11–25
DOI 10.1007/s10641-006-9089-1

ORIGINAL PAPER

The New River, Virginia, muskellunge fishery: population dynamics, harvest regulation modeling, and angler attitudes

Travis O. Brenden · Eric M. Hallerman · Brian R. Murphy · John R. Copeland · Joseph A. Williams

Received: 7 September 2005 / Accepted: 20 June 2006 / Published online: 25 July 2006

Abstract Although muskellunge, *Esox masquinongy*, fisheries in northern US states and Canadian provinces are increasingly being managed by introduction of restrictive harvest regulations (e.g. 1370-mm (54″) minimum length limits), many southern US muskellunge fisheries continue to be managed with comparatively liberal regulations (e.g. 762-mm (30″) minimum length limits) that are implemented statewide. We studied the population dynamics of the New River, Virginia, muskellunge fishery and used predictive modeling to determine whether restrictive harvest regulations also might prove beneficial for this southern latitude fishery. A creel survey was also conducted to learn more about angler attitudes to the New River muskellunge fishery. Muskellunge grew quickly, with fish reaching harvestable lengths (762 mm, 30″) in 2–3 years. Muskellunge fishing pressure, harvest rates, and voluntary release rates were low compared with reports for more northern areas. Most anglers, irrespective of how often they fished for muskellunge, defined "trophy" muskellunge to be approximately 1050–1100 mm (41–43″) in length. Although angler support for restrictive harvest regulations was low, abundance of memorable-length (≥1070 mm, 42″) muskellunge was predicted to increase under all evaluated length limits. Muskellunge yield would remain static at 914-mm (36″) and 1016-mm (40″) length limits, because of the rapid growth of fish, but yield would decline dramatically with a 1143-mm (45″) length limit, because male muskellunge rarely exceeded 1100 mm (43″). Because of rapid growth and low release rates, implementation of higher length limits (e.g. 965–1067 mm, 38–42″) may indeed prove beneficial for augmenting "trophy" muskellunge production on the New River. Angler support for higher minimum length limits might be increased by educating anglers about the rapid growth rates of muskellunge and the expected size structure changes that will result from a length-limit increase. Size structure changes resulting from an increase in the minimum length limit may be difficult to detect because of potential increases in fishing pressure or reduced

T. O. Brenden (✉) · E. M. Hallerman · B. R. Murphy
Department of Fisheries and Wildlife Sciences, Virginia Polytechnic Institute and State University, 100 Cheatham Hall, Blacksburg, VA 24061-0321, USA
e-mail: tbrenden@umich.edu

J. R. Copeland · J. A. Williams
Virginia Department of Game and Inland Fisheries, 2206 S. Main St., Suite C, Blacksburg, VA 24060, USA

T. O. Brenden
Quantitative Fisheries Center, Department of Fisheries and Wildlife, Michigan State University, 13 Natural Resources Building, East Lansing, MI 48824-1222, USA

fish growth as a result of competition for food resources. Long-term monitoring of muskellunge growth and angling pressure may therefore be needed to ensure that new regulations are indeed benefitting the fishery.

Keywords *Esox masquinongy* · Dynamic pool model · Creel survey · Minimum length limit · Exploitation · Voluntary release angling

Introduction

Although anglers often associate muskellunge, *Esox masquinongy*, with water bodies in northern US states and Canadian provinces, muskellunge fisheries do exist in southern US regions (e.g. North Carolina, Tennessee, Virginia, and West Virginia). Whereas many northern muskellunge fisheries are managed with harvest regulations tailored specifically for an individual system based on factors such as stock abundance, harvest rate, and growth rate (Simonson and Hewett 1999, OMNR 2005[1]), southern US muskellunge fisheries continue to be managed primarily with statewide limits that are substantially less restrictive than the regulations enforced in some northern water bodies. For example, all muskellunge fisheries in Virginia and North Carolina are managed with 762-mm (30″) minimum length (this and all lengths reported herein are total length measurements) and two-fish per day creel limits (NCWRC 2005[2], VDGIF 2005[3]). In comparison, length limits enforced in places such as Wisconsin and Ontario can be as high as 1270 and 1371 mm (50 and 54″) (Simonson and Hewett 1999, OMNR 2005[1]). Considering that most muskellunge fisheries are managed as "trophy" fisheries (Hanson et al. 1986), this difference in management approaches raises a question regarding the appropriateness of statewide, liberal harvest regulations in southern regions.

Although liberal harvest regulations can be beneficial for increasing production of "trophy" fish by reducing fish densities and preventing competition for food resources (Simonson and Hewett 1999), the paucity of information about muskellunge biology and ecology in southern systems probably is one of the major reasons why southern-latitude muskellunge fisheries are managed with statewide limits. Low length and high creel limits were once common for muskellunge fisheries even in northern areas, but regulations have become more restrictive as interest in muskellunge sportfishing has increased and the impact of overharvest has become more evident (Crossman 1986; Graff 1986). Because little research addressing muskellunge fisheries has been conducted in southern regions, biologists may have lacked data regarding muskellunge growth and mortality rates. Without such information, predicting the impact and justifying alternative harvest regulations for southern muskellunge would indeed prove difficult.

The purpose of our research was to determine whether the New River, Virginia, muskellunge fishery would benefit from implementation of a minimum length limit that differed from the current statewide limit. Very little is known about muskellunge in the New River, because fish are not frequently collected by Virginia Department of Game and Inland Fisheries (VDGIF) personnel during regular fish sampling. The fishery is managed for the production of "trophy" fish and is stocked annually with fingerling (~100 mm, 4″) fish. In recent years, some anglers have become concerned that Virginia's statewide harvest regulations do not afford sufficient protection for fish and that overharvest may be limiting "trophy" fish production. Specific objectives for this research were to estimate important population dynamic rate functions for the New River muskellunge fishery, to predict the effect that alternative harvest regulations would have on the fishery, and to measure angler attitude to the fishery.

[1] Ontario Ministry of Natural Resources. 2005. 2005 Fishing Regulations Survey. OMNR web site address: http://www.mnr.gov.on.ca/MNR/pubs/pubmenu.html#fish

[2] North Carolina Wildlife Resources Commission. 2005. North Carolina Inland Fishing, Hunting and Trapping

[3] Virginia Department of Game and Inland Fisheries. 2005. Freshwater Fishing in Virginia. VDGIF web site

Materials and methods

Study area

The New River originates in northwest North Carolina and flows northward through southwest Virginia and into West Virginia (Fig. 1). The New River merges with the Gauley River in West Virginia to form the Kanawha River, which is a tributary of the Ohio River. Naturalized expansion of muskellunge into the New River was possibly prevented by Kanawha Falls, a waterfall located several kilometers below the confluence of the New and Gauley rivers (Jenkins and Burkhead 1993). Stocking of the New River with muskellunge by VDGIF began in the early 1960s.

The Virginia portion of the New River is 245 kilometers (152 miles) long and the river is impounded by five dams—the Buck, Byllesby, Claytor, Fields, and Fries dams (Fig. 1). Muskellunge are currently stocked in two sections of the river, from the North Carolina-Virginia border to Fries Dam (upper section) and from Claytor Dam to the Virginia-West Virginia border (lower section). Approximately 5,000 fingerling muskellunge are annually stocked in these river sections in early to mid-summer.

Data collection

From 1998 to 2003 muskellunge were sampled by angling and by boat electrofishing at sites located primarily in the lower section of the river. The electrofishing unit consisted of two drop-wire boom-mounted anodes and a Type VI-A electrofisher (Smith Root, Vancouver, Washington, USA). Electrofishing was conducted in the fall, winter, and spring months using pulsed-DC output at 4 A and a frequency of 60 Hz. Muskellunge was the only species targeted when sampling. Angling occurred throughout the year.

Lengths of captured fish were measured to the nearest mm and fish were weighed, depending on size, to the nearest 0.001, 0.01, or 0.1 kg. Mus-

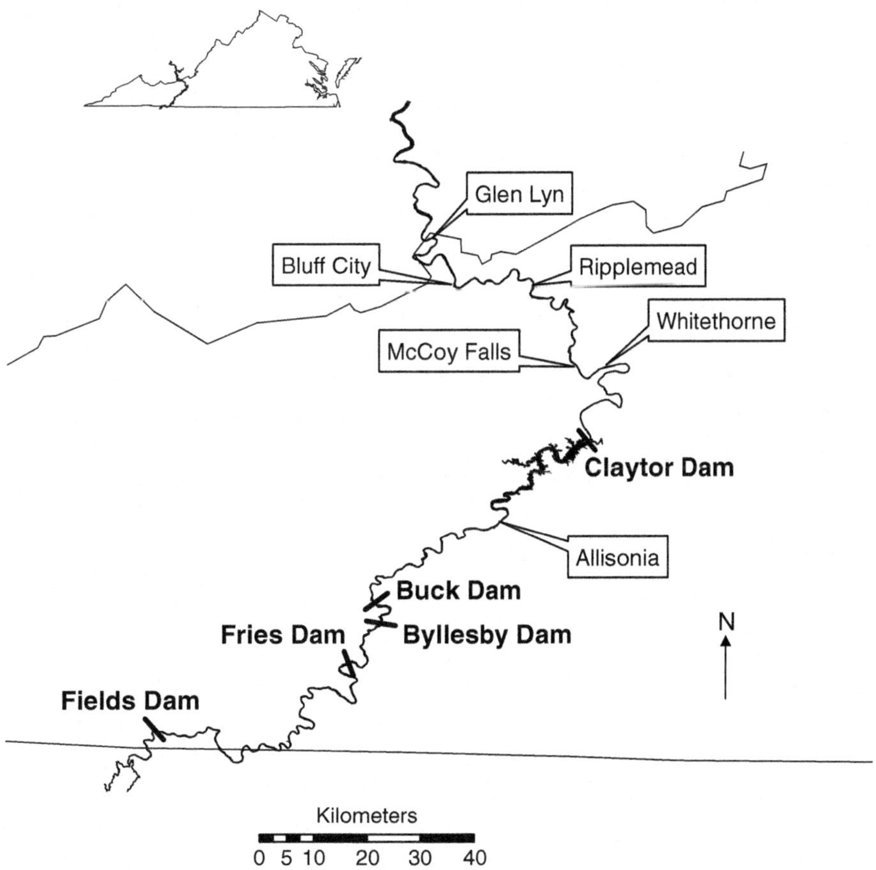

Fig. 1 Map of the New River in southwestern Virginia

kellunge that were captured by electrofishing were sexed using the method of LeBeau and Pageau (1989). A pelvic fin ray from most fish captured in 2000, 2001, 2002, and 2003 was clipped with wire cutters as close to the body as possible and stored in coin envelopes for aging—fin rays were prepared and fish ages were estimated using methods described by Brenden et al. (2006).

Forty-seven muskellunge >600 mm (23″) in length captured below Claytor Dam and at Whitethorne in 2000, 2001, and 2002 were equipped with internal trailing-whip antenna radiotelemetry transmitters (Advanced Telemetry Systems, Isanti, Minnesota, USA) so that more detailed information about sources of fish mortality could be collected. Transmitters listed contact information for VDGIF personnel and a notice that a reward would be given to those who harvested a tagged fish and returned the tag. Weekly attempts to locate telemetered fish were conducted from March 2000 to June 2003. Surveys were typically conducted only in the areas where muskellunge were originally tagged and released, although surveys throughout the entire lower section of the New River, including some West Virginia portions of the river, also were completed to detect whether any fish had moved beyond the areas regularly surveyed.

Angler attitude to the New River muskellunge fishery was measured through a creel survey conducted from 16 March to 8 November 2002 (Copeland 2005). The survey was stratified temporally by season (spring, summer, fall), day type (weekday, weekend/holiday), and river section. The angler intercept method depended on which section of the river was sampled. Region A (Virginia–North Carolina border to Byllesby Reservoir) was sampled by means of a combination of access point and road roving survey, whereby one clerk stayed at individual access sites while another clerk conducted a roadside roving survey. Regions B (Byllesby Reservoir to Allisonia) and C (Claytor Dam to Whitethorne) were sampled with access point surveys. Region D (McCoy Falls to Ripplemead) was sampled with an on-water roving survey. Region E (Bluff City to Glen Lynn) was sampled with a bus-route survey. Survey questions pertaining to the muskellunge fishery that were asked included how many days per year anglers fished exclusively for muskellunge, how many muskellunge had anglers harvested in the past three years, what length of muskellunge did anglers consider a trophy, what length of muskellunge would anglers consider harvesting, whether anglers believed that harvest regulations should be changed, and whether anglers belonged to specialized fishing clubs. Creel clerks also asked about targeted species and catch/harvest for the current fishing trip.

Data analysis

The size structure of the New River muskellunge population was assessed by use of length–frequency histograms and size-structure indices. Length–frequency histograms were constructed for each method of capture to evaluate differences in sampling effectiveness. Proportional stock density (PSD) and relative stock density (RSD) were used to summarize stock size structure. The length classes used to calculate these indices were stock (510–759 mm, 20–30″), quality (760–969 mm, 30–38″), preferred (970–1,069 mm, 38–42″), memorable (1,070–1,269 mm, 42–50″), and trophy (≥1,270 mm, 50″) lengths (Gabelhouse 1984). Lengths and weights of muskellunge were $\log_{10}$-transformed and regressed using least-squares simple linear regression. Differences between regression slopes among years of capture, capture method, and fish sex were tested by use of analysis of covariance (ANCOVA). Growth of male and female muskellunge was described by use of von Bertalanffy growth models, as described in Brenden et al. (2006). Differences between growth of the sexes were tested with sum-of-squares reduction F tests (Quinn and Deriso 1999).

Two separate methods were used to estimate muskellunge mortality. First, pooled catch-at-age data were used to construct catch curves for the New River muskellunge stock. On the basis of the appearance of the plot of $\log_e$ catch against fish age (Results section), we elected to fit two catch-curve regressions to the data. The first catch-curve regression was fit to the $\log_e$ catch values of age-1 to age-4 fish and the second catch-curve regression was fit to the $\log_e$ catch values of age-4 and older fish. On the basis of estimated fish growth and the 762-mm (30″) minimum length limit that was in effect, the first catch-curve regression primarily estimated natural mortality whereas the second catch-curve regression incorporated both fishing and natural mortality.

The telemetry method of Hightower et al. (2001) was the other approach used to estimate muskellunge mortality. This method estimates mortality on the basis of the number of telemetered fish located during each tracking event relative to the number of at-large telemetered fish during previous time periods. Although telemetered muskellunge were relocated weekly, we used a monthly time-step for estimating mortality. A fish was considered located in a particular month if it was located at least once during that month, unless it was discovered that the fish had been harvested or died from natural causes. Fish that were repeatedly found at the same location were assumed to have died from natural causes (although the possibility of hooking mortality cannot be discounted). Only those fish that were confirmed to be alive at least four months after transmitter implantation were included in the analysis to prevent surgery-related mortality from biasing the estimates. The temporal scale of the mortality analysis was reduced by scaling all fish relocation data to a common time span that began in November and lasted until the May nineteen months later. Thus, fish that were equipped with transmitters during the same month of different years were considered to have been released at the same time. The fate of fish after the 19-month time frame was ignored. Mortality rates were estimated using several different models in which mortality remained constant or varied quarterly or monthly (Hightower et al. 2001). Akaike's information criteria corrected for small sample sizes (AIC_c) and Akaike weights were constructed for the candidate models. Model-average estimates of fishing and natural mortality were computed from the mortality estimates and Akaike weights for the candidate models (Hightower et al. 2001).

Creel survey results were summarized according to the number of angling parties interviewed (as opposed to the total number of anglers), because angling parties formed the sampling unit for the survey questions. Angler subgroups were classified according to how often anglers fished for muskellunge. Response differences to individual questions among angler subgroups were tested using multi-response permutation procedures (Mielke and Berry 2001). The multi-response permutation procedure (MRPP) is a distance-based analytical method for testing differences among groups that uses randomization theory to determine the probabilities of test statistics (Cade & Richards 2005[4]). One of the main advantages of using MRPP as a statistical testing method is it makes few distributional assumption about the data. Detection of overall significant differences among angler subgroups were followed by Bonferroni-corrected pairwise MRPP comparisons between subgroups. The MRPP tests were conducted with Blossom (Cade and Richards 2005[4]) and *P*-values for the tests were obtained by using a Pearson Type-III approximation to the permutation distribution (Cade and Richards 2005[4]). All statistical tests were conducted with a Type-I error rate of 0.05. The Bonferroni-corrected error rate for pairwise comparisons of angler subgroup responses was 0.0083.

The effect of alternative length limits on the New River muskellunge fishery was assessed by use of a dynamic pool model as implemented in the Fishery Analyses and Simulation Tools software package (Slipke and Maceina 2000). Evaluated regulations were 914-mm (36″), 1016-mm (40″), and 1143-mm (45″) minimum length limits. Because of different fish-growth rates, responses of male and female muskellunge were modeled separately, but the results were combined for final evaluation. We assumed an initial population size of 2,000 fish per sex and that recruitment to the population occurred at age 1. Harvest regulations were modeled with conditional annual fishing and natural mortality rates ranging from 5 to 30%. The effects of alternative harvest regulations were evaluated on the basis of the change (%) in population abundance of memorable-length (≥1070 mm, ≥42″) muskellunge, the change (%)

[4] Cade, B. S. & J. D. Richards. 2005. User Manual for BLOSSOM Statistical Software. U.S. Geological Survey

in the total number of fish harvested, and the change (%) in yield-per-recruit (kg ha^{-1} year^{-1}).

Results

Population dynamics

From 1998 to 2003, 523 muskellunge were collected from the New River, 236 by electrofishing and 285 by angling. Two other fish were found dead in the river. A total of 445 muskellunge were seen, but not necessarily captured, when electrofishing. The median electrofishing catch rate for all sampled sites was 1.36 fish per electrofishing hour. Of the captured muskellunge for which sex was identified, the male-to-female ratio was approximately equal (female 49%; male 51%).

Lengths of captured muskellunge ranged from 358 to 1270 mm (14–50″). Comparison between sampling methods suggested that angling captured larger fish more effectively than electrofishing (Fig. 2). This different sampling effectiveness in turn affected size structure estimates. Muskellunge PSD, RSD-P, and RSD-M for fish captured by electrofishing equaled 64, 25, and 5, respectively. Muskellunge PSD, RSD-P, and RSD-M for fish captured by angling equaled 100, 45, and 17. Muskellunge PSD, RSD-P, and RSD-M combined across both sampling methods equaled 87, 38, and 12, respectively.

Significant differences in $\log_{10}$-transformed length–weight relationships by method (ANCOVA: F=2.36, df=1,511, P=0.125) and year (ANCOVA: F=0.90, df=1,503, P=0.482) of capture were not detected. There was, however, a significant difference between the slopes of the length–weight relationships for the different sexes (ANCOVA: F=4.73, df=1,142, P=0.0312). The $\log_{10}$-transformed length–weight relationships for females and males were:

$$\log_{10} W_f = -6.496 + 3.459 \cdot \log_{10} L_f$$

and

$$\log_{10} W_m = -5.991 + 3.280 \cdot \log_{10} L_m,$$

where W_f and W_m are female and male weights (g), and L_f and L_m are female and male lengths (mm), respectively. Estimated von Bertalanffy growth equations for female and male muskellunge were:

$$L_f(t) = 1,300 \cdot \left[1 - \exp^{-0.3169 \cdot (t+0.2743)}\right]$$

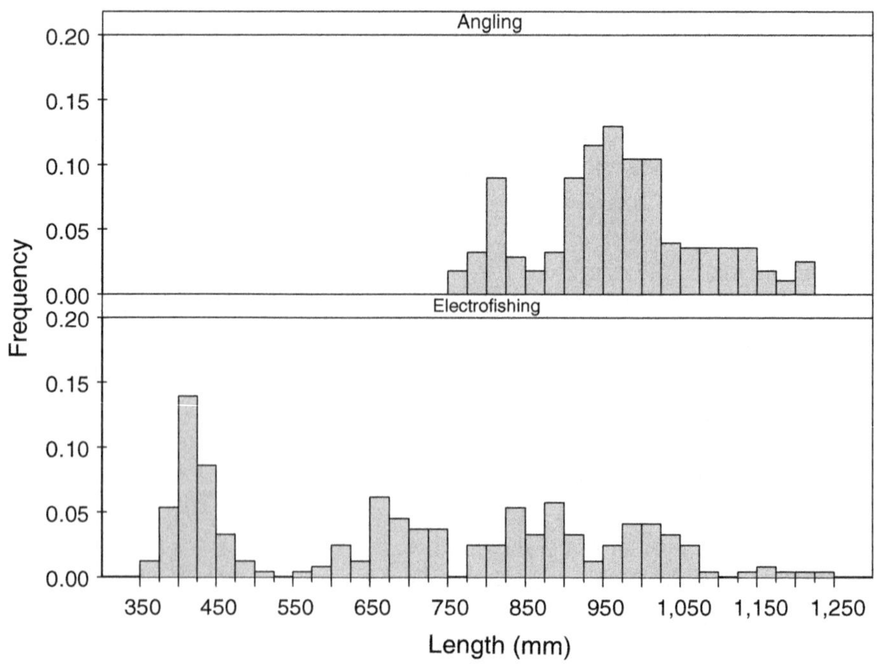

Fig. 2 Length–frequency histograms for muskellunge captured by angling and electrofishing in the New River, Virginia

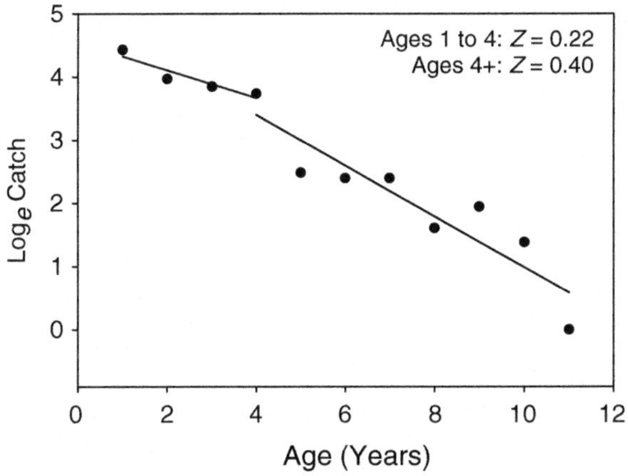

Fig. 3 Relationship between $\log_e$ catch and age data and the catch-curve regression fits for ages 1–4 and ages 4+ fish. On the basis of the scatterplot of $\log_e$ catch against age, it was assumed that fish were recruited to the sampling gear at age 1

and

$$L_m(t) = 1,100 \cdot \left[1 - \exp^{-0.4263 \cdot (t+0.1536)}\right],$$

where $L_f(t)$ and $L_m(t)$ are predicted lengths at age (t) for females and males, respectively (Brenden et al. 2006). Significant differences were detected among the parameter estimates of the sex-specific growth equations (F test: $F = 75.98$, df=2,173, $P < 0.0001$).

Total instantaneous mortality estimates from catch-curve regressions equaled 0.22 for age-1 to age-4 fish and 0.40 for age-4 and older fish (Fig. 3). Assuming that total mortality of age-1 to age-4 fish consisted primarily of natural mortality and that this rate remained constant for older fish, instantaneous natural and fishing mortality for muskellunge were estimated to be 0.22 and 0.18, respectively. The telemetry method of Hightower et al. (2001) yielded model-average instantaneous natural and fishing mortality to be 0.05 and 0.15, respectively. Averaging the mortality from the two methods yielded instantaneous natural and fishing mortality estimates of 0.135 and 0.165, respectively. Conditional annual fishing (cf) and natural (cm) mortality were estimated to be 15.2 and 12.6%, respectively. The annual exploitation rate for New River muskellunge was estimated to be 14.2%.

Angler attitudes

Seven-hundred-and-seventy angling parties (hereafter referred to as anglers) were interviewed during the New River creel survey. Twelve percent (n=91) of the interviews were repeat interviews, meaning that those interviewed had been previously interviewed at least once before. When conducting a repeat interview, clerks only asked questions relating to the current angling trip (e.g. targeted species, number of fish caught) and did not repeat questions about angler opinion. Muskellunge was identified as the targeted species by 2.2% (n=17) of anglers. Thirty-five percent (n=6) of those targeting muskellunge had caught a muskellunge during that particular fishing trip (total number caught=6), with one fish approximately 1200 mm (47″) long having been harvested. In comparison, only one percent (n=7) of anglers that were not targeting muskellunge had caught a muskellunge during that particular fishing trip (total number caught=8), with one fish approximately 920 mm (36″) long having been harvested.

Interviewed anglers fished for muskellunge an average of 4 days per year. For the purpose of differentiating angler attitudes, angler subgroups were defined as follows:

- non-muskellunge anglers—those who fished 0 days per year for muskellunge;
- occasional muskellunge anglers—those who fished 1–5 days per year;
- moderately-dedicated muskellunge anglers—those who fished 6–25 days per year; and
- dedicated muskellunge anglers—those who fished >25 days per year.

More than 80% (n=555) of those interviewed were classified as non-muskellunge anglers. Of those anglers who fished at least once a year for muskellunge, 50% (n=61) were occasional, 30% (n=37) were moderately dedicated, and 20% (n=25) were dedicated muskellunge anglers. Only one interviewed angler belonged to a muskellunge fishing club. That angler was a member of a West Virginia Muskies Inc. Chapter and fished frequently enough to be regarded as a dedicated muskellunge angler.

No significant differences in opinion of the length of muskellunge considered a "trophy" were found among the angler subgroups (Table 1; MRPP: δ=0.261, P=0.322). Most anglers defined "trophy" muskellunge lengths to be approximately 1050–1100 mm (41 to 43"). Seventy-four percent of anglers (n=302) with an opinion about "trophy" lengths of muskellunge believed that a "trophy" muskellunge was at least 1016 mm (40″) and 11% (n=22) believed that a "trophy" muskellunge was at least 1270 mm (50″) (Fig. 4).

Significant differences were found among the angler subgroups about the length of muskellunge they would consider harvesting (Table 1; MRPP: δ=15.58, $P<0.0001$). Approximately 35% of non-muskellunge anglers indicated they would consider harvesting a muskellunge approximately 800 mm (31″) in length whereas most dedicated muskellunge anglers indicated they would only consider harvesting a muskellunge 1200 mm (47″) or larger (Fig. 4). In terms of differences between angler subgroups, non-muskellunge anglers identified minimum harvestable lengths that were significantly smaller than lengths identified by occasional (MRPP: δ=4.676, P=0.005), moderately-dedicated (MRPP: δ=4.343, P=0.006), and dedicated (MRPP: δ=22.882, $P<0.0001$) muskellunge anglers. Dedicated muskellunge anglers identified minimum harvestable lengths that were larger than lengths identified by occasional (MRPP: δ=6.314, P=0.001) and moderately-dedicated MRPP: δ=6.019, P=0.001) muskellunge anglers. None of the other differences between angler subgroups was significant (P>0.0083).

When asked about minimum harvestable lengths at least a few anglers from all subgroups indicated that all muskellunge caught would be voluntarily released (Fig. 4). Although results from the overall test to discover whether angler subgroups differed in the percent of anglers that specified voluntary release of muskellunge would be practiced were significant (MRPP: δ=2.54; P=0.026), none of the pairwise MRPP comparisons between angler subgroups showed differences were significant (P>0.0083). Overall, 40% (n=240) of anglers with opinions about minimum harvestable length of muskellunge specified that voluntary release would always be practiced.

A significant difference in the percent of anglers who had harvested a muskellunge in the past 3 years (calculated from those that had caught a muskellunge in the past 3 years) was found among the angler subgroups (MRPP: δ=2.17; P=0.039). Pairwise comparisons indicated that a higher percentage of muskellunge anglers than non-muskellunge anglers had harvested a muskellunge (MRPP: δ=4.45; P=0.0065). None of the other pairwise comparisons was significant (P>0.0083).

Table 1 Responses of angler subgroups to questions asked during a creel survey conducted on the New River in 2002 regarding the muskellunge fishery. Anglers that had no opinions were excluded from the analysis

Point addressed	Angler Subgroups			
	Non-muskellunge	Occasional	Moderately-Dedicated	Dedicated
Trophy length (mm)	1036	1051	1071	1074
Minimum harvestable length (mm)	858	936	922	1109
Previously harvested a muskellunge (%)[a]	20	22	48	35
Average length of harvested fish (mm)	946	1023	916	995
Support a regulation change (%)	10	28	27	56
Support a creel limit change (%)[b]	83	76	80	79
Support a lower creel limit (%)[c]	31	77	75	82
Support a length limit change (%)[b]	33	76	50	71
Support a higher length limit (%)[d]	32	92	40	80
Suggested length limit (mm)[e]	527	973	550	897

[a] Percentages are calculated out of those anglers who indicated a muskellunge had been caught

[b] Percentages are calculated out of those anglers who believed regulations should be changed

[c] Percentages are calculated out of those anglers who believed creel limit should be changed

[d] Percentages are calculated out of those anglers who believed length limit should be changed

[e] Suggested length limits are weighted averages

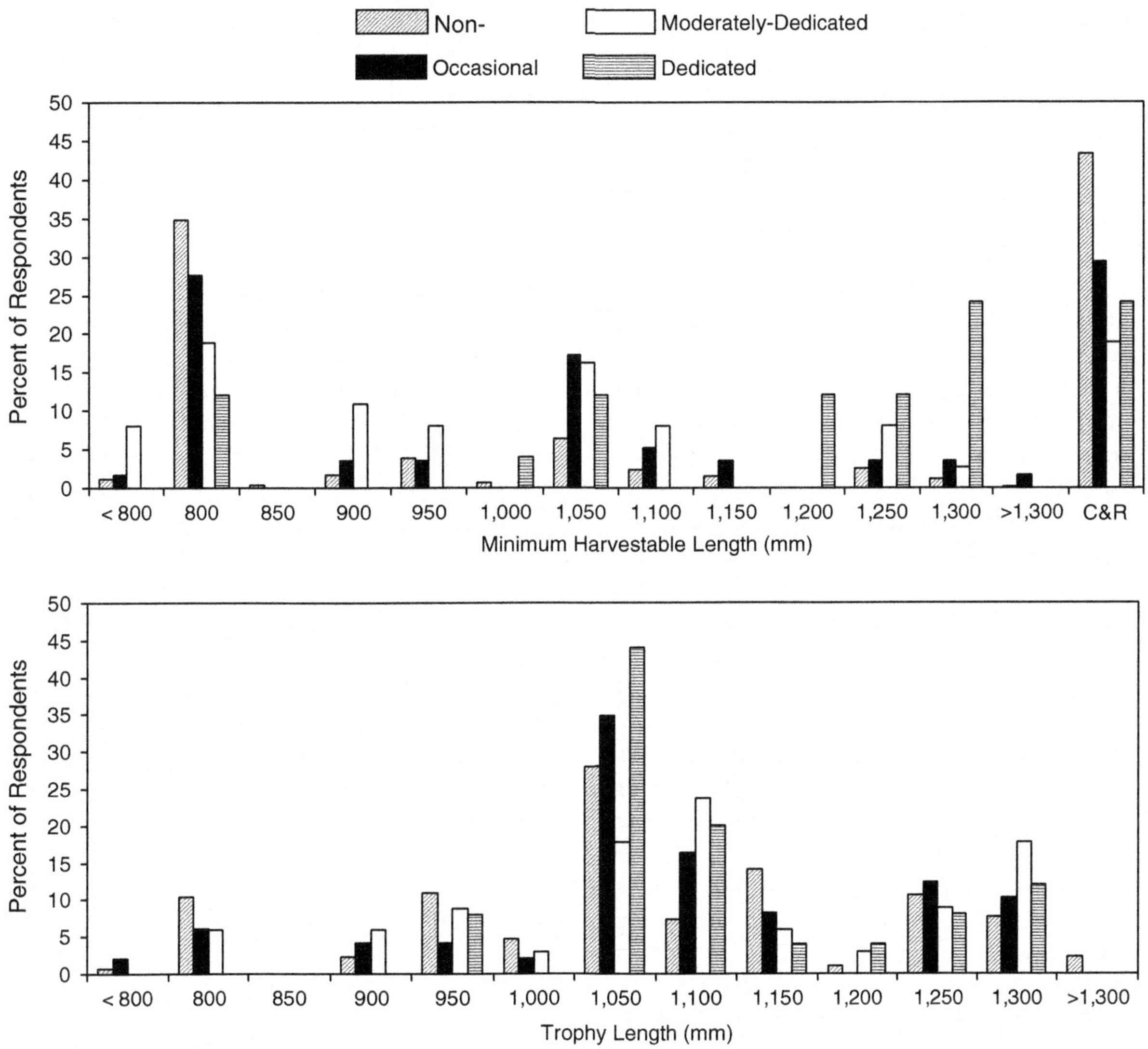

Fig. 4 Lengths of muskellunge regarded as a "trophy" (**bottom**) and as the minimum harvestable length (**top**) by non-muskellunge anglers, occasional muskellunge anglers, moderately-dedicated muskellunge anglers, and dedicated muskellunge anglers. The percent of anglers that specified all muskellunge would be released are under the category marked *C&R* (catch and release)

The percentage of anglers who believed muskellunge harvest regulations should be changed was significantly different among the angler subgroups (MRPP: δ=20.21; $P<0.0001$). A lower percentage of non-muskellunge anglers believed regulations needed to be changed than occasional (MRPP: δ=9.71; P=0.00012), moderately-dedicated (MRPP: δ=5.22; P=0.004), or dedicated muskellunge (MRPP: δ=30.17; $P<0.0001$) anglers. Of the anglers that thought harvest regulations for muskellunge should be changed, 86% (n=85) believed the creel limit should be changed and 56% (n=55) believed the size limit should be changed. Of those anglers who believed the creel limit should be changed, the percentage that wanted a lower creel limit was significantly different among the angler subgroups (MRPP: δ=5.67; P=0.0005). A lower percentage of non-muskellunge anglers believed a lower creel limit should be implemented than occasional (MRPP: δ=5.43; P=0.003) and dedicated (MRPP: δ=5.90; P=0.002) muskellunge anglers.

Of the anglers interviewed who thought the size limit for muskellunge should be changed, the percentage that believed a higher minimum length limit should be implemented was significantly different among the angler subgroups (MRPP: δ=4.56; P=0.002). A lower percentage of non-muskellunge anglers than occasional muskellunge anglers believed a higher minimum length limit should be implemented (MRPP: δ=7.40; P=0.0006). None of the other differences among the angler subgroups was significant (P>0.0083). Of the anglers who believed a higher minimum length limit should be implemented, the mean length specified (irrespective of angler subgroup) was 988 mm (39″) and ranged from 863 to 1219 mm (34–48″). The weighted-average size limits proposed by the angler subgroups, which were found by multiplying the percentages of anglers that wanted lower and higher size limits by the mean size limits they specified, ranged from 527 to 973 mm (21–38″) (Table 1).

Alternative length-limit evaluations

Alternative length-limit modeling indicated that increasing the minimum length limit to 914 mm (36″) would increase the population abundance of memorable-length (≥1070 mm, ≥42″) muskellunge by approximately 25%, on the basis of current mortality estimates (Fig. 5). For other mortality levels, the expected abundance increase could be between 5 and 50%. Total number of harvested fish would be expected to decline by approximately 20% on the basis of current mortality estimates, although as a much as a 40% decrease could be expected with lower natural and higher fishing mortality estimates (Fig. 5). If the minimum length was 914 mm (36″), yield of muskellunge would actually increase, albeit only slightly, on the basis of current mortality estimates (Fig. 5). The increase in yield if the minimum length limit were 914-mm (36″) would be expected to be even larger in a low natural mortality and high fishing mortality scenario (Fig. 5).

With a 1016-mm (40″) minimum length limit, the population abundance of memorable-length (≥1070 mm, ≥42″) muskellunge was predicted to increase by 50% at current mortality (Fig. 5). Expected abundance increases could be as high as 130% with low natural mortality and high fishing mortality, but also as low as 10% with low fishing mortality. On the basis of current mortality estimates, harvest of muskellunge would decline by 40% if the minimum length limit was 1016-mm (40″) (Fig. 5). Yield would be expected to decline by approximately 0.25 kg ha^{-1} $year^{-1}$ (0.22 lbs $acre^{-1}$ $year^{-1}$) per recruit, on the basis of current mortality estimates, but could also increase by as much as 1.25 kg ha^{-1} $year^{-1}$ (1.12 lbs $acre^{-1}$ $year^{-1}$) per recruit given a higher estimate of fishing mortality (Fig. 5).

An increase in the minimum-length limit to 1143 mm (45″) was predicted to nearly double the population abundance of memorable-length (≥1070 mm, ≥42″) muskellunge on the basis of current mortality estimates, although the total number of harvested fish was also predicted to decline by as much as 80% (Fig. 5). Unlike the other size limits that were considered, yield under a 1143-mm (45″) minimum length limit was predicted to decline, irrespective of the mortality rates considered in the analysis (Fig. 5). The decline in yield was predicted to be as much as 2.00 kg ha^{-1} $year^{-1}$ (1.78 lbs $acre^{-1}$ $year^{-1}$) per recruit.

Discussion

New River muskellunge population dynamics

The New River muskellunge fishery was found to consist of fast-growing individuals that become harvestable at young ages under Virginia's current 762-mm (30″) minimum length limit. Nearly all New River muskellunge exceeded 762 mm (30″) by 3 years of age, and fish could exceed 1016 mm (40″) by age 5. This rate of growth is among the fastest reported for any muskellunge population. In northern regions, muskellunge typically do not reach 762 mm (30″) until 4–6 years of age, and fish from some populations may not reach this length until nine to ten years of age (Johnson 1971; Harrison and Hadley 1979; Casselman et al. 1999). In a study of 15 northern Wisconsin lakes, muskellunge took an average of 11.5 years to reach 1016 mm (40″) (Margenau and AveLallemant 2000). Growth of New River muskellunge also was rapid compared with

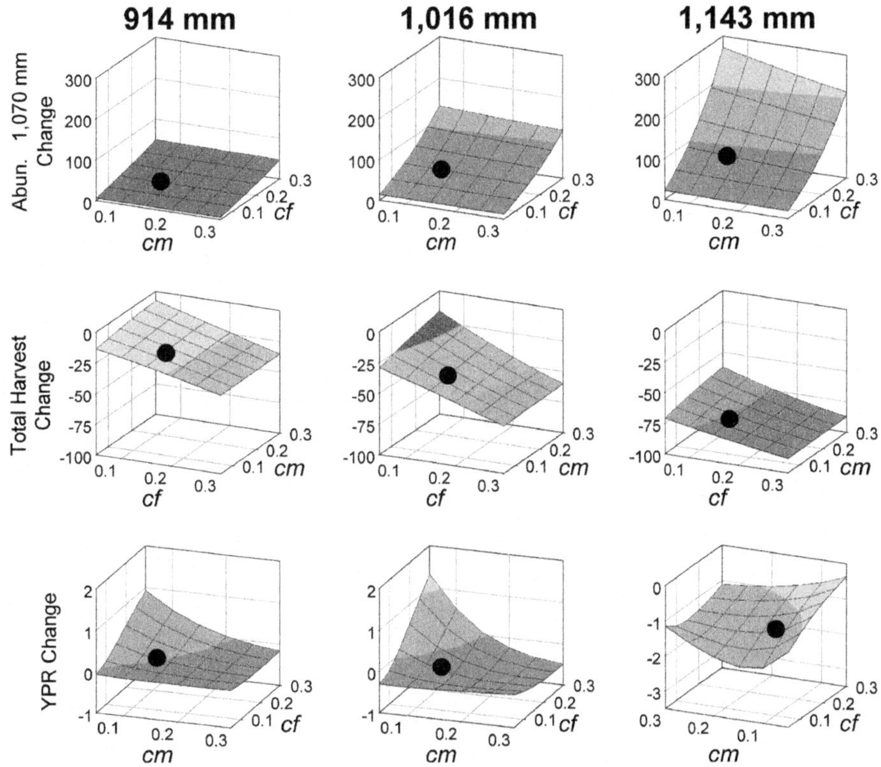

Fig. 5 Predicted changes in population abundance of memorable length (≥1070 mm) fish (% change), total harvest (% change), and yield per recruit (kg ha^{-1} year^{-1} change) for the New River muskellunge fishery on implementation of a 914-mm (36″), 1,016-mm (40″), or 1143-mm (45″) minimum length limit. Changes were predicted with annual conditional natural (*cm*) and fishing (*cf*) mortality ranging from 5–30%. Expected changes on the basis of current mortality estimates (cm=12.6%, cf=15.2%) are indicated by *filled circles*

growth rates from other southern populations. In North Carolina's French Broad River, muskellunge did not attain 762 mm (30″) until 4–5 years of age (Monaghan and Borawa 1988). Likewise in Tennessee, West Virginia, and Kentucky muskellunge did not reach 762 mm (30″) until 5–6 years of age (Parsons 1959; Miles 1978; Axon and Kornman 1986).

An abundance of appropriate-size prey and suitable thermal conditions are two factors probably contributing to the rapid growth of muskellunge in the New River. Fast growth of riverine muskellunge has previously been linked to availability of cyprinids as prey (Harrison and Hadley 1979), and, indeed, New River muskellunge have been found to forage extensively on species such as telescope shiners, *Notropis telescopus*, and white shiners, *Luxilus albeolus*, until approximately 600 mm (24″) in length (Brenden et al. 2005). New River muskellunge also feed primarily on catostomids later in life (Brenden et al. 2005); this also has been correlated with faster growth (Hanson 1986). Water temperature in the New River remains within 20–30°C (68–86°F) throughout much of the summer; this is the temperature range regarded as optimum for muskellunge feeding and growth (Bevelhimer et al. 1985). Further, the water temperature dips below 5°C (41°F), the temperature at which muskellunge feeding stops (Bevelhimer et al. 1985), from early December until early February only. Thus muskellunge in the New River may experience a nearly 10-month feeding and growing season, which may explain the faster growth rates for this stock.

The 14.2% annual exploitation level estimated for New River muskellunge stock was low compared with levels reported elsewhere. Although exploitation ranging from 5.4 to 15% has been reported for muskellunge fisheries in New York (Bimber and Nicholson 1981) and Wisconsin (Hanson 1986), exploitation in excess of 25% has been reported for muskellunge fisheries in West Virginia, Kentucky, and Wisconsin (Miles 1978; Brewer 1980; Axon 1981; Brege 1986; Hanson 1986; Hoff and Serns 1986; Cornelius and Margenau 1999). Exploitation in excess of 25% is believed to be too high for managing a quality muskellunge fishery (Hanson 1986); it thus seems unlikely that overharvest of New River muskellunge fishery is problematic. Suggested regulation changes would therefore be intended primarily to enhance trophy fish production and to prevent overharvest from possibly occurring in the future.

Angler attitudes

The New River angler-attitude survey provided an interesting contrast to surveys conducted in more northern areas. Between 11.8 and 67.3% of anglers fishing on eight northern Wisconsin lakes were found to specifically target muskellunge (Hanson 1986), which is substantially greater than the 2.2% of anglers that were targeting muskellunge on the New River. Between 7.5 and 14.0% of anglers targeted muskellunge on Cave Run Lake, Kentucky (Axon 1981), so percentages of New River anglers targeting muskellunge were more comparable with, albeit still lower than, the percentages for that fishery. New River anglers fished for muskellunge an average of 4 days per year, substantially less than when Wisconsin anglers fished for muskellunge ($\bar{x}$=13 days per year; Margenau and Petchenik 2004).

Lengths of "trophy" muskellunge identified by New River anglers were smaller than the lengths identified by Wisconsin anglers. Eighty-nine percent of general anglers surveyed in Wisconsin believed a "trophy" muskellunge should be greater than 1016 mm (40″) and 44% believed the fish should be greater than 1270 mm (50″) (Margenau and Petchenik 2004). In comparison, 74% of New River anglers believed that "trophy" muskellunge should be greater than 1016 mm (40″) and 11% believed that muskellunge should be greater than 1270 mm (50″). Lengths of "trophy" fish identified by muskellunge anglers also differed between Wisconsin and the New River. Ninety-eight percent of anglers in Wisconsin who were members of a muskellunge fishing club believed muskellunge should be greater than 1016 mm (40″) to be a "trophy" and 62% believed a fish should be at least 1270 mm (50″) (Margenau and Petchenik 2004). In comparison, 92% of New River anglers that fished most often for muskellunge believed a "trophy" muskellunge should be at least 1016 mm (40″), and 13% believed a "trophy" muskellunge should be at least 1270 mm (50″).

Although New River anglers that fished less often for muskellunge identified smaller minimum harvestable lengths and were more likely to suggest that higher creel or lower size limits should be enacted than anglers that fished more often for muskellunge, there is little evidence to indicate that harvest by non-muskellunge anglers is negatively affecting the New River muskellunge stock. Although incidental catch of muskellunge on the New River may not necessarily be a rare occurrence, non-muskellunge anglers seem just as likely as muskellunge anglers to release caught fish. Rather, our results indicate that overharvest of New River muskellunge would be more likely to occur because of the greater inclination of all anglers toward harvest of muskellunge. Only 40% of all interviewed anglers indicated that all caught muskellunge would be released, and the estimated release of muskellunge on the basis of the creel survey results was 86%. Release rates for New River muskellunge calculated from Virginia's Angler Recognition program, a program that gives formal recognition to anglers for catching "trophy" fish, was estimated to be 46% (J. Williams, unpublished data). In comparison, ninety percent of general anglers in Wisconsin voluntarily released muskellunge, and release rates for muskellunge have been estimated to be as high as 99.9% (Fayram 2003; Margenau and Petchenik 2004).

One factor possibly contributing to a greater tendency for muskellunge harvest in the New River and other southern fisheries is the lack of specialized muskellunge angling clubs (e.g. Muskies Inc.) in southern regions. Such clubs often promote voluntary-release angling both within the club and to the general angling public (Gasbarino 1986; Oehmke et al. 1986). This peer pressure to release muskellunge may be a significant factor shaping angler behavior in locales where the clubs operate. In areas without active clubs there may be less incentive for anglers to release fish. Other factors possibly contributing to higher muskellunge harvest rates include the novelty aspect associated with muskellunge fishing in Virginia or the possibility of regionally different angling release rates.

Alternative harvest regulations

At current estimated mortality levels, abundance of memorable-length (≥1070 mm, ≥42″) was predicted to increase by 25, 50, and 100% on implementation of 914-mm (36″), 1016-mm (40″), and 1143-mm (45″) minimum length limits. Although the total number of harvested fish would also decrease if these length limits were implemented, yield per recruit would occasionally increase (914-mm (36″) minimum length limit) or decrease only slightly (1016-mm (40″) minimum length limit). This is primarily because of the rapid growth of muskellunge—addition of fish biomass will offset the decrease in overall catch. A 1143-mm (45″) minimum length limit would result in the largest decrease in yield, because of the inability of male muskellunge to reach these larger lengths. Thus, 1143-mm (45″) would result in a "female only" muskellunge fishery. Although "female only" muskellunge fisheries are not a major concern to fishery biologists in northern-latitudes (Simonson and Hewett 1999), in an introduced southern latitude fishery where anglers, perhaps, view muskellunge fishing as a novelty, the yield of the fishery may need to be a greater concern to biologists.

Although it might seem unequivocal that higher length limits will lead to increases in the abundance of larger fish, this is not always true. Cornelius and Margenau (1999) found that abundance of large muskellunge in Bone Lake, Wisconsin initially remained stable after the lake's size limit was increased from 762 to 864 mm (30–34″). Similarly, Margenau and AveLallemant (2003) did not find the abundance of large muskellunge to have increased with implementation of a 1016-mm (40″) minimum length on seven northern Wisconsin lakes. One factor in particular that can prevent regulation changes from affecting muskellunge size structure is an increase in fishing pressure and harvest rate coincidental with the regulation change (Cornelius and Margenau 1999). Given that all other muskellunge fisheries in Virginia are managed with the statewide 762-mm (30″) minimum length limit, an increase in the New River muskellunge length limit may indeed lead to increased fishing pressure as anglers may be attracted to the fishery because of to the uniqueness of the regulation.

Management implications

Intraspecific or interspecific competition for food resources resulting in slow growth, physiological inability to reach large sizes, and high voluntary release of fish by anglers are a few possible reasons why low minimum length limits can be useful for muskellunge fisheries (Simonson and Hewett 1999). On the basis of our results, however, none of these factors seems applicable to the New River muskellunge fishery. As a result, an increase in the New River minimum length limit to approximately 965–1067 mm (38–42″) may prove beneficial for increasing the production of "trophy" muskellunge, which is the stated management goal for the fishery. Although the increase in abundance of "trophy" muskellunge would be even more substantial with a length limit change to 1143 mm (45″) or greater, such a drastic change in the size limit seems unnecessary given New River angler opinions about "trophy" lengths of muskellunge. Our observation that a small minority of even the most dedicated muskellunge anglers rarely identified "trophy" muskellunge to be greater than 1270 mm (50″) means that few anglers may be supportive of length limits of that order of magnitude. Even winning support for a 965-mm to 1067-mm (38–42″) size limit might prove difficult given the small number of anglers who believed it was necessary to change the minimum length limit. Public support for a length limit change could possibly be increased by publicizing the rapid growth rates of muskellunge and by explaining how higher limits should result in a greater abundance of "trophy" fish.

Negative consequences that could result from a change in the length limit include reduced growth rates of muskellunge as a result of competition for food resources, increased predatory impacts on other New River fish species, and increased fishing pressure on the muskellunge stock (Cornelius and Margenau 1999). To ensure that muskellunge growth is not affected, growth rates of fish, particularly during early life stages, should be monitored subsequent to harvest regulation changes.

If it becomes evident that muskellunge growth rates have been reduced, it may be necessary to return the harvest regulation to the original limits or to reduce stocking densities. In terms of increased predatory effects on other fish species, muskellunge consumption per initial abundance of 100 age-1 muskellunge has been estimated to be 0.63 kg ha^{-1} $year^{-1}$ (0.56 lb $acre^{-1}$ $year^{-1}$) for catostomids, 0.31 kg ha^{-1} $year^{-1}$ (0.28 lb $acre^{-1}$ $year^{-1}$) for cyprinids, 0.18 kg ha^{-1} $year^{-1}$ (0.16 lb $acre^{-1}$ $year^{-1}$) for smallmouth bass, *Micropterus dolomieu*, and 0.43 kg ha^{-1} $year^{-1}$ (0.38 lb $acre^{-1}$ $year^{-1}$) for Lepomis spp. (Brenden et al. 2005). With enactment of a 1067-mm (42″) minimum length limit, consumption of catostomids, cyprinids, smallmouth bass, and Lepomis spp. would increase by 51, 8, 44, and 18%, respectively (T. Brenden, unpublished data). Monitoring of muskellunge growth and the concurrent abundance assessment of these food items should be conducted to ensure that elevated consumption rates do not affect fish stocks. Future creel or angler surveys also should be used to determine whether a concurrent increase in muskellunge fishing pressure occurs as a result of any regulation change.

This research was initiated to determine whether a southern-latitude muskellunge fishery might benefit from more restrictive harvest regulations. Although the results from the study do indicate that the New River would benefit from a higher length limit, it remains to be seen whether these results are typical among other southern latitude muskellunge fisheries. Growth of muskellunge in other populations may not be nearly as fast, nor may fish be able to achieve lengths as large as those from the New River population. The one factor that might remain similar among fisheries is a greater tendency of anglers to harvest muskellunge, given the lack of specialized muskellunge angling clubs in southern regions. Before length limits are changed in other fisheries, biologists should ensure that conditions within the fishery warrant the change. If biologists lack suitable data or resources for assessing these conditions, collaboration with experienced muskellunge anglers may provide a convenient method for collecting the information required.

Acknowledgments Funding for this research was provided by the Acorn Alcinda Foundation and the Virginia Department of Game and Inland Fisheries through Federal Aid in Sportfish Restoration Project F-122-R. The authors thank G. Gardener and A. Scott for their angling collections, S. Malvestuto for his assistance with the creel survey design, S. Arthur, J. Hightower, and J. Hines for their assistance with the telemetry mortality model, and the numerous clerks who interviewed anglers for the creel survey. Reviews by V. Dicenzo, D. Isermann, D. Orth, E. Smith, D. Stauffer, J. Younk, and several anonymous reviewers aided the preparation of this manuscript.

References

Axon JR (1981) Development of a muskellunge fishery at Cave Run Lake, Kentucky, 1974–1979 North Am J Fish Manage 1:134–143

Axon JR, Kornman LE (1986) Characteristics of native muskellunge streams in eastern Kentucky. In: Hall GE (ed) Managing Muskies. American Fisheries Society, Special Publication 15, Bethesda, MD, pp 263–272

Bevelhimer MS, Stein RA, Carline RF (1985) Assessing significance of physiological differences among three esocids with a bioenergetics model. Can J Fish Aquat Sci 42:57–69

Bimber DL, Nicholson SA (1981) Fluctuations in the muskellunge (*Esox masquinongy* Mitchill) population of Chautauqua Lake, New York. Environ Biol Fish 6:207–211

Brege DA (1986) A comparison of muskellunge and hybrid muskellunge in a southern Wisconsin lake. In: Hall GE (ed) Managing Muskies. American Fisheries Society, Special Publication 15, Bethesda, MD, p 203–207

Brenden TO, Murphy BR, Hallerman EM (2005) Predatory impact of muskellunge on New River, Virginia, smallmouth bass. Proc Southeastern Assoc Fish Wildl Agencies 58:12–22

Brenden TO, Hallerman EM, Murphy BR (2006) Sectioned pelvic fin ray ageing of muskellunge *Esox masquinongy* from a Virginia river: comparisons among readers, with cleithrum estimates, and with tag-recapture growth data. Fish Manage Ecol 13:31–37

Brewer DL (1980) A study of native muskellunge populations in Eastern Kentucky streams. Fishery Bulletin 64. Kentucky Department of Fish and Wildlife Resources, Frankfort, p 107

Casselman JM, Robinson CJ, Crossman EJ (1999) Growth and ultimate length of muskellunge from Ontario water bodies. North Am J Fish Manage 19:271–290

Copeland JR (2005) New river creel survey key findings. Virginia Department of Game and Inland Fisheries, Performance Report F-111-R-14, Richmond. 5 pp

Cornelius RR, Margenau TL (1999) Effects of length limits on muskellunge in Bone Lake, Wisconsin. North Am J Fish Manage 19:300–308

Crossman EJ (1986) The noble muskellunge: a review. In: Hall GE (ed) Managing Muskies. American Fisheries Society, Special Publication 15, Bethesda, MD, pp 1–13

Fayram AH (2003) A comparison of regulatory and voluntary release of muskellunge and walleyes in northern Wisconsin. North Am J Fish Manage 23:619–624

Gabelhouse DW Jr (1984) A length-categorization system to assess fish stocks. North Am J Fish Manage 4:273–285

Gasbarino P (1986) Catch and release of muskellunge—philosophy and methods. In: Hall GE (ed) Managing Muskies. American Fisheries Society, Special Publication 15, Bethesda, MD, pp 300–308

Graff DR (1986) Musky management—a changing perspective from past to present. In: Hall GE (ed) Managing Muskies. American Fisheries Society, Special Publication 15, Bethesda, MD, pp 195–199

Hanson DA (1986) Population characteristics and angler use in eight northern Wisconsin lakes. In: Hall GE (ed) Managing Muskies. American Fisheries Society, Special Publication 15, Bethesda, MD, pp 238–248

Hanson DA, Axon JR, Casselman JM, Haas RC, Schiavone A, Smith MR (1986) Improving musky management: a review of management and research needs. In: Hall GE (ed) Managing Muskies. American Fisheries Society, Special Publication 15, Bethesda, MD, pp 335–341

Harrison EJ, Hadley WF (1979) Biology of muskellunge (*Esox masquinongy*) in the upper Niagara River. Trans Am Fish Soc 108:444–451

Hightower JE, Jackson JR, Pollock KH (2001) Use of telemetry methods to estimate natural and fishing mortality of striped bass in Lake Gaston, North Carolina. Trans Am Fish Soc 130:557–567

Hoff MH, Serns SL (1986) The muskellunge fishery of Escanaba Lake, Wisconsin under liberalized angling regulations, 1946–(1981). In: Hall GE (ed) Managing Muskies. American Fisheries Society, Special Publication 15, Bethesda, MD, pp 249–256

Jenkins RE, Burkhead NM (1993) Freshwater fishes of Virginia. American Fisheries Society, Bethesda, MD pp 1,079

Johnson LD (1971) Growth of known-age muskellunge in Wisconsin and validation of age and growth determination methods. Wisconsin Department of Natural Resources, Technical Bulletin 49, Madison pp 24

LeBeau B, Pageau G (1989) Comparative urogenital morphology and external sex determination in muskellunge, *Esox masquinongy* Mitchill. Can J Zool 67:1053–1060

Margenau TL, AveLallemant SP (2000) Effects of a 40-inch minimum length limit on muskellunge in Wisconsin. North Am J Fish Manage 20:986–993

Margenau TL, Petchenik JB (2004) Social aspects of muskellunge management in Wisconsin. North Am J Fish Manage 24:82–93

Mielke PW Jr, Berry KJ (2001) Permutation methods: a distance function approach. Springer-Verlag, New York, pp 352

Miles RL (1978) A life history study of the muskellunge in West Virginia. In: Kendall RL (ed) Selected coolwater fishes of North America. American Fisheries Society, Washington D. C, pp 140–145

Monaghan JP Jr, Borawa JC (1988) Recovery of riverine muskellunge populations in North Carolina. Proc Annu Conf Southeastern Assoc Fish Wildl Agencies 40(1986):258–265

Oehmke AA, Stange D, Ogden K, Addis JT, Mooradian SR (1986) The role of anglers and private organizations in muskellunge management. In: Hall GE (ed) Managing Muskies. American Fisheries Society, Special Publication 15, Bethesda, MD, pp 323–334

Parsons JW (1959) Muskellunge in Tennessee streams. Trans Am Fish Soc 88:136–140

Quinn TJ II, Deriso RB (1999) Quantitative fish dynamics. Oxford University Press, New York, pp 542

Simonson TD, Hewett SW (1999) Trends in Wisconsin's muskellunge fishery. North Am J Fish Manage 19:291–299

Slipke JW, Maceina MJ (2000) Fishery Analyses and Simulation Tools (FAST). Auburn University, Auburn, AL

Environ Biol Fish (2007) 79:27–36
DOI 10.1007/s10641-006-9132-2

CROSSMAN

Population dynamics of muskellunge in Wisconsin waters of Green Bay, Lake Michigan, 1989–2005

Kevin L. Kapuscinski · Brian J. Belonger · Steven Fajfer · Terrence J. Lychwick

Received: 10 September 2005 / Accepted: 21 July 2006 / Published online: 30 August 2006

Abstract Muskellunge, *Esox masquinongy*, were an important component of the Green Bay ecosystem prior to mid 1900s, but were extirpated by over-fishing, pollution, habitat degradation, and the introduction of exotic species. The Green Bay ecosystem improved after the passage of the Clean Water Act, and the Wisconsin Department of Natural Resources (WI DNR) started a muskellunge reintroduction program in 1989. Monitoring the results of reintroduction efforts is necessary to achieve the program goal of establishing a self-sustaining population. We used available data to provide a 2005 spawner abundance estimate for a Green Bay tributary, estimates of contributions to that spawning stock from fall fingerling and yearling stocking, a weight–length relationship, a growth analysis, and a description of size and age at maturity. Our results indicate that stocking efforts have been successful in producing an adult population, with yearlings contributing to the spawning stock at a higher proportion than fingerlings (14.69:1). Our weight–length and growth analyses suggest that Green Bay muskellunge are unlikely to reach record length, but that it is possible for females to achieve record weight. The rapid growth of Green Bay muskellunge results in their maturing at larger sizes than other stocks, but the relationship between age and maturity is not well understood. Reintroduction efforts in Green Bay have created stocked populations capable of supporting trophy fisheries, but evidence of successful natural reproduction has not been observed. Future research should focus on the reproductive requirements of muskellunge reintroduced into altered habitats.

K. L. Kapuscinski (✉)
Wisconsin Department of Natural Resources, 2984 Shawano Avenue, Box 10448, Green Bay, WI 54307, USA
e-mail: kevin.kapuscinski@dnr.state.wi.us

B. J. Belonger
N 3231 Kozuzek Road, Peshtigo, WI 54157, USA

S. Fajfer
Wisconsin Department of Natural Resources, Wild Rose Fish Hatchery, N5871 State Road 22, Wild Rose, WI 54984, USA

T. J. Lychwick
N9099 Spruce Road, Casco, WI 54204, USA

Keywords Growth · Maturity · Native species · Rehabilitation · Stocking

Introduction

Muskellunge, *Esox masquinongy*, were historically an important component of the Green Bay ecosystem and near-shore fishery (Smith and Snell 1891; Greene 1935), but were extirpated due to ecosystem changes that began during 1800s. Over-fishing, pollution, habitat degradation, and the introduction of exotic species caused the decline or extirpation of many native fish species by mid-1900s (Lake Michigan Fisheries Team 2004). Habitat degradation was caused by agricultural runoff, damming of spawning streams, pollution from industrial and domestic sources, and filling in marshes used as spawning and nursery grounds (Jude and Leach 1993). Remnant populations of muskellunge persisted into 1930s, but likely became extirpated shortly thereafter (Lake Michigan Fisheries Team 2004).

Green Bay's water quality and fishery improved after the passage of the 1972 Clean Water Act, and fishery surveys during mid-1970s and 1980s indicated a shift toward a wider variety and number of gamefish species, although carp abundance remained high (Lee Meyers, Wisconsin Department of Natural Resources, personal communication). The Wisconsin Department of Natural Resources (WI DNR) started a walleye, *Sander vitreus*, reintroduction program in 1977, and walleye were self-sustaining by 1984. Biologists wanted to further improve the Green Bay fishery and the stability of the fish community by reintroducing the once-native muskellunge, and the Lake Michigan Fisheries Management Plan and Lower Green Bay Remedial Action Plan supported the proposed reintroduction (Lake Michigan Fisheries Team 2004; WI DNR 1986). The WI DNR biologists drafted a three phase plan to reintroduce muskellunge of Great Lakes origin to Green Bay: (1) locate an egg source, obtain eggs, and successfully hatch, rear, and release fish, (2) develop a captive broodstock population in Wisconsin, and (3) develop a self-sustaining population. Reintroduction efforts began in 1989 with strong financial support from local muskellunge angling clubs and The Musky Clubs Alliance of Wisconsin. Phases one and two of the reintroduction plan have been completed, and a total of 113,130 fall fingerlings and 2,762 yearlings were released into Green Bay and its tributaries during 1989–2005 (Table 1, Fig. 1).

Stocking efforts were successful in producing muskellunge populations that support trophy fisheries of growing popularity; however, the abundance of these populations is unknown and

Table 1 Number of fall fingerling and yearling muskellunge stocked into Green Bay during 1989–2005

Year	Number stocked	
	Fall fingerlings	Yearlings
1989	5,261	–
1990	1,274	9
1991	2,624	–
1992	2,107	152
1993	1,394	215
1994	–	237
1995	1,803	–
1996	3,135	247
1997	1,842	130
1998	4,311	278
1999	3,305	294
2000	2,451	295
2001	1,854	176
2002	9,281	140
2003	33,107	103
2004	20,772	161
2005	18,609	325
Total	113,130	2,762

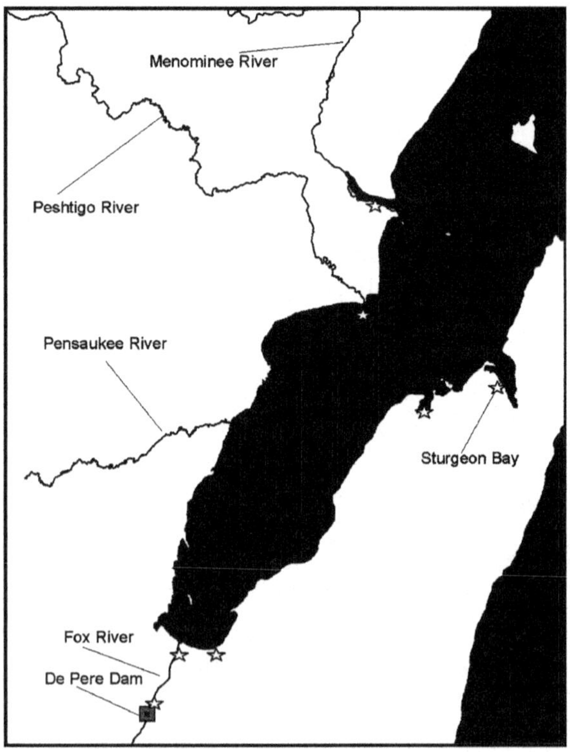

Fig. 1 Southern Green Bay, Lake Michigan, and selected tributaries. Stars indicate stocking sites during 1989–2005

fishery surveys have not found evidence of successful natural reproduction in Green Bay or its tributaries. Understanding the dynamics of Green Bay muskellunge populations will help guide management plans and evaluate management actions, and is essential for successful reintroduction (Van Den Avyle 1993). In the remainder of this paper, we will use available data to present: (1) a 2005 spawner abundance estimate for a Green Bay tributary, (2) estimates of contributions to the spawning stock from fall fingerling and yearling stocking, (3) a weight–length relationship, (4) growth estimates, and (5) a description of size and age of mature muskellunge.

Methods

Muskellunge data was collected incidentally during spring, summer, and fall walleye surveys in 1989–2004 on southern Green Bay and the Fox River downstream from the De Pere Dam. Fall electrofishing surveys targeting muskellunge were conducted on the Fox River during 2001–2004, a brief broodstock collection effort was completed during 2004, and a spring survey targeting spawning muskellunge was completed during 2005. Four fyke nets (1.52 m frame, 5.08 cm mesh) were set perpendicular to shore and fished overnight for one night during the 2004 broodstock collection effort, and seven fyke nets were fished overnight for 14 nights during the 26 April–24 May 2005 spring survey. In all surveys, captured muskellunge were measured for total length, and examined for Floy tags and fin-clips. Muskellunge captured during the spring were only assigned a sex and considered mature if they expelled gametes upon examination; we did not consider sex and maturity data from examination of morphometric characteristics reliable enough for our analyses. Most muskellunge captured during 1990–2005 were Floy tagged and many were weighed (muskellunge captured during 1989 were not tagged). Most muskellunge captured during 2005 were Passive Integrated Transponder (PIT) tagged. Additional weight and length data were collected from anesthetized hatchery-reared yearling muskellunge prior to release during most years. We only included records of known age fish in our age-based analyses. Ages were calculated (to the nearest 0.01 year) using tag-recapture records of hatchery-reared muskellunge tagged prior to stocking and recaptured from Green Bay. Descriptions of sample size, gears used to collect data, and the years and seasons of data collection for each analysis are shown in Table 2.

We used Chapman's adjustment to the Schnabel multiple-census mark-recapture population estimation method (Ricker 1975) to estimate the 2005 muskellunge spawner abundance in the Fox River downstream from the De Pere Dam. We calculated ~95% confidence limits by considering the number of recaptures as a Poisson variable (Ricker 1975). Only samples of mature muskellunge tagged and recaptured during 26 April–24 May 2005 were used to estimate spawner abundance.

We used mature muskellunge of known origin (stocked as a fingerling or yearling, determined via fin-clip and Floy tag) captured in our 2005 spring survey to calculate contributions to the 2005 Fox River spawning stock from 1989 to 2002

Table 2 Sex of muskellunge included, sample size (*N*), gears used to collect data, and the years and seasons of collection for data used in each analysis of our study

Analysis	Sex	N	Gears used	Years	Seasons
Spawner abundance	Both	63 (three recaptures)	Fyke nets	2005	Spring
Contributions	Both	50	Fyke nets	2005	Spring
Weight–length relationship	–	1,595	Fyke nets, electrofishing	1994–2005	Spring–fall
Growth	Females	62 (17 known age)	Fyke nets, electrofishing	1998–2005	Spring–fall
Growth	Males	820 (180 known age)	Fyke nets, electrofishing	1989–2005	Spring–fall
Length at maturity	Females	42	Fyke nets	2004–2005	Spring
Length at maturity	Males	51	Fyke nets	2004–2005	Spring
Age at maturity	Females	10	Fyke nets	2004–2005	Spring
Age at maturity	Males	6	Fyke nets	2004–2005	Spring

fall fingerling and yearling stocking events. We only considered muskellunge stocked in the Fox River and an adjacent site in extreme southern Green Bay. Muskellunge stocked at other locations were unlikely to be captured in the Fox River during spring based on tag-recapture observations (WI DNR, unpublished data). We only considered muskellunge stocked as fall fingerlings and yearlings during 1989–2002 because fall fingerlings stocked after 2002 would not have spawned during 2005 and no yearlings were stocked in or adjacent to the Fox River during 2003. We calculated the percent of recaptures that were stocked as fall fingerlings and yearlings. Next, we calculated the number of muskellunge contributed to the 2005 spawning stock from each stocking strategy:

$$NC_{\mathrm{FF}} = \%O^{*}_{\mathrm{FF}}A$$

or

$$NC_{\mathrm{Y}} = \%O^{*}_{\mathrm{Y}}A,$$

where NC_{FF} and NC_{Y} is the number of fall fingerlings and yearlings contributed to the 2005 Fox River spawning stock, $\%O_{\mathrm{FF}}$ and $\%O_{\mathrm{Y}}$ is the percent of mature muskellunge observed during the 2005 spring survey that were stocked as fall fingerlings and yearlings, and A is the 2005 Fox River spawning stock abundance estimate. Then we calculated the percent of all fall fingerlings and yearlings stocked during 1989–2002 that contributed to the 2005 Fox River spawning stock:

$$\%C_{\mathrm{FF}} = N_{\mathrm{FF}}/NC^{*}_{\mathrm{FF}}100$$

or

$$\%C_{\mathrm{Y}} = N_{\mathrm{Y}}/NC^{*}_{\mathrm{Y}}100,$$

where $\%C_{\mathrm{FF}}$ and $\%C_{\mathrm{Y}}$ is the percent of all fall fingerling and yearling muskellunge stocked during 1989–2002 that contributed to the 2005 Fox River spawning stock, N_{FF} and N_{Y} is the number of fall fingerlings and yearlings stocked during 1989–2002, and NC_{FF} and NC_{Y} the number of fall fingerlings and yearlings contributed to the 2005 Fox River spawning stock.

We assumed the power function:

$$W = aL^{b},$$

where W is weight (g), L is total length (mm), and a and b are parameters, was an appropriate model for muskellunge weight as a function of length (Casselman and Crossman 1986; Anderson and Neumann 1996). We estimated parameters a and b by linear regression of logarithmically transformed weight–length data collected from hatchery-reared yearlings prior to release and muskellunge captured from the Fox River.

We estimated growth of muskellunge from data collected during all years and could not account for year effects because adequate annual sample sizes were not available. Similarly, sex-specific sample sizes were not adequate for estimation of male muskellunge growth, so growth was estimated for all muskellunge and then separately for females. Therefore, we could not describe sexually dimorphic growth of Green Bay muskellunge (Casselman and Crossman 1986; Casselman et al. 1999). We calculated mean length-at-age from a sub-sample with known ages, and we used an age-length key to expand mean length-at-age calculations from the sub-sample to the sample. We then used a Gauss–Newton least-squares method (SYSTAT 11, SYSTAT Software Inc., step halving = 8, tolerance = 0.0001, loss convergence = 1e-6, parameter = 1e-5) to estimate the parameters and 95% confidence intervals of the additive error von Bertalanffy growth model:

$$L_t = L_{\infty}(1 - \mathrm{e}^{-K(t-t_0)}) + \varepsilon_i,$$

where L_t is length at age t, L_{∞} is the asymptotic average maximum length (starting value = 1,270 mm), K describes the rate at which L_t approaches L_{∞} (starting value = 0.2), t is age (years), and t_0 is hypothetical age when L_t is zero (starting value = –0.1 years, Quinn and Deriso 1999). For comparative purposes, we used the above method to estimate growth parameters from mean length-at-age data reported in the literature from Lake St. Clair muskellunge captured during 1969–1974 (Haas 1978). We also plotted our growth model for female Green Bay muskellunge against growth estimates reported in the literature for other Great Lakes stocks (Georgian Bay, Lake St. Clair, and the St. Lawrence River, Casselman et al. 1999). Next, we estimated the asymptotic average maximum weight W_∞ for all Green Bay muskellunge and females separately as:

$$W_\infty = 10^{\alpha+\beta*\log_{10} L_\infty},$$

where α and β are the parameters we estimated from the weight–length relationship, and L_∞ the asymptotic average maximum length we estimated from the von Bertalanffy growth model. We calculated 95% confidence intervals for asymptotic average maximum weight W_∞ by substituting the lower and upper 95% confidence limits of asymptotic average maximum length L_∞ into the above equation.

Finally, we calculated mean length and age of mature muskellunge for each sex. We estimated age for mature female muskellunge of unknown age with our sex-specific growth model; we did not estimate age of mature male muskellunge because we did not have a sex-specific growth model.

Results

A total of 63 mature muskellunge were captured during the 2005 Fox River spring survey; three of these were recaptures previously tagged during the survey. We estimated that 425 (~95% CI 173–1,061) muskellunge contributed to the Fox River spawning stock during 2005.

A total of 50 mature muskellunge of known origin were captured during the 2005 Fox River spring survey; 30 (60%) were originally stocked as fall fingerlings and 20 (40%) as yearlings. Therefore, we extrapolated that of the estimated 425 spawning muskellunge present in the Fox River during 2005, 255 were originally stocked as fall fingerlings and 170 as yearlings. About 0.71% of the fall fingerlings and about 10.43% of the yearlings stocked during 1989–2002 contributed to the 2005 spawning stock.

We developed a weight–length relationship from 1,595 observations of captive yearling and Fox River survey-recaptured hatchery-reared muskellunge as:

$$\log_{10} W = -6.197 + 3.360 * \log_{10}(L),$$

where W is weight (g) and L is total length (mm, Fig. 2).

We estimated the von Bertalanffy growth model for Green Bay muskellunge from 820 observations (length-age sub-sample = 180) as:

$$L_t = 1,217(1 - e^{-0.312\,(t+0.246)}).$$

We estimated the model for female muskellunge from 62 observations (length-age sub-sample = 17) as:

$$L_t = 1,355\,(1 - e^{-0.170\,(t+2.705)}).$$

We estimated the model for Lake St. Clair muskellunge captured during 1969–1974 from 329 observations as:

$$L_t = 1,293\,(1 - e^{-0.249\,(t-0.754)}).$$

A broad range of ages was used to estimate growth for Green Bay muskellunge (1.35–11.98 years, mean 3.57 years, Fig. 3), female Green Bay muskellunge (2.44–11.98 years, mean = 6.67 years, Fig. 4), and Lake St. Clair muskellunge (3–10 years, mean = 6.9 years).

Estimated growth of Green Bay muskellunge was more rapid than growth of Lake St. Clair muskellunge, but the asymptotic average maximum length of muskellunge was smaller for Green Bay compared to Lake St. Clair (Fig. 3, Table 3). However, the 95% confidence intervals for both parameters overlapped (Table 3). Growth of female Green Bay muskellunge was more rapid than growth of females from Georgian Bay, Lake St. Clair, and the St. Lawrence River (Fig. 4), but asymptotic average maximum length was smaller for female Green Bay muskellunge compared to the St. Lawrence River. The upper limit of the asymptotic average maximum length L_∞ that we estimated for Green Bay muskellunge indicates that they are unlikely to attain record length.

We estimated asymptotic average maximum weight W_∞ at 14,783 g (95% CI 12,072–17,896 g)

Fig. 2 Weight (g)–length (mm) data and the curvilinear relationship between weight and length from 1,595 captive yearling and Fox River survey-recaptured muskellunge sampled during 1994–2005. The equation and coefficient of determination (R^2, observed versus predicted) of the curvilinear relationship are given

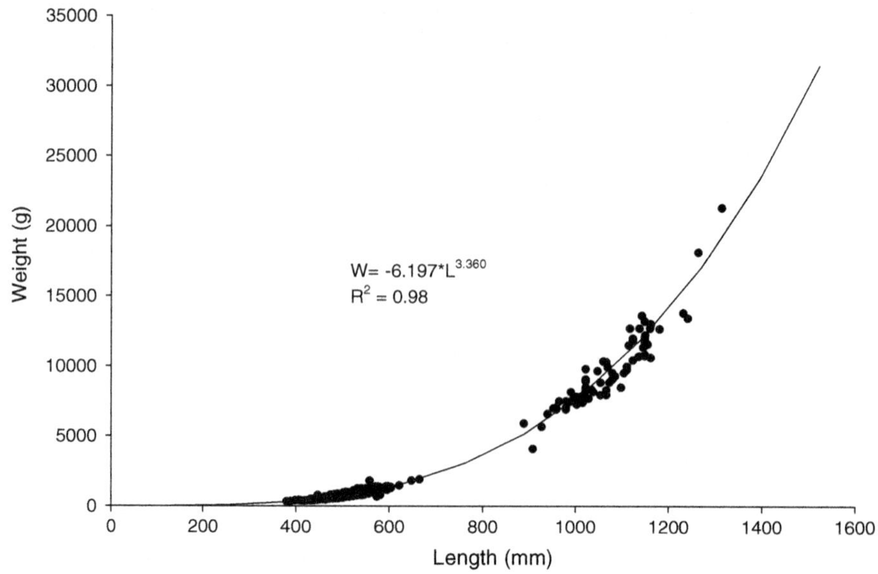

Fig. 3 Length-at-age for muskellunge captured from Green Bay (1989–2005) and Lake St. Clair (1969–1974). The observed length-at-age data from the Green Bay population are indicated by *closed circles*, and von Bertalanffy growth curves (*solid line*-Green Bay, *dashed line*-Lake St. Clair), equations, and coefficients of determination (R^2, observed versus predicted) are provided

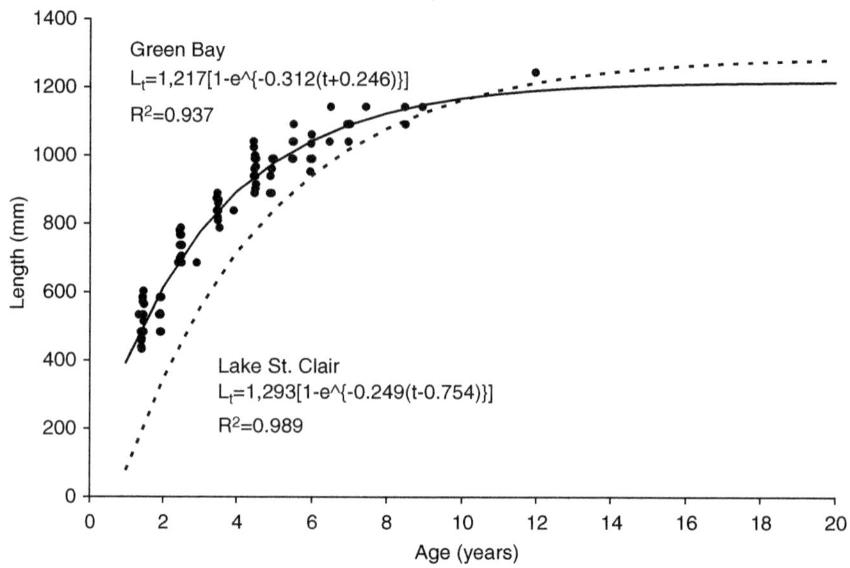

for all Green Bay muskellunge and 21,188 g (95% CI 13,125–32,219 g) for females (Table 3). The upper limit of the asymptotic average maximum weight W_∞ that we estimated for females suggests that some individuals may attain record weight.

A total of 93 mature muskellunge were captured during 2004–2005 spring surveys; 51 males and 42 females. Mean length of males was 1,013 mm (range 815–1,215 mm, 95% CI 989–1,038 mm) and mean length of females was 1,115 mm (range 980–1,264 mm 95% CI 1,093–1,137 mm). Mean age of mature males was 5.80 (n = 6, range 4.98–6.01 years, 95% CI 5.38–6.23 years) and 6.78 years (n = 10, range 4.98–11.98, 95% CI 5.36–8.20) for mature females. We estimated age for 32 mature females of unknown age via our sex-specific growth model. Mean estimated age of females was 8.10 years (range 5.56–13.16 years, 95% CI 7.43–8.77 years).

Discussion

Understanding the abundance of muskellunge spawning in Green Bay is essential to reintroduction efforts. The estimated 425 muskellunge spawning in the Fox River suggests that stocked muskellunge are surviving, maturing, and spawn-

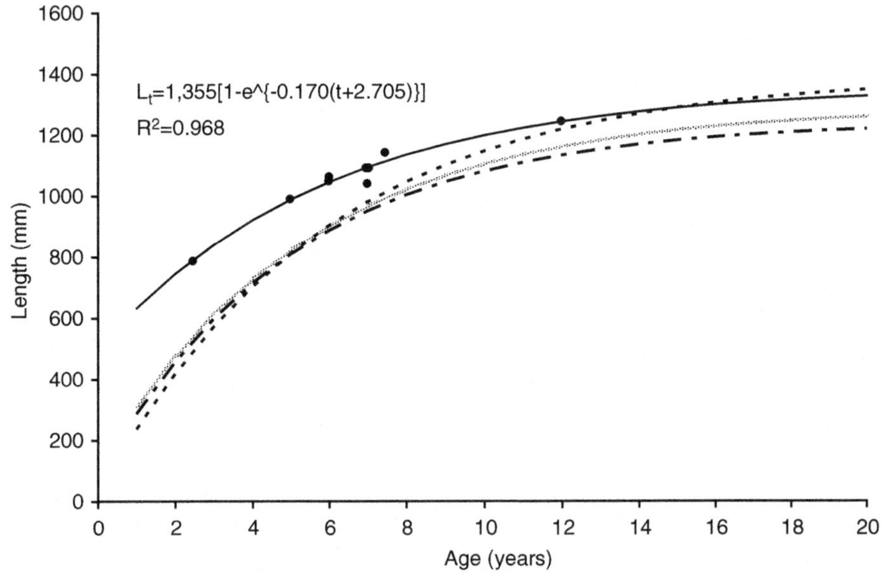

Fig. 4 Length-at-age for female muskellunge captured from Georgian Bay (1977–1992), Green Bay (1989–2005), Lake St. Clair (1980–1996), and the St. Lawrence River (1964–1983). The observed mean length-at-age data from the Green Bay population are indicated by *closed circles*, von Bertalanffy *growth curves* (*solid line*-Green Bay, *dashed line*-St. Lawrence River, *grey line*-Georgian Bay, *dash/long dash line*-Lake St. Clair) and the equation and coefficient of determination (R^2, observed versus predicted) for the Green Bay population are provided

ing in habitats adjacent to historical stocking sites. Unfortunately, the precision of this estimate was limited by few recaptures, and estimates for other areas of Green Bay are not calculable because muskellunge surveys have not been conducted outside of the Fox River. The Fox River survey should be repeated annually to monitor abundance near this most heavily stocked and sport-fished location. Surveys targeting spawning muskellunge should be conducted near all stocking sites to monitor the status of spawning stocks, and abundance estimation methods allowing for the examination of data across years and sites should be explored.

The ratio of yearling to fall fingerling muskellunge stocked during 1989–2002 and observed as 2005 spawners was 2:3. However, the ratio expressed as the percent originally stocked and contributing to the 2005 spawning stock was about 14.69:1 yearling to fall fingerling. We also observed that muskellunge stocked as yearlings contributed more (per fish stocked) to the spawning stock in Long Lake, Waushara County, Wisconsin than muskellunge stocked as fingerlings (WI DNR unpublished data). Yearling muskellunge have been stocked in Green Bay during the spring or summer. The cost of rearing spring yearlings is less than twice the cost of

Table 3 Summary of von Bertalanffy growth parameters and 95% confidence limits (in parentheses) for muskellunge captured from Green Bay (1989–2005) and Lake St. Clair (1969–1974)

	Green Bay		Lake St. Clair
	All	Females	All
N	820 (180)	62 (17)	329
L_∞	1,217 (1,146, 1,289)	1,355 (1,175, 1,535)	1,293 (1,102, 1.483)
K	0.312 (0.249, 0.375)	0.170 (0.074, 0.265)	0.249 (0.114, 0.383)
t_0	–0.246 (–0.543, –0.051)	–2.705 (–4.968, –0.443)	0.754 (–0.295, 1.803)
W_∞	14,783 (12,072, 17,896)	21,188 (13,125, 32,219)	–

N is the number of samples used in the analysis (known-age sub-sample size in parentheses)

rearing fingerlings (Larson et al. 2006), so yearlings are more cost effective (at surviving and returning as spawning adults) than fingerlings in Green Bay. Margenau (1992) found that spring yearlings survived better to age 18 months (19%) than fall fingerlings (4%) in three different Wisconsin lakes, and spring yearlings were 1–4 times more cost-effective at surviving through 18 months than fall fingerlings. In a continuation of Margenau's 1992 study, Margenau (1999) found that for four Wisconsin lakes stocked with spring yearling and fall fingerling muskellunge, at least 50% of the survey captured muskellunge were stocked as spring yearlings; generally this number was more than 80%. Similarly, Johnson (1982) reported that muskellunge stocked at 246–297 mm survived 12.6% better than those stocked at 191–241 mm in 20 Wisconsin lakes, and Wahl (1999) concluded that survival increased with size at stocking. Beyerle (1981), however, recommended that tiger muskellunge *Esox lucius* × *E. masquinongy* fingerlings be stocked at 89 mm rather than a larger size in lakes with low-predator densities. Hatchery-reared muskellunge stocked into Green Bay as yearlings survived better, were more cost effective, and most importantly, contributed more to the spawning stock compared to those stocked as fall fingerlings. However, spawning by stocked fingerlings and yearlings has not resulted in successful reproduction. Muskellunge that are hatchery-reared for extended lengths of time may not behave similarly to muskellunge stocked as fry or born in the wild. Fry stocking has been shown to be successful in areas of the Great lakes with suitable habitat (Farrell and Werner 1999). Survival to adulthood of stocked muskellunge is an important consideration for reintroduction efforts, but more emphasis and resources should be directed at comparing the reproductive potential of muskellunge stocked as fry, fingerlings, and yearlings in available habitats.

The weight–length relationship we calculated describes the body shape of hatchery-reared muskellunge that are stocked into the wild and live in or near the Fox River downstream from the De Pere Dam. Muskellunge stocked into other areas of Green Bay may have different weight–length relationships due to differences in factors such as diet and water temperature. We estimated the parameter b at 3.360, which suggests that Green Bay muskellunge become more rotund as length increases (Anderson and Neumann 1996). Casselman and Crossman (1986) calculated parameter b at 3.135 for 495 trophy muskellunge from across their North American range, and Haas (1978) calculated parameter b at 3.016 for 296 muskellunge sampled from Lake St. Clair. While we calculated the Fox River muskellunge weight–length relationship from 1,595 observations, significant data gaps existed for fish between 600–950 mm and greater than 1,160 mm, and for muskellunge from other areas in Green Bay. The weight–length relationship for Fox River muskellunge should be updated as more data become available, and relationships for muskellunge at other locations should be calculated to monitor the condition of reintroduced stocks.

Growth of Green Bay muskellunge was very rapid during early ages, and length-at-age was greater than that observed for Lake St. Clair muskellunge through age 10. However, length-at-age of Green Bay muskellunge was surpassed by Lake St. Clair muskellunge by age 11. Growth of female Green Bay muskellunge was similarly rapid, and greater than that reported for Georgian Bay and Lake St. Clair for all ages, and the St. Lawrence River through age 14 (Casselman et al. 1999). Again, length-at-age of female Green Bay muskellunge was surpassed by St. Lawrence River muskellunge by age 15. The rapid early growth of Green Bay muskellunge was the result of being hatchery-reared for 4–15 months, and exceptional growth was maintained in the wild for many years following stocking. The continued exceptional growth rate in the wild may be attributed to high productivity of southern Green Bay (Johnson 1971), low muskellunge densities relative to other stocks, and availability of gizzard shad, *Dororsoma cepedianum*, as a prey source (Quist et al. 2003). The rapid growth of Green Bay muskellunge allows them to achieve large sizes at relatively young ages, but also appears to compromise their ability to achieve record lengths. This growth pattern is consistent with observations of Becker (1983) who noted that record muskellunge have been taken from Wisconsin stocks with average growth, but no records have been taken from stocks with excellent growth. Similarly, Casselman et al. (1999) concluded that the largest muskellunge are produced by stocks exhibiting moderate and steady early growth that continues for many years. While Green Bay muskellunge may be incapable of achieving record lengths, our results indicate that it is possible for females to achieve record weight.

Green Bay muskellunge mature at longer lengths and possibly older ages than other muskellunge stocks. The mean length of mature male and female Green Bay muskellunge was greater than the mean lengths reported for muskellunge in Big Spider Lake, Bone Lake, and Lac Court Oreilles, WI, USA (Johnson 1971), and Grace and Waconia lakes, MN, USA (Younk and Strand 1992). The mean length of mature male muskellunge from Green Bay was also greater than the mean length reported for males from Lake St. Clair (Haas 1978), but the mean length of female muskellunge was similar for both systems. Furthermore, we did not capture any mature male muskellunge at age 3, which is inconsistent with the findings of Johnson (1971) and Younk and Strand (1992), but consistent with the results of Haas (1978). For example, Johnson (1971) reported that some male muskellunge from Bone Lake, Lac Court Oreilles, and Big Spider Lake matured at age 3, Younk and Strand (1992) found mature male muskellunge at age 3 in Grace and Waconia Lakes, but Haas (1978) did not report any mature male muskellunge at age 3 in Lake St. Clair. We also found very few mature female muskellunge less than age 5 (two of ten known age, 0 of 32 estimated age), which is inconsistent with the findings of Johnson (1971), but consistent with the findings of Haas (1978) and Younk and Strand (1992). If Green Bay muskellunge are maturing later than muskellunge in inland waters of Wisconsin, they may have more potential to grow to trophy size because of a longer somatic growth period (Younk and Strand 1992). However, the low-sample size of muskellunge from Green Bay makes this conclusion tenuous. For example, if the muskellunge stocked as yearlings that would have shown up as ages 3 and 4 spawners during 2005 had poor survival, they may not have been represented in our low sample size, thereby biasing our conclusions about maturity. Similarly, high tag shedding rates could have limited our observations. More data must be collected and analyzed to accurately describe the maturation schedule and growth potential of Green Bay muskellunge.

Reintroduction efforts have been successful in producing muskellunge populations in Green Bay that support trophy fisheries. We have documented spawning in some areas and assume it occurs in others, but we have no evidence that spawning has successfully produced young muskellunge. We know that stocked fingerling and yearling muskellunge can survive to adulthood and grow rapidly in Green Bay, and therefore we assume that the main impediment to establishing a self-sustaining population is the lack of successful natural reproduction. Most muskellunge stocked in Green Bay have been released into the two largest, most altered tributaries, so reproduction of stocked muskellunge is likely limited by the available habitat. Reproduction may be further limited by stocking only fall fingerling or yearling muskellunge. Future research efforts should continue to build existing data sets, but efforts that identify suitable spawning and rearing habitats and promote natural reproduction should be the highest priority. This will require habitat examination in near-shore areas and tributary streams, comparisons of the reproductive potential of muskellunge stocked as different life stages and spawning in various habitats, and surveys targeting young-of-year and juvenile muskellunge to gauge natural reproduction in areas where spawning is observed or assumed to take place. Low densities of adult muskellunge may also be limiting reproduction, so stocking efforts should focus on increasing densities where successful natural reproduction is documented or most expected to occur.

Acknowledgments Many WI DNR and Michigan DNR personnel have contributed to reintroduction efforts and collected data used in this study. Dave's Musky Club, Packerland Musky Club, The Musky Clubs Alliance of Wisconsin, Muskies Unlimited, and the Lower Fox River and Green Bay Area Natural Resources Damage Assessment Fund provided major financial support for reintroduction efforts. George Boronow, David Clapp, James Diana, Steve Hogler, Matt Mangan, Mike Toneys, and an anonymous reviewer provided valuable comments on drafts of this manuscript. Kyle Burton provided technical support during figure creation.

References

Anderson RO, Neumann RM (1996) Length, weight, and structural indices. In: Murphy BR, Willis DW (eds) Fisheries techniques, 2nd edn. American Fisheries Society, Bethesda, Maryland, pp 447–482

Becker GC (1983) Fishes of Wisconsin. The University of Wisconsin Press, Madison, Wisconsin

Beyerle GB (1981) Comparative survival and growth of 8.9- and 17.8-cm (3.5- and 7.0-inch) tiger muskellunge planted in a small lake with forage fishes. Michigan Department of Natural Resources, Fisheries Division, Fisheries Research Report No. 1894, p 7

Casselman JM, Crossman EJ (1986) Size, age, and growth of trophy muskellunge and muskellunge-northern pike hybrids—the cleithrum project, 1979–1983. In: Hall GE (ed) Managing muskies, a treatise on the biology and propagation of muskellunge in North America. American Fisheries Society Special Publication, pp 93–110

Casselman JM, Robinson CJ, Crossman EJ (1999) Growth and ultimate length of muskellunge from Ontario water bodies. North Am J Fish Manage 19:271–290

Farrell JM, Werner RG (1999) Distribution, abundance, and survival of age-0 muskellunge in upper St. Lawrence River nursery bays. North Am J Fish Manage 19:309–320

Greene CW (1935) The distribution of Wisconsin fishes. Wisconsin Conservation Commissio, WI, USA, p 235

Haas RC (1978) The muskellunge in Lake St. Clair. Am Fish Soc Special Publ 11:334–339

Johnson LD (1971) Growth of known-age muskellunge in Wisconsin: and validation of age and growth determination methods. Wisconsin Department of Natural Resources Technical Bulletin 49, Madison, WI, p 22

Johnson LD (1982) Factors affecting short-term survival of stocked muskellunge fingerlings in Wisconsin. Wisconsin Department of Natural Resources Report No. 117, Madison, WI, p 24

Jude DJ, Leach J (1993) The Great Lakes fisheries. In: Kohler CC, Hubert WA (eds) Inland fisheries management in North America. American Fisheries Society, Bethesda, Maryland. pp 517–551

Lake Michigan Fisheries Team (2004) Lake Michigan integrated fisheries management plan 2003–2013. Wisconsin Department of Natural Resources, Bureau of Fisheries Management and Habitat Protection, Administrative Report No. 56, Madison, WI

Larson TR, Blankenheim JE, Fuller DB (2006) An evaluation comparing survival of fall fingerling versus spring yearling stocked musky in southern Wisconsin lakes. Wisconsin Department of Natural Resources, Bureau of Fisheries Management, Madison, WI

Margenau TL (1992) Survival and cost-effectiveness of stocked fall fingerling and spring yearling muskellunge in Wisconsin. North Am J Fish Manage 12:484–493

Margenau TL (1999) Muskellunge stocking strategies in Wisconsin: the first century and beyond. North Am J Fish Manage 19:223–229

Quinn TJ II, Deriso RB (1999) Quantitative fish dynamics. Oxford University Press, New York

Quist MC, Guy CS, Schultz RD, Stephen JL (2003) Latitudinal comparisons of walleye growth in North America and factors influencing growth of walleyes in Kansas Reservoirs. North Am J Fish Manag 23:677–692

Ricker WE (1975) Computation and interpretation of biological statistics of fish populations. Department of the Environment Fisheries and Marine Service, Ottawa

Smith HH, Snell M (1891) Review of the fisheries of the Great Lakes in 1885, with introduction and description of fishing vessels and boats by J. W. Collins. Report of Commissioner of Fish and Fisheries for 1887, US Commission of Fish and Fisheries, p 333

Van Den Avyle MJ (1993) Dynamics of exploited fish populations. In: Kohler CC, Hubert WA (eds) Inland fisheries management in North America. American Fisheries Society, Bethesda, Maryland, pp 105–135

Wahl DH (1999) An ecological context for evaluating the factors influencing muskellunge stocking success. North Am J Fish Manage 19:238–248

Wisconsin Department of Natural Resources (1986) Lower Green Bay remedial action plan. Wisconsin Department of Natural Resources, Bureau of Fisheries Management and Habitat Protection, Madison, WI

Younk JA, Strand RF (1992) Performance evaluation of four muskellunge *Esox masquinongy* strains in two Minnesota lakes. Minnesota Department of Natural Resources, Investigational Report 418, St. Paul, Minnesota

Environ Biol Fish (2007) 79:37–47
DOI 10.1007/s10641-006-9133-1

SPECIAL ISSUE – CROSSMAN

Quantification of native muskellunge nursery habitat: influence of body size, fish community composition, and vegetation structure

Brent A. Murry · John M. Farrell

Received: 11 September 2005 / Accepted: 18 July 2006 / Published online: 29 August 2006

Abstract Habitat utilization of native young-of-the-year (YOY) muskellunge, *Esox masquinongy*, was quantified by the type and density of vegetation present, water depth, and fish communities associated with their presence and abundance in nursery bays of the Upper St. Lawrence River. We completed 441 seine hauls and captured 400 YOY muskellunge in 11 bays that were sampled each July and August over a 3-year period (2002–2004). We hypothesized a change in habitat utilization related to increasing body size, as YOY muskellunge doubled in total length from July to August. Fine-leafed submerged and emergent macrophytes and prey availability (cyprinids, *Notropis* sp., banded killifish, *Fundulus diaphanous* and tessellated darter, *Etheostoma olmstedi*), were positively related to muskellunge use in July, while in August coverage of broad-leafed submerged macrophytes and increased overall vegetation density were the best habitat descriptor. In both months, muskellunge were associated with moderate (20–60%) vegetation coverage and density, however, captures were in areas of significantly greater vegetation coverage and density than was generally available. A negative relation of muskellunge occurrence with water depth, yellow perch, *Perca flavescens*, and stonewort, *Chara vulgaris*, was observed in both months. The negative relationship between muskellunge and depth, plus their strong linkages to nearshore submerged vegetation and forage fish that inhabit the nearshore areas, highlights the importance of protecting the ecological integrity of nearshore habitats. Our findings should assist managers in protecting native stocks, planning restoration and enhancement initiatives, and in regulating riparian and nearshore development.

B. A. Murry (✉) · J. M. Farrell
Faculty of Environmental Forest Biology,
College of Environmental Science and Forestry,
State University of New York, 104 Illick Hall,
1 Forestry Drive, Syracuse, NY 13210, USA
e-mail: bamurry@syr.edu

Keywords Ontogenetic habitat utilization · Submerged macrophytes · Predator-prey relationships

Introduction

Compared to most other sport fish, basic biological data related to muskellunge habitat requirements are lacking (Dombeck 1986). This is especially true of nursery habitat for muskellunge young-of-the-year (YOY). The few studies that have addressed their nursery habitat use are generally descriptive in nature (Osterberg 1985; Craig and Black 1986; Bendig 1995). Specific nursery habitat conditions, such as the density, structure, and composition of the fish and plant communities may greatly influence recruitment during the summer nursery period.

The nursery period, from swim-up to the first winter is a critical period for year class formation. Several studies suggest that relative year class strength in large predatory fishes is determined during the first year of life (May 1974; Juanes et al. 1994; Garvey et al. 1998). Recruitment of northern pike, *Esox lucius*, is influenced by the timing of spawning (Farrell et al. 2006), which affects later match/mismatch of optimal temperature, prey availability, and vegetative cover during the nursery period (Cushing 1990; Juanes et al. 1994). Similarly, rapid growth during the nursery period, which is directly linked to prey availability, is also critical to first winter survival, as larger fish have greater over-winter survival rates (Garvey et al. 1998). These observations highlight the importance of the nursery period and habitat conditions in mediating annual recruitment in predatory fishes such as the muskellunge.

Wahl (1999) expressed the importance of fish community relationships to the success of muskellunge stocking. Predator abundance can influence the survival of stocked esocids, though it was suggested that high abundance of alternative prey might mediate the impacts. Prey availability also influences foraging success, growth, and survival of young muskellunge (Szendrey and Wahl 1996; Wahl 1999). Muskellunge stomach contents suggest negative selection of centrarchids as prey, but a preference for soft-rayed fusiform prey such as cyprinids (Szendrey and Wahl 1996; Farrell 1998). Further, Wahl and Stein (1988) showed that muskellunge grew more rapidly on clupeid and cyprinid diets than a centrarchid diet. Thus, the fish community composition (assembly and relative densities) is likely an important component of YOY muskellunge success and should be included in analyses of habitat utilization along with more traditional physical habitat parameters.

The vegetative community is an important component of physical habitat structure in near-shore lakes and large rivers. Qualitatively, YOY muskellunge have been found associated with low to moderate densities (and avoided high densities) of submerged and emergent macrophytes (Craig and Black 1986; Werner et al. 1990; Clapsadl 1993; Bendig 1995), but were negatively associated with dense *Chara* beds (Clapsadl 1993). Although YOY muskellunge tended to avoid areas of dense macrophytes, they have been associated with tall vegetation that reaches the water's surface (Jonckheere 1994). Catch of YOY muskellunge has also been found to be inversely related to water depth (Farrell and Werner 1999).

Physical habitat composition and fish community structure likely interact in their influence on YOY muskellunge. Patterns of habitat use may change with increasing muskellunge size and different muskellunge habitat relationships are expected through their first year ontogeny. For instance, vegetative structure (cover) and predator abundance are expected to be of primary importance earlier in the nursery period when muskellunge are small and more susceptible to predation. This relationship may be dampened at larger body sizes with reduced predation risk (Werner and Gilliam 1984). Later in the summer, as predation risk diminishes with increasing body size and the need to grow rapidly in preparation for winter increases (Garvey et al. 1998), we expect that relationships to prey fish abundances should drive habitat utilization. Whether YOY muskellunge respond to physical habitat and fish community factors independently or interactively is more complex. Different relationships may be observed if fish community and physical habitat variables are examined individually as opposed to collectively.

This paper is an effort to bridge the gap between autecological management and a more holistic community-level management approach. Successful habitat management depends on a more complete understanding of fish-habitat relationships. Our objectives were to (1) identify important fish community and vegetative habitat characteristics associated with YOY muskellunge presence and abundance as their body size increases, and (2) indirectly compare and contrast the importance of vegetative and fish community associations with muskellunge abundance. Identifying and quantifying significant relationships between particular habitat components (physical habitat and fish community) and muskellunge presence and abundance will be useful to managers as targets for the conservation of self-sustaining muskellunge populations. Relationships described in this study should also guide habitat enhancement programs and influence decisions regarding development within environmentally sensitive nearshore areas, where landowners disrupt the habitat by building docks, dredging, and applying herbicides (Craig and Black 1986).

Methods

Eleven bays of the St. Lawrence River between Cape Vincent and Alexandria Bay, NY were sampled in July and August of each year 2002–2004. Study sites are qualitatively representative of typical embayment habitats, are well distributed throughout the region, and are known muskellunge spawning bays (Farrell et al., this issue). Fish sampling was conducted using standardized seining protocols (Farrell and Werner 1999). The seine was slowly dragged 36.6 m (100 ft) and was "bagged" in the water by bringing the poles together and pulling the lead lines along the substrate until the bag is secure. Most seine hauls were conducted parallel to the shoreline in water less than 1.5 m (~4.5 ft) deep. Each study site was sampled once during each sampling period (July and August) each year between the 15th and 31st of each month. In July we used a fine mesh bag seine (9.14 m-long 1 mm-mesh, 10 yds-long ~1/32 in.-mesh), which sampled 223 m^2 (244 yds^2) and a large mesh seine (18.3 m-long 6.4 mm-mesh, 20 yds-long 1/4 in.-mesh) which sampled 447 m^2 (489 yds^2) was used in August. The fine mesh seine was used in July to ensure the capture of all muskellunge, including small individuals.

Captured fish were retained live in a cooler, muskellunge were held in a separate cooler. The total length (mm) and wet weight (0.01 g) were determined for every muskellunge and they were released in the area where they were captured. All other fish captured were identified, enumerated, and released. The total lengths of most other fish were recorded, but at times a random subsample of fish (>25 per species per haul) was measured. The fish community was not evaluated in August 2004. The number of hauls (6–12) made in each bay corresponded approximately to the seinable area of each bay and was consistent for all sampling periods.

To facilitate analyses fish species were placed into functional categories as they potentially relate to YOY muskellunge. Categories included potential predators, prey, competitors, and other fish that share similar habitat affinities (Table 1). The basic observational unit used to evaluate YOY muskellunge habitat was the seine haul, which for fish abundance was assumed to be equivalent to the catch per unit effort (CPUE).

Vegetative habitat was evaluated at the beginning, middle, and end point of each haul within a 1 m^2 (1,521 $in.^2$) floating grid. The water depth (m) and the height of the tallest plant (m) were measured to the nearest centimeter. Vegetation within each sampling quadrat was identified to species and analyzed in five structural groups (Table 2) including; fine-leafed submerged macrophytes, broad-leafed submerged macrophytes, emergent vegetation, floating vegetation, and macroalgae. The percent coverage of each group was estimated using a vegetation coverage score (VC-score, similar to the Braun–Blanquet scale) where 0–5% coverage = 0, 5–15% = 1, 15–25% = 2, 25–50% = 3, 50–75% = 4, and 75–100% = 5. The VC-score was also used to evaluate total vegetation coverage. Vegetation density in the water column, an index of the amount of water column space occupied by vegetation, was the product of the percent of the water column containing vegetation (vegetation height/depth) by the VC-score of the tallest vegetation group in each quadrat. The mean value of each parameter for each seine haul (n = 3) was utilized in statistical comparisons.

Statistical analyses

July and August data were analyzed separately because we were interested in evaluating habitat differences related to differences in body size. First we examined univariate patterns of habitat use with respect to YOY muskellunge presence/absence and abundance. We used a complete block design analysis of variance (ANOVA) to determine the absolute differences in each habitat parameter relative to YOY muskellunge presence/absence. Muskellunge

Table 1 Fish community variables used to differentiate YOY muskellunge habitat utilization

Fish groups	# Species	Dominant species	Month	% Total Catch	Mean total length (range, mm)	Mean # M YOY Present	Mean # M YOY Absent
YOY muskellunge	1	n/a	July	0.11	56 (31–122)	2.65	0
			August	0.50	118 (84–148)	1.78	0
Other Esocids[a,c]	2	Northern Pike	July	0.01	184 (46–662)	0.24	0.08
		Grass Pickerel	August	0.13	241 (103–643)	0.19	0.16
Bass[b,c]	2	Largemouth Bass	July	5.50	53 (30–191)	72	41
		Smallmouth Bass	August	6.66	64 (46–112)	11	7
Bottom[b]	5	White Sucker	July	0.34	77 (31–290)	4	3
		Bullheads	August	1.31	63 (32–179)	2	1
Cyprinids[b]	14	Bluntnose Minnow	July	14.06	52 (31–83)	228	75
		Golden Shiner	August	6.58	61 (48–106)	9	8
		Blackchin Shiner					
Fry[b] (all fish < 30 mm)		Bluntnose Minnow	July	67.13	17 (9–30)	529	750
		Rock Bass[d]	August	28.08[d]	29 (28–30)	46	30
Sunfish[b,c]	5	Pumpkinseed	July	3.65	82 (31–217)	58	21
			August	9.00	86 (31–165)	14	10
Killifish[b] and Tessellated Darter[b]	2	Killifish	July	5.86	51 (30–75)	76	45
		Tessellated Darter	August	4.57	63 (43–77)	4	7
Rock Bass[a]	1	n/a	July	1.18	73 (57–109)	16	9
			August	2.35	82 (38–129)	1	4
Yellow Perch[a]	1	n/a	July	2.10	93 (56–140)	17	23
			August	40.55	113 (94–147)	37	60
Gar[c]	1	n/a	July	0.01	77 (33–153)	0.28	0.04
			August	0.03	165 (125–205)	0.03	0.04
Other	9	Brook Silverside	July	0.05	134 (32–632)	0.42	0.48
		Sticklebacks	August	0.23	125 (37–790)	0.42	0.23

The '# species' is the number of species within each group and the 'dominant species' lists the species that are most abundant in each group. The right portion of the table provides catch and size data for each group. The fish groups, based on relative body size and assumed trophic niche are considered

[a] Potential predators

[b] Potential prey

[c] Possible competitors

[d] 100% of the fish in the fry group were YOY rock bass in August

YOY presence/absence was the main effect and data were blocked by year and site to account for repeated measures. Then we calculated Pearson correlation coefficients between YOY muskellunge abundance and each habitat parameter to examine direct independent linkages.

Principle component analysis (PCA) of the correlation matrix was used to examine patterns of variation within the fish community and physical habitat variables. Individual principle components (PCs) were subsequently correlated to YOY muskellunge CPUE. Each PC represented a unique gradient of variation within the total habitat components measured. The eigenvectors (loadings) and internal correlation matrix of each PC significantly correlated to YOY muskellunge abundance were examined to assess which individual habitat variables were the strongest determinants of each PC (gradient). This procedure was repeated three times for each month; first to examine the independent influence of the physical habitat and the fish community variables, and then third to assess the collective influence of both data sets.

PCA is the most widely used multivariate technique reported in aquatic science journals (Paukert and Wittig 2002), but it does have limitations. Organisms often exhibit unimodal distributions in response to long habitat gradients, which PCA cannot adequately represent due to its underlying linear structure. The habitat gradients reported here, however, were very short

Table 2 General description of physical habitat variables used to differentiate YOY muskellunge habitat utilization

Habitat Groups	# Species	Dominant Species	Month	Mean value M YOY present	Mean value M YOY absent
Depth (m)	n/a	n/a	July	0.82	0.87
			August	0.69	0.74
Total Cover	n/a	n/a	July	4.68	4.52
			August	4.45	4.62
Emergent	4	*Scirpus* sp.	July	0.25	0.13
			August	0.17	0.22
Fine-leafed Submerged Macrophytes	9	*Potamogeton pectinatus* *P. pucilus*	July	2.92	2.20
			August	1.21	0.90
Broad-leafed Submerged Macrophytes	6	*Vallsineara* sp. *Alsima* sp.	July	1.20	2.22
			August	2.31	1.69
Macroalgae	2	*Chara* sp. *Nitella* sp.	July	3.37	3.60
			August	3.21	3.98
Floating	3	*Nymphea* sp. *Nuphar* sp.	July	0.38	0.26
			August	0.35	0.32
Vegetation Diversity	Range: 0–8/quadrat	n/a	July	3.45	3.04
			August	3.18	2.82
Column Density	n/a	n/a	July	2.58	2.12
			August	2.55	2.08

The '# species' is the number of species within each group and the 'dominant species' lists the species that are most abundant in each group. The right portion of the table provides mean values for hauls within which YOY muskellunge were captured and absent

(average gradient length = 1.593 SD units, max = 2.560, where 2 SD units = 50% turnover and 4 SD units = 100% turnover) because sampling sites were within muskellunge nursery bays.

For our particular study PCA was chosen over detrended correspondence analysis (DCA) because it centers and standardizes the raw data, which is important because our habitat data set contains several different measurement scales (VC-scores, vegetation diversity counts, depth, and fish counts). The standardization procedure utilized in PCA (mean = 0, variance = 1) allows these different measurement scales to be compared without measurement scale bias.

Results

Mid-summer: July

In 263 seine hauls during July sampling we captured 286 YOY muskellunge. Up to 13 muskellunge were captured in a single haul and 41% of hauls contained at least one muskellunge. Muskellunge mean total length was 56 mm (2.2 in.) and ranged from 31 to 122 mm (1.2–4.8 in.). A significant difference was detected with ANOVA for 10 (50%) of the predictor variables between areas of YOY muskellunge presence and absence. There were significantly more longnose gar, *Lepisosteus osseus* ($F_{1,249} = 16.03$, $P < 0.0001$, Table 1) and other esocids ($F_{1,249} = 6.18$, $P = 0.0136$) captured in hauls containing muskellunge than those without, likely indicating shared habitat preferences. Hauls containing YOY muskellunge also contained significantly greater prey abundance. Over three times as many cyprinids (14% of total catch) were captured in hauls with muskellunge present ($F_{1,249} = 12.13$, $P = 0.0006$). Similarly, almost twice as many killifish, *Fundulus diaphanous*, and tessellated darters, *Etheostoma olmstedi*, were associated with muskellunge ($F_{1,249} = 4.67$, $P = 0.0316$). Other potential prey, such as YOY bass (majority were largemouth bass, *Micropterus salmoides*), bottom fishes, and fry (fish <30 mm TL) did not differ in abundance in association with muskellunge ($F_{1,249} < 1.20$, $P > 0.05$, in all cases). Yellow perch, *Perca flavescens*, and rock bass, *Ambloplites rupestris*, are both potential predators of YOY muskellunge and together represented 3.3% of the total catch. Yellow perch abundance did not, however, differ between hauls with YOY muskellunge present or absent ($F_{1,249} = 0.56$, $P = 0.4546$), but there were significantly more rock bass were caught in association with muskellunge ($F_{1,249} = 7.94$, $P = 0.0052$). The two remaining groups, the sunfishes, *Lepomis* sp., and the 'other' category, did not differ with YOY muskellunge presence/absence ($F_{1,249} < 1.20$, $P > 0.05$, in both cases).

The water depth of seine hauls where muskellunge were captured was significantly shallower than those hauls where they were absent ($F_{1,249} = 5.03$, $P = 0.0258$, Table 2). Muskellunge were captured in hauls containing significantly greater coverage of emergent vegetation ($F_{1,249} = 4.61$, $P = 0.0327$) and fine-leafed submerged vegetation ($F_{1,249} = 21.59$, $P < 0.0001$). Similarly, they were captured in areas of higher vegetation diversity ($F_{1,249} = 14.19$, $P = 0.0002$) and water column vegetation density ($F_{1,249} = 17.40$, $P < 0.0001$). Total vegetation coverage, broad-leafed submerged vegetation, macroalgae, and floating vegetation did not differ between hauls containing muskellunge and those without ($F_{1,249} < 1.20$, $P > 0.05$, in all cases).

Most of the fish community and habitat factors that were significantly different when compared in areas of muskellunge presence or absence were similarly correlated to YOY muskellunge abundance (catch per seine haul). Muskellunge abundance was positively correlated to other esocids ($r = 0.25$, $P < 0.0001$), bass ($r = 0.27$, $P < 0.0001$), cyprinids ($r = 0.21$, $P = 0.0007$), killifish and tessellated darters ($r = 0.27$, $P < 0.0001$), and rock bass ($r = 0.14$, $P = 0.0284$). Yellow perch were negatively related, but the relationship was non-significant ($r = -0.11$, $P = 0.0745$). Remaining correlations between YOY muskellunge and the fish groups were all non-significant ($r < 0.11$, $P > 0.05$ in all cases).

Muskellunge exhibited positive associations with fine-leafed submerged vegetation ($r = 0.26$, $P < 0.0001$), vegetation diversity ($r = 0.23$, $P = 0.0001$), and water column vegetation density ($r = 0.16$, $P = 0.0091$). Though the coverage of macroalgae, *Chara* sp., did not differ between areas of muskellunge presence and absence, muskellunge abundance showed a significant negative correlation ($r = -0.16$, $P = 0.0102$). Correlations with the remaining physical habitat variables were all non-significant ($r < 0.11$, $P > 0.05$ in all cases).

The PCA for the fish group independent of the physical habitat data resulted in two PCs that were significantly correlated to YOY muskellunge abundance (PC1: $r = 0.37$, $P < 0.0001$; PC7: $r = 0.18$, $P = 0.0043$; Table 3). The first PC accounted for 16.15% of the total fish community variance and described prey abundance, being strongly correlated to cyprinid, killifish, and tessellated darter abundance. The PC7 described 9.97% of the remaining fish community variance and described a fish community gradient contrasting bass in the positive direction to other esocids and yellow perch in the negative direction.

The physical habitat PCA (independent of the fish community) also resulted in two PCs that were significantly related to YOY muskellunge abundance (PC1: $r = 0.28$, $P < 0.0001$; PC4: $r = 0.13$, $P = 0.0304$, Table 3). The PC1 accounted for 26.2% of the total physical habitat variation and described a habitat preference gradient with fine-leafed submerged macrophytes and the water column density of vegetation on the positive side to macroalgae on the negative side. The PC4 accounted for 13.31% of the remaining variance and extended the habitat gradient in the positive direction with both emergent vegetation and vegetation diversity and negatively with depth.

When the fish community and physical habitat data sets were analyzed together two PCs were significantly related to muskellunge abundance (Table 4). PC1, which accounted for 15.40% of the total overall habitat variance, described the physical habitat and showed the strongest correlation to YOY muskellunge abundance ($r = 0.35$, $P < 0.0001$). PC1 was explained by fine-leafed submerged macrophytes and water column vegetation density contrasted to depth. PC5 ($r = 0.20$, $P = 0.0012$), which accounted for 6.67% of the total variance, described the fish community and contrasted yellow perch and killifish/tessellated

Table 3 The PCA loading coefficients for individual habitat components in July and August

Variable	July		August	
	PC1	PC7	PC2	PC4
Esocids	0.3253	**–0.3629**	–0.2569	**0.6262**
Bass	0.2212	**0.7053**	0.4930	**0.5134**
Bottom	0.0487	0.1233	**0.5993**	0.4161
Cyprinids	**0.7967**	–0.2048	0.4253	–0.1555
Fry	0.1412	–0.1362	0.4570	–0.1892
Sunfish	–0.1219	0.0454	0.3337	**–0.3266**
Killi and tess	**0.6755**	0.1047	–0.1175	–0.0982
Rock bass	0.4122	0.1959	–0.0572	–0.1620
Yellow perch	0.0202	**–0.3756**	**–0.5373**	0.1878
	PC1	PC4	PC1	PC2
Depth (m)	–0.1817	**–0.3771**	0.3235	0.4597
Total cover	0.4125	–0.1027	0.0537	**0.9196**
Emergent	0.0091	**0.8803**	–0.1609	–0.0018
Fine-leafed	**0.8922**	–0.1281	**0.7709**	0.0400
Broad-leafed	0.2933	–0.0682	**0.8610**	0.0684
Macroalgae	**–0.2557**	–0.0237	**–0.6460**	**0.6583**
Floating	0.3949	0.0162	–0.0288	**–0.2177**
Veg. diversity	0.6050	**0.4471**	**0.7579**	0.1408
Column density	**0.8280**	–0.2209	**0.6736**	0.0359

Bold loading coefficients were used in interpretation

Table 4 The PCA loading coefficients for all habitat variables in July and August

Variable	July		August	
	PC1	PC5	PC1	PC2
Esocids	0.3619	0.1073	0.1520	–0.0182
Bass	0.1896	0.3383	0.3589	–0.3714
Bottom	0.1307	–0.2870	0.0846	**–0.6555**
Cyprinids	**0.5144**	0.1648	0.3318	–0.1216
Fry	0.1731	**–0.4765**	0.0528	–0.2557
Sunfish	0.0746	–0.3981	0.3952	0.0116
Killi and tess	0.3558	**0.4821**	–0.1982	0.0758
Rock bass	0.2135	0.1165	0.0996	0.3016
Yellow perch	–0.1644	**0.4645**	–0.1843	0.3014
Depth (m)	**–0.3748**	0.1039	0.2175	**0.6084**
Total cover	0.4435	0.1320	–0.0718	**0.7284**
Emergent	0.0295	0.1024	–0.1474	–0.0593
Fine-leafed	**0.8280**	–0.0971	**0.7166**	0.1970
Broad-leafed	0.0364	0.1251	**0.8174**	0.2255
Macroalgae	–0.1425	0.1274	**–0.7047**	0.3674
Floating	0.4529	–0.1375	0.0551	–0.3350
Veg. diversity	0.4410	0.1058	**0.7305**	0.1232
Column density	**0.8023**	–0.0887	**0.6611**	0.0823

Bold loading coefficients were used in interpretation

darter abundance to fry (all fish <30 mm) abundance.

Late-summer: August

In 178 seine hauls completed during August sampling, we caught 114 YOY muskellunge. There were up to nine YOY muskellunge caught in a single haul and overall 36% of hauls contained at least one muskellunge. Although total CPUE decreased from July to August (1.09 in July, 0.64 in August), the total length increased to 118 mm (4.7 in.) on average (range: 84–148 mm, 3.3–5.8 in.). Habitat utilization specificity based on presence/absence declined in August. We observed no differences in the abundance of any of the fish groups between hauls with and without muskellunge ($F_{1,165} < 2.60$, $P > 0.05$ in all cases, Table 1) and the only physical habitat variable that differed was vegetation density in the water column ($F_{1,165} = 4.71$, $P = 0.0315$ and $F_{1,165} < 2.60$, $P > 0.05$ in all other cases, Table 2), which was denser where muskellunge were captured.

The YOY muskellunge abundance, however, was positively correlated with bass ($r = 0.19$, $P = 0.0103$) and bottom fish ($r = 0.22$, $P = 0.0031$) abundance and negatively associated with both rock bass ($r = -0.18$, $P = 0.0140$) and yellow perch ($r = -0.17$, $P = 0.0266$) abundance. All other correlations with fish groups were non-significant ($r < 0.16$, $P > 0.05$ in all cases). The YOY muskellunge abundance showed a positive association with broad-leafed submerged macrophytes ($r = 0.17$, $P = 0.0268$) and a negative association with macroalgae ($r = -0.19$, $P = 0.0136$). All other correlations between YOY muskellunge abundance and physical habitat variables were non-significant ($r < 0.16$, $P > 0.05$ in all cases).

The PCA of the fish community independent of the physical habitat data resulted in two PCs that were significantly correlated to YOY muskellunge abundance (PC2: $r = 0.23$, $P = 0.0019$; PC4: $r = 0.21$, $P = 0.0060$, Table 3). The second PC accounted for 16.39% of the variance in the fish community and contrasted bottom fishes in the positive direction with yellow perch in the negative direction. The PC4 described 11.85% of the variance and described fish co-occurrence. Other esocids and bass were the strongest positive variables, while sunfish was the negative component.

The PCA of the physical habitat variables also resulted in two PCs significantly related to muskellunge abundance (PC1: $r = 0.17$, $P = 0.0279$; PC2: $r = -0.15$, $P = 0.0400$, Table 3). The first two PCs collectively accounted for nearly 50% of the total physical habitat variance (32.39 and 17.39%, respectively). The first PC described; broad and fine-leafed submerged macrophytes, vegetation diversity, and water column vegetation density. Macroalgae vegetation was the opposite gradient extreme. The PC2 identified total cover and macroalgae as important factors contrasted against coverage of floating vegetation.

When the fish community and physical habitat data sets were analyzed together the first two PCs were related to muskellunge abundance (PC1: $r = 0.19$, $P = 0.0093$; PC2: $r = -0.23$, $P = 0.0023$, Table 4). The first two PCs described 18.00 and 11.66% of the overall habitat variance, respectively. The first PC contrasted fine-leafed submerged macrophytes, broad-leafed submerged macrophytes, vegetation diversity, and water column density of vegetation with macroalgae. Water column density of vegetation was a secondary factor on the positive side of the gradient. The second PC contrasted total vegetation coverage and depth with the abundance of bottom fishes.

Discussion

"Typical" muskellunge nursery habitat, defined by their presence and abundance, was linked to both vegetation structure and fish community composition. In general, physical habitat variables, primarily positive associations with submerged macrophyte density and negative associations with depth and stonewort, *Chara vulgaris*, showed stronger linkages than fish community variables to YOY muskellunge abundance. The fish community data did, however, provide very useful insight, expanding our understanding of YOY muskellunge habitat utilization, particularly in July when YOY muskellunge presence was positively associated with prey abundance, especially cyprinids (*Pimephales* sp. and *Notropis* sp.), banded killifish, and tessellated darters, while in both months was negatively associated with yellow perch.

Differences in YOY muskellunge habitat utilization between July and August were evident in their relationship to the fish community, and although they were contrary to our expectations, may have still been related to ontogenetic changes in body size. Prey fish, especially cyprinids, banded killifish, and tessellated darter were significant factors in all July analyses of YOY muskellunge presence and abundance. We had expected strong negative associations with predator abundance in July because the average total length of YOY muskellunge in July was 56 mm

(2.2 in.), two thirds the size of the average rock bass (73 mm, 2.9 in.) and half the size of the average yellow perch (93 mm, 3.7 in.), making muskellunge highly susceptible to predation from larger rock bass and yellow perch. In late August predation risk from these potential predators likely declined as muskellunge more than doubled their mean total length to 118 mm (4.7 in.), slightly larger than the average total length of yellow perch (113 mm, 4.5 in.) and 36 mm (1.4 in.) longer than the average rock bass (82 mm, 3.2 in.). Although not directly measured in this study, mortality was substantial as our overall YOY muskellunge CPUE in August was half of that in July (0.64 and 1.09, respectively), despite the fact that twice the area was sampled (due to gear size). The sharp decline in CPUE and the size differences between YOY muskellunge and potential predators provide support for a functional linkage to the strong relationship between YOY muskellunge and dense vegetation. In contrast to muskellunge: prey linkages observed in July, relationships between muskellunge abundance and the fish community in August were nearly non-existent, consisting primarily of associations with bottom fishes, which were numerically rare.

In relation to body size and muskellunge habitat selection, the high growth exhibited by the muskellunge (doubling their mean total length in a 3–4 week period) was likely related to the high abundance of prey. In fact, the strongest relationships between muskellunge and fish variables were the positive associations with cyprinids and the killifish/ tessellated darter group in July. Farrell (1998) studied YOY muskellunge diet from 1994 to 1996, through all years examined, killifish, tessellated darters, and cyprinids, *Notropis* sp., were the three most dominant prey items. *Lepomis* sp. were consumed, but always represented less than 20% of their diet and in 2 of 3 years were less than 5%. In 1994 the majority of prey (47%) in YOY muskellunge stomachs were tessellated darters and in 1995 and 1996 over 50% of their diet consisted of killifish. Similarly, in the present study YOY muskellunge occurrence coincided with three times as many cyprinids and about twice as many killifish and tessellated darters than in hauls where muskellunge were absent. Similarly, fry (fish less than 30 mm (1.2 in.)), were highly abundant in all areas, exceeding 500 individuals per seine haul on average. The lack of correlation in July between muskellunge and potential predators as well as the strong positive associations with prey fish strongly suggests that habitat use is dictated by foraging opportunities rather than predation risk, though the strong relationships to submerged macrophyte coverage is suggestive of both predator avoidance and foraging cover.

Correlations and significant differences in abundance observed in this study, however, cannot be interpreted as cause and effect relationships. Observed relationships could be the result of similar microhabitat utilization, an alternative that can only be evaluated through a full community analysis. The correlative findings presented here describing habitat selection based on prey abundance are supported by experimental work conducted by Szendrey and Wahl (1996) who observed negative selection of centrarchids and preference for cyprinids by young muskellunge. Wahl and Stein (1988) also documented increased growth of muskellunge on a cyprinid/ clupeid diet over a centrarchid diet. The only relationship we observed between muskellunge and the sunfish group was in the August 'fish only' PCA where sunfish were present on the negative side of a secondary PC gradient. Cyprinids, on the other hand, were a dominant factor in differentiating muskellunge habitat. In a similar, qualitative survey of YOY muskellunge nursery habitat in South Georgian Bay, Lake Huron, Craig and Black (1986) observed a somewhat opposite pattern, where pumpkinseed were the most common co-habiting species found in 90% of hauls with YOY muskellunge and bluntnose minnows, *Pimephales notatus*, and blackchin shiners, *Notropis heterodon* (dominant in this study) were found co-existing in only 40% of bays with muskellunge. They did not, however, separate fishes by hauls as we did here, thus on a finer scale, it is not clear what the relationships may have been.

The vegetative community of the nursery bays we studied changed from fine-leafed submerged macrophyte dominance (over extensive *Chara*) in July to dominance of broad-leafed submerged macrophytes (again over extensive *Chara*) in

August. In each month the dominant macrophyte type was the most important positive factor in each analysis. Similar to past qualitative studies (Osterberg 1985; Craig and Black 1986; Bendig 1995), peak YOY muskellunge abundance occurred at low to moderate densities of each vegetation type (~20–60% substrate coverage). It should be pointed out, however, that in our study YOY muskellunge were captured in areas of greater macrophyte coverage than was generally available. Muskellunge were also associated with emergent vegetation. One area in particular, Peos Bay near Cape Vincent, NY, consistently produced the greatest YOY muskellunge catch per haul and was characterized by over 50% emergent coverage directly adjacent to the shoreline, while other stands of extensive emergent vegetation not associated with the shoreline rarely yielded YOY muskellunge.

Chara vulgaris (macroalgae) was the overall dominant cover type in nearly all the study sites, yet muskellunge were negatively related to it, differentially utilizing areas with less *Chara*. Likewise, muskellunge were negatively associated with depth. We captured more muskellunge in shallower water adjacent to shorelines than in deeper water away from shore. Farrell and Werner (1999) also observed a negative relationship between muskellunge abundance and depth within these sites.

The data presented here linking YOY muskellunge presence and abundance to specific physical and fish community habitat components are valuable for evaluating suitable from non-suitable areas to guide population enhancement programs, habitat restoration, and riparian management decisions. These data are the first known to us that connect abundances of other fish species to muskellunge abundance and support the growing interest in and importance of community-level management. Further, because of the overall importance of nearshore vegetation as spawning substrate (Farrell et al. 1996; Farrell 2001) and nursery habitat shared by multiple species, we are in full agreement with Craig and Black (1986) who stress the importance of biologically based riparian land management, as well as, and perhaps more importantly, the regulation of dock building, dredging, and herbicide use in sensitive, biologically important nearshore nursery areas. We recommend that further research evaluate site-level differences in YOY muskellunge production and the factors related to high productivity. Scaling-up to the site-level will likely enlighten us with a new suite of ecological factors such as site exposure, substrate composition, and important patch size information that will certainly improve our understanding of factors influencing muskellunge recruitment processes.

Acknowledgments We would like to thank the Hoosier Muskie Hunters Chapter of Muskies Inc., the Esocid Technical Committee, and Jim Diana for organizing and sponsoring the Ed Crossman Muskellunge Symposium. We also thank our crew at the Thousand Island Biological Station for field assistance and Molly Ramsey and four anonymous reviewers that certainly improved this manuscript. This work was funded through a Federal Aid in Sportfish Restoration Grant # FA-5-R. This is a research contribution of the State University of New York, College of Environmental Science and Forestry, Thousand Islands Biological Station.

References

Bendig A (1995) 1994 St. Lawrence River muskellunge nursery habitat inventory. St. Lawrence River Subcommittee Report to the Lake Ontario Committee, Great Lakes Fishery Commission, pp 83–91

Clapsadl MD (1993) An examination of habitat characteristics influencing the survival of stocked muskellunge, *Esox masquinongy*, in four St. Lawrence River Embayments. Master's thesis, College of Environmental Science and Forestry, State University of New York, Syracuse, NY

Craig RE, Black RM (1986) Nursery habitat of muskellunge in southern Georgian Bay, Lake Huron, Canada. Am Fish Soc Spec Publ 15:79–86

Cushing DH (1990) Plankton production and year-class strength in fish populations: an update of the match/mismatch hypothesis. Adv Mar Biol 26:249–294

Dombeck MP (1986) Muskellunge habitat with guidelines for habitat management. Am Fish Soc Spec Publ 15:208–215

Farrell JM (1998) Population ecology of sympatric age-0 northern pike and muskellunge in the St. Lawrence River. PhD dissertation, College of Environmental Science and Forestry, State University of New York, Syracuse, NY, USA

Farrell JM (2001) Reproductive success of sympatric northern pike and muskellunge in an upper St. Lawrence River Bay. Trans Am Fish Soc 130:796–808

Farrell JM, Klindt R, Cassellman JM, LaPan SR, Werner RG, Schiavone A (this issue) Development, implementation, and evaluation of an international muskellunge management strategy for the upper St. Lawrence River. Environ Biol Fish 00:000–000

Farrell JM, Meade JV, Murry BA (2006) Protracted spawning of St. Lawrence River northern pike (Esox lucius): simulated effects on survival, growth, and production. Ecol Freshw Fish 15:169–179

Farrell JM, Werner RG (1999) Distribution, abundance, and survival of age-0 muskellunge in upper St. Lawrence River nursery bays. N Am J Fish Manag 19:309–320

Farrell JM, Werner RG, LaPan SR, Claypoole KA (1996) Egg distribution and spawning habitat of northern pike and muskellunge in a St. Lawrence River marsh, New York. Trans Am Fish Soc 125:127–131

Garvey JE, Wright RA, Stein RA (1998) Overwinter growth and survival of age-0 largemouth bass (*Micropterus salmoides*): revisiting the role of body size. Can J Fish Aquat Sci 55:2414–2424

Jonckheere BV (1994) Production and survival of juvenile muskellunge (*Esox masquinongy*) in the St. Lawrence River, and the habitat characteristics affecting distribution. Master's thesis, College of Environmental Science and Forestry, State University of New York, Syracuse, NY

Juanes F, Buckel JA, Conover DO (1994) Accelerating the onset of piscivory: intersection of predator and prey phenologies. J Fish Biol 45(Suppl A):41–54

May RC (1974) Larval mortality in marine fishes and the critical period concept. The early life history of fishes. Springer-Verlag, New York

Osterberg DM (1985) Habitat partitioning by muskellunge and northern pike in the international portion of the St. Lawrence River. NY Fish Game J 32:158–166

Paukert CP, Wittig TA (2002) Applications of multivariate statistical methods in fisheries. Fisheries 27(9):16–22

Szendrey TA, Wahl DH (1996) Size-selective survival and growth of stocked muskellunge: effects of predation and prey availability. N Am J Fish Manag 16:395–402

Wahl DH (1999) An ecological context for evaluating the factors influencing muskellunge stocking success. N Am J Fish Manag 19:238–248

Wahl DH, Stein RA (1988) Selective predation by three esocids: the role of prey behavior and morphology. Trans Am Fish Soc 117:142–151

Werner EE, Gilliam JF (1984) The ontogenetic niche and species interactions in size-structured populations. Ann Rev Ecol Syst 15:393–425

Werner RK, LaPan SR, Klindt R, Farrell JM (1990) St. Lawrence River Muskellunge Investigations Phase I—Final Report: Identification of muskellunge spawning and nursery habitat. SUNY-ESF and NYSDEC, Albany, NY

Environ Biol Fish (2007) 79:49–60
DOI 10.1007/s10641-007-9187-8

SPECIAL ISSUE CROSSMAN

Using movements and diet analyses to assess effects of introduced muskellunge (*Esox masquinongy*) on Atlantic salmon (*Salmo salar*) in the Saint John River, New Brunswick

R. Allen Curry · Chad A. Doherty · Timothy D. Jardine · Steven L. Currie

Received: 15 September 2005 / Accepted: 10 December 2006 / Published online: 26 January 2007

Abstract The muskellunge was introduced in the Saint John River system from stockings in a headwater lake in the 1970s. They have migrated down the system as far as the river's first dam, Mactaquac Hydroelectric Facility, at Fredericton and appear to have established several reproducing populations along the river. This exotic invader represents a potential threat to the severely depleted Atlantic salmon stocks in the river. We radio-tracked muskellunge over a 2-year period in the middle reaches. Home ranges extended to ~100 km in both riverine and lacustrine areas, including 78% of individuals translocated upstream of the dam making their way back through the dam successfully. Downstream of the dam, home ranges were <25 km. No spawning areas were detected. An isotope analyses of diet indicated that the large sub-adults and adults had established the greatest proportion of their biomass in a more ^{15}N depleted environment typical of areas farther upstream. Isotope mixing models could not accurately determine the proportion of Atlantic salmon smolts that may have been consumed by muskellunge, but anadromous salmon had ≤7% probabilities of being in the diet. A bioenergetics model suggested ≤5% of the annual food intake by muskellunge occurs during the smolt out-migration period. For the Saint John River, the impacts of growing numbers of muskellunge are multi-faceted creating a complex management challenge. Muskellunge appear to minimally increase predation risk for Atlantic salmon smolts while their increasing numbers are creating a growing recreational fishery and potential threat to the native fish community and ecosystem.

Keywords Radio and acoustic tracking · Stable isotopes · Modeling diets

R. A. Curry (✉) · C. A. Doherty · T. D. Jardine
Canadian Rivers Institute, New Brunswick Cooperative Fish and Wildlife Research Unit and Stable Isotopes in Nature Laboratory, Biology Department, University of New Brunswick, E3C 6E1 Fredericton, NB, Canada
e-mail: racurry@unb.ca

S. L. Currie
New Brunswick Department of Natural Resources, Region 3, E3E IG3 Islandview, NB, Canada

Introduction

Introduction of non-native species can result in significant changes in ecosystem structure and functions if the introduced species becomes pervasive, displacing and eliminating native species. There are many examples of introduced species becoming invasive in aquatic ecosystems. The Great Lakes of North America have experienced some dramatic changes in structure and function of their ecosystems from introductions of sea lamprey, *Petromyzon marinus*, zebra mussels, *Dreissena polymorpha*, and round gobies, *Neogobious melanostomas*, as described for Lake Michigan (Madenjian et al. 2002). We are learning that effects can begin at any trophic level and then permeate through the entire food web of the ecosystem.

Introductions of large, predatory fish such as *Esox* spp. can significantly impact resident fish communities (e.g., He and Kitchell 1990; Wynne 1995). Introductions of large predators such as muskellunge, *Esox maquinongy*, are generally perceived to create dramatic alterations of ecosystems because their impacts typically occur in the realm of recreational fisheries, i.e., where humans are directly affected. For example, smallmouth bass, *Micropterus dolomieui*, introductions into northeastern North American waters have been correlated with changes in lake trout, *Salvelinus namaycush*, population structure (VanderZanden et al. 1999). The invasion of sea lampreys into lakes Erie, Huron, Michigan, and Superior are strongly correlated with the declines in both recreational and commercial fisheries for lake trout (e.g., Krueger et al. 1995). But introductions are not always negative. Species such as muskellunge, smallmouth and striped bass, *Morone saxatilis*, and Pacific salmon, *Oncorhynchus* spp., are highly prized and can support socially and economically acceptable recreational fisheries after their introduction whether introduced through planned management decisions or illegally (Neal et al. 1999; Simonson and Hewett 1999; Tanner and Tody 2002).

The introduction of muskellunge into a headwater lake of the Saint John River (SJR; Fig. 1) was an unintended expansion of a planned management introduction in the province of Quebec (Stocek et al. 1999). The species is now present throughout approximately 500 km of mainstem and tributary rivers and lakes upstream of Fredericton, NB. Increasing numbers of individuals are being captured in the first fish passage system on the river (Mactaquac Hydroelectric Facility, 20 km upstream of Fredericton; see Fig. 2) and by anglers in the head pond stretching 100 km upstream from there (see for example, http://www.muskiesnb.ca).

The species' impact on the river ecosystem is unknown. The river once supported a large population of wild Atlantic salmon, *Salmo salar*, but numbers have declined to <3,000 returning adults and the population is on the verge of being declared endangered (Jones et al. 2004). With so few salmon in the river, any threat to their survival has raised concerns by angler organizations and agencies responsible for managing the salmon fishery. These groups consider muskellunge to be significant predators of salmon, especially out-migrating smolts and small salmon or grilse returning after one sea winter (Canadian Department of Fisheries and Oceans—DFO and Saint John River Management Advisory Commission, personal communication). There is only one report of *Esox* spp. feeding on out-migrating salmonids in reservoirs (Jepsen et al. 1998). DFO presently destroys all muskellunge captured in their fish traps at all dams along the river.

The continued expansion of muskellunge in the SJR system is a dilemma for fisheries managers. No *Esox* spp. is native to the system (the smaller chain pickerel, *Esox niger*, was introduced in the 1800s) and therefore the muskellunge represents a large, exotic predator with the potential to alter the ecosystem (Townsend 2003). In upstream areas, the ecosystem functions being influenced by muskellunge are unknown, but populations are growing in the river and in several lakes where annual surveys capture age 0+ and 1+ muskellunge (E. LeBlanc, New Brunswick Department of Natural Resources, personal communication). Muskellunge diet, as reviewed by Scott and Crossman (1973), typically consists of three main forage fish: yellow perch, *Perca flavescens*, brown bullhead, *Amerius nebulosus*, and sucker species, *Catostomus* spp., all of which occur in abundance in the SJR (Curry and Munkittrick 2005). Across the muskellunge range, they coexist with many salmonid species, but significantly different habitats most probably restricts interspecific interactions. In the SJR, there is a greater potential for contact between muskellunge and Atlantic salmon because of the salmon's upstream and downstream migrations through cool- to warm-water runs, pools, and reservoirs that are not holding habitats for salmon, but where muskellunge reside. How much contact occurs and

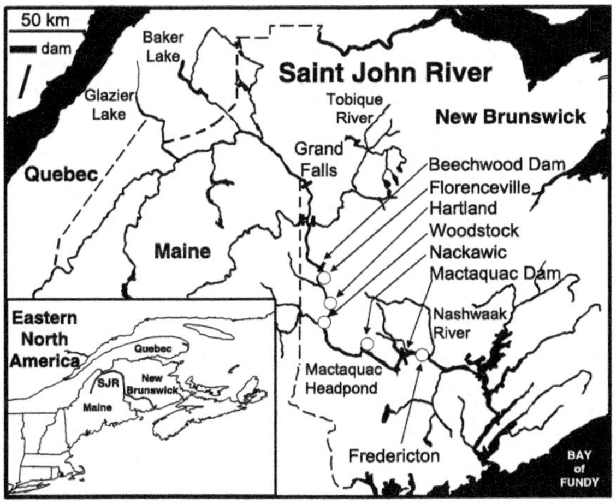

Fig. 1 The Saint John River and landmarks related to muskellunge movements in the middle reach from Florenceville to Fredericton, NB

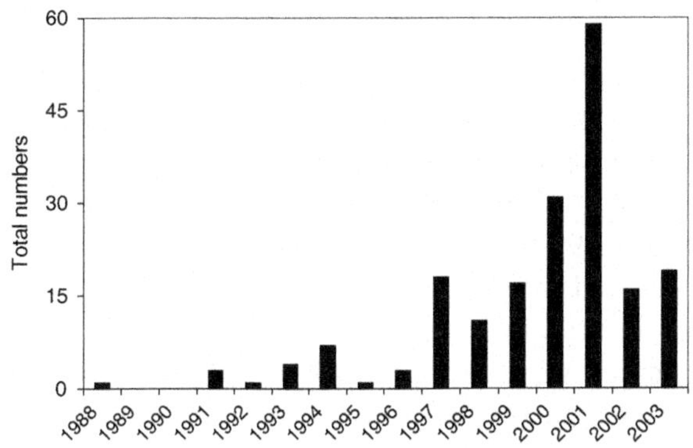

Fig. 2 Total numbers of muskellunge captured annually in the fish trap at the base of the Mactaquac Dam in the Saint John River

whether muskellunge exploit the opportunity to predate Atlantic salmon grilse or smolts is unknown.

Our long-term goal is to determine the impact of muskellunge on the SJR ecosystem. In this study, we continue developing our understanding of their ecology in the river by examining movements and diet patterns in the middle reach of the river where Atlantic salmon smolt production is greatest (Jones et al. 2004). We concentrated on the area downstream of the Mactaquac Hydroelectric Facility, the first dam on the river. Here, the species may have the greatest contact because muskellunge are known to reside in the spillway area of the dam, Atlantic salmon smolts are concentrated and potentially stressed in this area where they have either fallen over the dam or passed through the turbines (there is no fishway), and Atlantic salmon grilse moving upstream are held-up below the dam. Muskellunge predation on Atlantic salmon is potentially greatest in this area.

Study area background

The SJR is >650 km in length with portions of its >55,000 km^2 drainage in Maine, Quebec and New Brunswick (Fig. 1). The fish community consists of >35 species with the greatest diversity downstream of the Mactacquac Dam (Curry and Munkittrick 2005). Muskellunge were stocked as fingerlings multiple times during the 1970s in Lac Frontiere, Quebec (a planned management decision). Muskellunge were first recorded in the system in Maine during the mid-1980s followed by confirmation of the species at the Mactaquac Dam in New Brunswick in 1988 (Stocek et al. 1999). Populations supporting recreational fisheries are established in Glazier and Baker lakes, and the recreational fishery is rapidly expanding at Woodstock (Fig. 1) at the head of the Mactaquac Dam's head pond.

The area used by muskellunge in this study stretched from Fredericton upstream to Hartland, NB (Fig. 1). Fredericton is ~25 km downstream of the Mactaquac Dam where the river is ~1 km wide and up to 3 m deep. From 5 to 15 km downstream of the dam there is a series of large islands creating multiple channels across a 1 km wide channel, but there is only one principle flow channel along the west bank of the river (1–3 m deep). The dam has deep water in its immediate spillway (~3 m) inhabited by many fishes and where a fish trap at the base of the dam is located and operated by Fisheries and Oceans Canada and New Brunswick Power Corporation.

The Mactaquac Dam's head pond stretches 65 km upstream to Woodstock, NB, but the most lacustrine environment extends 30 km upstream to Nackawic, NB (Fig. 1). Upstream of Woodstock is a riverine environment with channel widths from 100 to 200 m and depths up to 2 m. River flow is regulated by the Beechwood Dam upstream of Florenceville, NB, which operates as a daily peaking facility.

Methods

Radio-tagging

Sub-adult to adult muskellunge were taken from the fish trap at the Mactaquac Dam in 2000 (5–24 July) and 2001 (4 July–20 August) and surgically implanted with radio transmitters (Lotek 3-volt micro beeper; 16 mm × 46 mm, 6.7 g). Transmitters, 30 beeps per minute, emitted a frequency of 150–151 MHz, with each tag separated by 20 KHz (minimum battery life of 342 days). Transmitter weight represented less than the 2% of fish mass standard in fish telemetry studies. Fish length ranged from 69.7 to 93.5 cm averaging 83.7 ± 10.5 cm (±1 SD, $n = 18$); weight ranged from 3,000 to 8,450 g averaging 5,470 ± 2,343 g; 50% of muskellunge were females (the largest individuals), 25% males, and 25% immature.

Surgery followed procedures outlined in McKinley and Power (1992) and Curry et al. (2002, 2006). Fish were anesthetized using a clove oil solution of 50 ppm and once loss of equilibrium was reached, a moist V-notched sponge-operating platform was used during tag implantation, to prevent excessive mucus loss. Water flow over the gills was maintained throughout the surgical procedure. An incision, slightly larger than tag diameter was made in the abdomen of the fish approximately mid-way between the pectoral and pelvic girdles. The tag was gently inserted internally into the body cavity, with an external whip antenna fed through an 18-gauge needle, posterior to the incision. Two to four evenly spaced sutures using either a 3–0 or 4–0 needle with non-absorbable black silk were used to close the incision. To ensure incision closure and to help prevent infection, VetBond glue was applied followed by Polysporin antiseptic gel. Three of 6 (2000) and 6 of 12 (2001) tagged fish were released in the reservoir above the dam (≤5 km upstream) and the remainder in the river below the dam (3.5 km downstream).

Attempts to locate fish were conducted on approximately a weekly basis, using either a boat- or truck-mounted multi-element antenna with a Lotek SRX 400 receiver. Detailed notes on fish location in reference to easily distinguishable landmarks were made for calculations of distance traveled. When individuals moved to new locations, we determined exact positions by triangulating signals using different locations. Only individuals tracked successfully for >14 days were used in analyses.

Our tracking data set was not complete enough to produce fine-scale home ranges. We examined locations of individuals in the spring, summer, fall, and winter to determine preferred locations (habitats) within this reach of the river system in distances from the release point (Mactaquac Dam). We define home range as the distance between maximum upstream and downstream locations observed.

Stable isotope analysis

Thirteen muskellunge captured at the Mactaquac Dam during the week of 9 July 2001 and destined for removal from the river were sampled for length, weight, sex (internal verification), and white muscle tissue which provides a good overall representation of assimilated diet (e.g., Hesslein et al. 1993). Tissue samples (~1 g) from the sacrificed muskellunge were excised from above the lateral line and anterior to the dorsal fin. Samples were stored in 95% ETOH until later analyses in the laboratory. Six individuals of all other species captured in the trap were sacrificed and a white muscle sample excised. The samples were supplemented with electrofishing at the dam tailrace to increase sample sizes and species captured.

Tissue samples were dried at 50°C for 48 h and ground into a fine powder with a mortar and pestle. Approximately 0.2 mg aliquots were packed into 3 × 5 mm tin cups. Samples were combusted to gas using a Thermoquest NC 2500 elemental analyzer, and gases were submitted via a continuous flow of helium to a Finnigan MAT Delta Plus isotope-ratio mass spectrometer. Results of ^{13}C:^{12}C and ^{15}N:^{14}N isotope ratios were calculated as parts per thousand by the formula

$$\delta X = [(R_{\text{sample}}/R_{\text{standard}}) - 1] \times 1,000$$

where $X = {}^{15}N$ or ^{13}C and $R = {}^{15}N/{}^{14}N$ or $^{13}C/{}^{12}C$.

Values were said to be enriched when more positive than a comparison sample and depleted when more negative. Replicates of commercially available isotope standards yielded results that indicated sample runs were both accurate and precise (International Atomic Energy Agency). Ten percent of the samples were analyzed in duplicate. All analyses occurred at the Stable Isotope in Nature Laboratory, University of New Brunswick.

Stable isotope mixing model

We used the IsoSource mixing model developed by Phillips and Gregg (2003) to estimate dietary source contributions for muskellunge. We used diet-tissue fractionation factors (0.4‰ for δ^{13}C, 3.4‰ for δ^{15}N) from Post (2002) with six possible food sources. The sources were as follows: (1) stream fishes—a pooling of published data from multiple rivers in NB (Jardine et al. (2005)—Miramichi River Atlantic salmon smolts, Little River slimy sculpin, *Cottus cognatus*, Pointe Wolfe River brook trout, *S. fontinalis*; Kelly et al. (2006)—Miramichi River slimy sculpin; Doucett et al. (1996)—Miramichi River brook trout and blacknose dace, *Rhinichthys atratulus*; and, from this study Saint John River Atlantic salmon smolts); (2) adult anadromous fishes at the Mactaquac Dam including alewives, *Alosa pseudoharengus*, blueback herring, *A. aestivalis*, American shad, *A. sapidissima*, and sea lamprey; (3) forage fishes from the vicinity of the Mactaquac dam, including yellow perch, white sucker, *Catostomus commersoni*, and young-of-the-year alosids; (4) forage fishes (yellow perch and white sucker) from Woodstock 60 km upstream from the dam; (5) forage fishes (yellow perch and white sucker) from Hartland 98 km upstream from the dam, and (6) forage fishes (yellow perch and white sucker) from Florenceville 120 km upstream from the dam. These groups were chosen to represent the existing forage fish community within the limits of muskellunge home ranges as estimated from telemetry. We pooled isotope data for species within categories following Phillips et al. (2005), who recommended pooling to reduce the number of sources for modeling provided the pooled sources were isotopically and functionally similar. This was the case in our study, e.g., our mean values for NB stream fishes (δ^{13}C = –27.1 ± 3.3‰, δ^{15}N = 9.6 ± 0.7‰) were consistent with values reported in literature surveys by France (1995a) for lotic fish δ^{13}C (modal value –28‰) and France (1995b) for freshwater fish δ^{15}N (9 ± 3‰). We also examined additional models with smolts as a separate, one-species category. The model generates feasible combinations of dietary sources that satisfy isotopic mass balance, but does not create unique solutions (Phillips and Gregg 2003).

To assess the isotope mixing model, we modeled the annual muskellunge dietary requirements for the Saint John River (Table 1) using Fig. 3 of Bevelhimer et al. (1985) that predicts food intake (mass of individual) at varied water temperatures. We used seasonal water temperatures in the SJR (RAC, unpublished data). For individuals >4,000 g, we assumed they were mature, spawning, and therefore not feeding for a 2-week period in May when water temperatures were ~10°C (Scott and Crossman 1973). We next calculated the potential quantity of Atlantic salmon smolts that could be consumed and their dietary contribution based on an average smolt size of 40 g and out-migration period of 21 days in May and early June at the Mactaquac Dam (Jones et al. 2004; T. Goff, Fisheries and Oceans Canada, Mactaquac Biodiversity Centre, unpublished data). In these analyses, we calculated the maximum potential rates.

We used estimates of smolt out-migration for the SJR tributaries from Jones et al. (2004). All salmon reproduction is assumed to occur in the tributaries. The Tobique River out-migration is 3,600–9,500 smolts annually (average = 5,700/year) and Nashwaak River out-migrations averages 17,000 smolts/year (average = 8,500/year). There are four similar sized tributaries where Atlantic salmon may reproduce between Grand Falls, NB, and Fredericton, NB; therefore, the estimated total number of Atlantic salmon smolts that out-migrate past Fredericton is on the order of 60,000 (Table 1). Again, these estimates are maximum potential rates.

Table 1 The bioenergetics model of consumption of Atlantic salmon smolts by the muskellunge population downstream of Mactaquac Hydroelectricity Facility

Time period	Total days	Average °C[a]	g^{-1} g of muskellunge^{-1} day^{-1} [b]	Consumption per individual (of given size)		
				500 g	4,500 g	9,000 g
November–March	150	5	0.01	750	6,750	13,500
April	30	10	0.025	375	3,375	6,750
May	30	12.5	0.0375	562	2,531	5,063
June	30	15	0.05	750	6,750	13,500
July–September	90	25	0.125	5,625	50,625	101,250
October	30	15	0.05	750	6,750	13,500
Total yearly intake (g)				8,813	76,781	153,563
Diet during smolt migration period—last 2 weeks of May and first week of June (%)				5.3	3.9	3.9
Total # of 40 g smolts possibly consumed				12	74	148
Estimated annual diet of smolts (g) from isotope mixing model (average = 64%)				5,640	49,140	98,280
Estimated annual diet of smolts (#) from isotope mixing model				141.0	1,229	2,457
Estimates for a population of 100 muskellunge at base of dam (average size = 4,500 g)						
Estimated annual diet of smolts (#)—isotope mixing model				122,850		
Estimated annual diet of smolts (#)—bioenergetics model				7,383		
Estimate of # of smolts passing Mactaquac Dam[c]				~60,000		

[a] RAC, unpublished data

[b] from Bevelhimer et al. (1985)

[c] Jones et al. (2004)

Results

A total of 18 sub-adult to adult muskellunge were tracked over a 2-year period. They were not located on each tracking date, but the average number of dates when they were located was 15 ± 8. Total tracking periods averaged 293 ± 103 days. Their relative home ranges were 2.5–90.0 km and averaged 28.8 ± 25.2 km. The median and mode were 23 and 10 km, respectively.

Most muskellunge were located in the areas downstream of the Mactaquac Dam (Fig. 3). Nine individuals were trans-located upstream of the Mactaquac Dam and seven of these traveled back over or through the dam facility. The majority of muskellunge downstream of the dam exhibit skin lacerations and fin damage consistent with passage through a sluiceway turbine. Most were located in the islands along the east bank ~15 km from the dam during May when river temperatures were 5–10°C and Atlantic salmon smolts were passing through the reach (see Jones et al. 2004). Some remained and some dispersed upstream to the dam area and the main river channel by late May and into June. In winter, they preferred the slow, deeper reaches from the dam downstream to Fredericton.

Upstream of the dam, there were fewer fish to track (45 of our 268 total data points when fish were located = 17%), but these nine individuals traveled in proximity of the dam for short periods after being released and then either dispersed upstream to the Nackawic area (lacustrine) or upstream to Woodstock and Hartland (riverine) and seven eventually traveled downstream past the dam (Fig. 3). In winter, locations used upstream of the dam were lacustrine environments. One muskellunge, 89.5 cm fork length and 6,417 g wet weight (estimated age 5+; Stocek et al. 1999) traveled past the dam downstream to Fredericton after its release in the summer, and then up the Nashwaak River 26 km from the dam where it resided for the summer. Its "home pool" was 30 m long, 15 m wide, and <2 m deep.

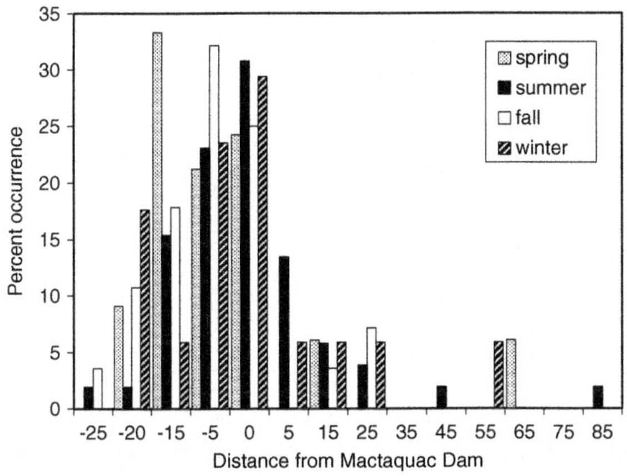

Fig. 3 Locations of radio-tagged muskellunge as % occurrence of tagged individuals in each of four seasons in the Saint John River, July 2000 to July 2003. Distances are from the Mactaquac Dam

Muskellunge captured at the base of the Mactaquac Dam had average $\delta^{13}C = -25.8 \pm 1.7$ and $\delta^{15}N = 13.9 \pm 1.0$ ($n = 12$). Values for other species and groups are given in Table 2. The values for the muskellunge adjusted for diet-tissue fractionation (0.4‰ for $\delta^{13}C$, 3.4‰ for $\delta^{15}N$; Post 2002) place them within the dietary source polygon for the categories modeled (Fig. 4).

According to the mixing model, stream fishes were the most important item in the diet of muskellunge, comprising 26–64% of the diet (Fig. 5). Forage fishes from the Mactaquac Dam were also important, comprising 2 to 48% (Fig. 5). All other groups were less important, with 0% contributions as the most feasible value (Fig. 5). When Atlantic salmon smolts were added separately as a dietary component, they were predicted to be an important item for muskellunge, comprising 38–80% of the diet.

The bioenergetics modeling of dietary requirements for muskellunge in the SJR downstream of the Mactaquac Dam suggests the total intake is 9–170 kg/annum for individuals ranging from 0.5 to 10.0 kg total body weight. For the 3-week period at the end of May and beginning of June (the period of muskellunge spawning and out-migration of Atlantic salmon smolts), total food intake ranged from 4 to 5% of the total annual intake for an individual muskellunge, which represents 12–164 Atlantic salmon smolts consumed if they were the only diet item (for 0.5–10.0 kg muskellunge, respectively). The isotope mixing model suggested the muskellunge diet ranged from 38 to 80% Atlantic salmon smolts. At 38% of a muskellunge's diet ranges from 33 kg or 84 smolts to 65 kg or 1,621 smolts if they were the entire diet (for 0.5 and 10.0 kg muskellunge, respectively). A smolt diet of 80% ranges from 71 kg or 176 smolts to 137 kg or 3,415 smolts (for 0.5 and 10.0 kg muskellunge, respectively). Assuming the average size for muskellunge downstream of the Mactaquac Dam in the Saint John River was 4.5 kg (average for tagged and isotope-sampled muskellunge), 100 individuals inhabit the reach (the most observed in a year was 58; Fig. 2), and only smolts were consumed, then the bioenergetics model predicts that 7,400 smolts would be consumed and the mixing model predicts 73,000–154,000 smolts would be consumed annually by the muskellunge downstream of the dam.

Table 2 Stable isotope levels for selected fishes captured at the base of the Mactaquac Dam, Saint John River, 1999 (average, 1 SD, and sample size)

Muskellunge		Atlantic salmon smolts		Forage fishes[a]		Smallmouth bass		Anadromous fishes[b]	
$\delta^{13}C$	$\delta^{15}N$	$\delta^{13}C$	$\delta^{15}N$	$\delta^{13}C$	$\delta^{15}N$	$\delta^{13}C$	$\delta^{15}N$	$\delta^{13}C$	$\delta^{15}N$
–25.8	13.9	–26.4	9.8	–25.6	11.5	–26.0	14.3	–19.1	12.6
1.7	1.0	1.9	0.9	1.7	0.8	0.8	1.4	1.2	0.9
12	12	19	19	97	97	21	21	71	71

[a] Forage fishes = *Alosa* spp. (young-of-the-year), brown bullhead, banded killifish (*Fundulus diaphanous*), creek chub (*Semotilus atromaculatus*), common shiner (*Luxilus cornutus*), golden shiner (*Notemigonus chrysoleucas*), lake chub (*Couesius plumbeus*), pumpkinseed sunfish (*Lepomis gibbosus*), white perch (*M. americana*), white sucker, and yellow perch

[b] Anadromous fishes = *Alosa* spp. (adults), sea lamprey, and striped bass

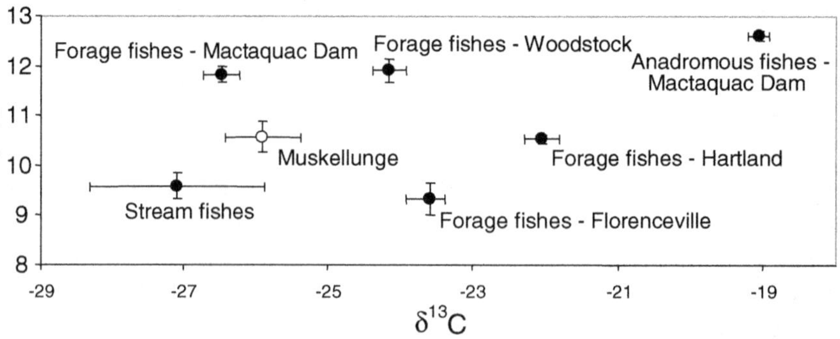

Fig. 4 δ^{15}N and δ^{13}C isotope values (average and 1 SE) for the fish community along the Saint John River and for a composite sample of stream fishes from New Brunswick rivers. Muskellunge values were adjusted by diet-tissue fractionation factors of 0.4‰ for δ^{13}C and 3.4‰ for δ^{15}N (Post 2002)

Discussion

Our tracking of muskellunge captured at the Mactaquac Dam indicates that there is a population of muskellunge in the middle reaches of the SJR from about Florenceville, NB, to Fredericton, NB. Large sub-adults and adults readily traveled up to 100 km upstream of the Mactaquac Dam in both lacustrine and river environments and passed over or through the dam. The apparent home ranges were variable in size extending from a few kilometers to 90+ km traveled along the river within a 2-year period, which is consistent with other populations (Strand 1986; LaPan et al. 1996). SJR muskellunge used a diversity of broad-scale habitats within their range, e.g., lacustrine, large river, and a small tributary. If they had a preferred habitat, it was not readily apparent; however, it would appear that a general, downstream movement by these large muskellunge was the trend given that 78% of individuals that were trans-located upstream of the dam made their way back downstream of the dam.

The muskellunge diet inferred from the stable isotope analyses suggests individuals have not been at the dam or in the downstream reaches for much of their lives. Rather, the isotope analyses support an origin upstream of Florenceville, or perhaps from one or more tributaries along the river. The difference in δ^{15}N and δ^{13}C between muskellunge and its reported principle forage of yellow perch and white sucker (Scott and Crossman 1973) for all sites along the main stem of the river was less than the 3.4‰ and 0.4‰, respectively, which is the typical difference between fish predators and their prey (Post 2002). Only the pooled stream fishes category appeared to align with muskellunge as an important dietary item. White muscle tissue reflects the amalgamated diet history of individual ectotherms over the longer-term of several years (Hesslein et al. 1993). If muskellunge in the middle and lower reaches arrive from upstream at age ≥ 2+, then they have already added a large proportion of their biomass in a more ^{15}N-depleted environment upstream (observed by Curry and Munkittrick 2005), i.e., their δ^{15}N should fit best in an upstream food web. Alternatively, reproduction and juvenile life history periods may occur within a tributary to the main stem because forested tributary environments also tend to be depleted in ^{15}N relative to larger, more developed main stems of rivers (Anderson and Cabana 2005). Atlantic salmon smolts are principally recruited from the tributary streams (Jones et al. 2004) and their average δ^{15}N = 9.8 ± 0.9‰ (from captures at the Mactaquac Dam). A top predator in their tributary streams is predicted to have δ^{15}N ~ 13.3‰ and the Mactaquac Dam muskellunge have an average δ^{15}N = 13.9 ± 1.0‰. There was no observational evidence suggesting muskellunge reproduce and juveniles rear in any of the middle and lower reach tributaries to the SJR, including no reports from the numerous anglers in these tributaries (one tracked muskellunge probably

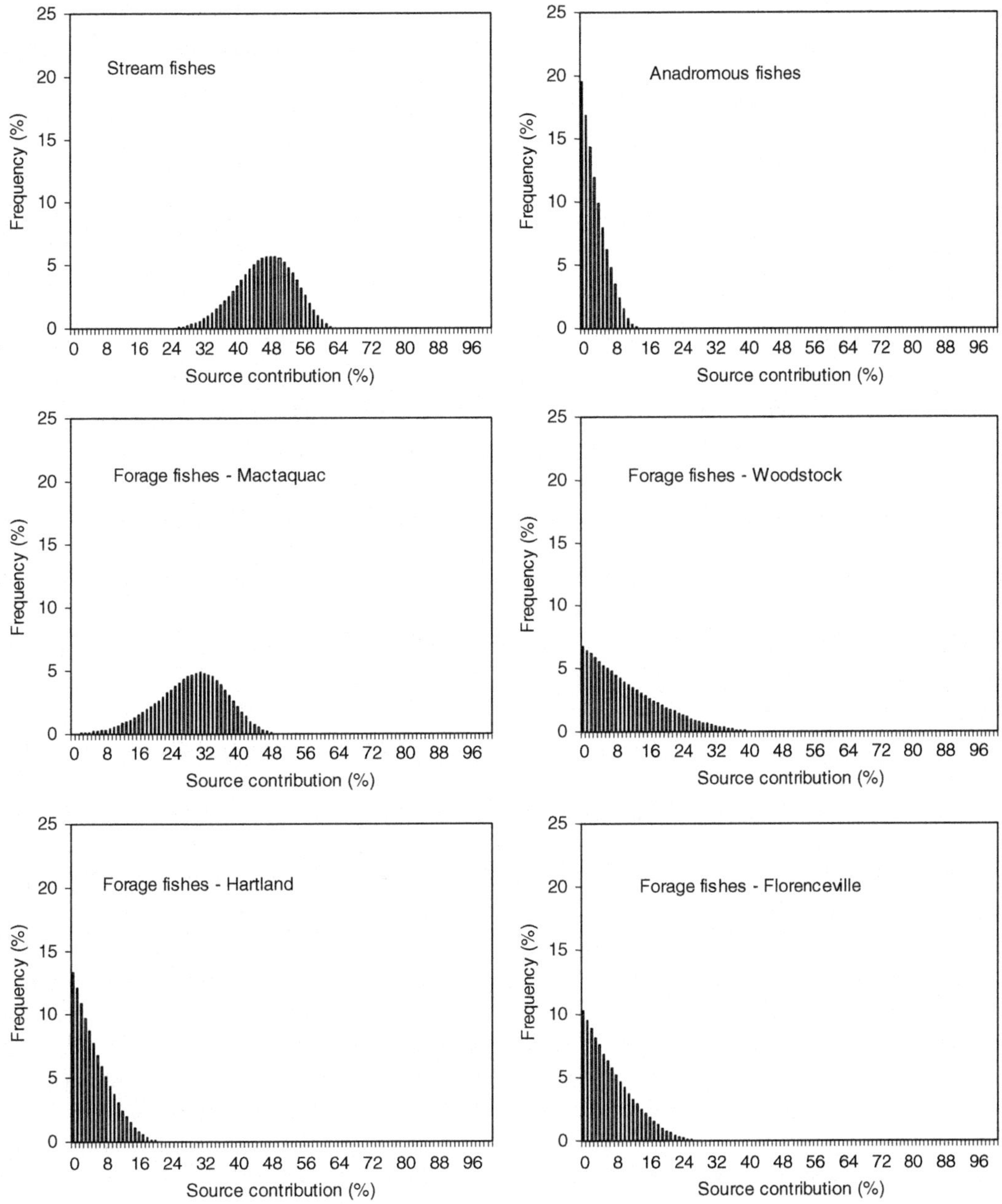

Fig. 5 Proportional estimates of feasible source contributions of various food items to the diet of muskellunge in the Saint John River estimated from a stable isotope mixing model

age 5+ moved into and resided in the Nashwaak River at Fredericton for a summer). Our evidence indicates there is a detachment between muskellunge and the food webs in the middle and lower reaches.

The disconnect causes problems for interpreting isotope analyses and resultant mixing models to predict muskellunge diet. The mixing model output suggested a large contribution to muskellunge diets by Atlantic salmon smolts (38–80%) or "stream fishes", with minor contributions from forage fishes and negligible contributions from anadromous fishes. Application of the bioenergetics models of Bevelhimer et al. (1985) contradicts the isotope mixing model for the smolt. The mixing model predicts smolt consumption could be 73,000–154,000, the bioenergetics models predict 7,400 smolts could be consumed, and total annual out-migration of smolts was estimated to be ~60,000. While the isotope mixing models suggest that muskellunge may be a major threat to Atlantic salmon populations in the river via consumption of smolts (but not grilse as the isotope signatures indicate), the smolt consumption estimate is quite doubtful.

Several factors in the isotope analysis must be considered as we attempt to revolve the complex interactions between muskellunge and their prey. First, mixing models are highly sensitive to diet-tissue fractionation for ^{13}C and ^{15}N (Connolly et al. 2005). In our analysis, we used mean values from the literature (0.4‰ for C, 3.4‰ for N—Post 2002); however, there was a large degree of uncertainty associated with those estimates, as fractionation is known to vary with nutritional state, age, etc. (McCutchan et al. 2003; Vanderklift and Ponsard 2003). We considered using fractionations for more narrowly defined species, but the literature contains no estimates of fractionation for species in the Esocidae family. Running the model using a lower fractionation estimate of 2.0‰ from McCutchan et al. (2003) returned similar dietary mixtures. Second, the isotopic signature of Atlantic salmon smolts was most similar to tributary food webs, which suggests and supports the hypothesis that muskellunge are recruited to the mainstem, either from upstream areas in general or tributaries. Finally, the abundance and biomass of Atlantic salmon smolts was extremely low relative to other potential prey for muskellunge in the system (Curry and Munkittrick 2005). Smolt availability for consumption is temporally limited to a brief pulse during their spring out-migration, which overlaps with muskellunge spawning. Availability is spatially limited because smolts move downstream rapidly and at night (Bakshtanskiy et al. 1980; Faangstam 1993) in the main channel of the river, while muskellunge are spawning in shallow, flooded, backwater areas (Scott and Crossman 1973; Dombeck 1986), e.g., the flooded islands upstream of Fredericton where they were located in spring, and feeding typically occurs in early morning and late evening at this time of year (Miller and Menzel 1986). Muskellunge most probably consumed some Atlantic salmon smolts as do esocids in other systems (Warner et al. 1968; Jepsen et al. 1998), but grilse or adult salmon have a low probability of being consumed based on the isotope signatures of anadromous fishes in the SJR. Many other fish predators that are significantly more abundant in the SJR (Curry and Munkittrick 2005) can consume smolts such as striped bass, *M. saxatilis* (Blackwell and Juanes 1998), smallmouth bass (Naughton et al. 2004), brown trout, *Salmo trutta* (Huntingford et al. 1988), and brook charr, *S. fontinalis* (Mohler et al. 2002). The best estimate of smolt consumption by muskellunge predicted a total of 7,400 smolts consumed downstream of the Mactaquac Dam if every muskellunge ate smolts only. Given these latter requirements, it is highly improbable that muskellunge have a significant effect on the salmon population in this reach of the SJR.

Since muskellunge began appearing at the Mactaquac Dam, only large sub-adults and adults have been captured. The number of captures has been increasing (although down in 2002 and 2003) and the population has grown to support a fishery and sport fishing association in the Woodstock, NB area (catch per unit effort = 0.8 muskellunge/angler-hour; C. Raynor, SJR Chapter, Muskies Canada, Inc., personal communication). Interestingly, all muskellunge recently captured in these reaches were larger individuals >3 kg. The smallest individual reported by the sport fishing association was 2.5 kg. Stocek et al. (1999) report two to five juveniles from the area and possibly one young-of-the-year from the islands at Fredericton (species identification was unconfirmed), but the majority were age 3+ and 4+ (63.8–93.4 cm total length, 1,850–6,275 g wet weight). We have conducted multiple fish surveys in the area from Florenceville to Fredericton, and specifically targeted juvenile esocids in the islands upstream of Fredericton where radio-tagged muskellunge were observed in spring during the time when spawning should have occurred; we have not yet located any juvenile muskellunge. Upstream of Grand Falls, NB, juveniles are regularly captured in fish community surveys and by local anglers (RAC, unpublished data; E. LeBlanc, personal communication). Given the absence of juveniles in the reach from Florenceville to Fredericton and the tendency for tracked individuals to move downstream, it may be that muskellunge are dispersing from reaches upstream of Florenceville where they are reproducing successfully, to the middle and downstream reaches. Our stable isotope results support this claim.

The complete effects of the introduced muskellunge in the SJR ecosystem will take time to resolve. Prior to the muskellunge's introduction, the chain pickerel, *E. niger* (1800s) and smallmouth bass (1940s) were introduced to the lower river (Livingstone 1951). These predators appear to have had no impact on Atlantic salmon in the river, which began to decline in population size in the 1980s. Regional fisheries managers have considered neither of these introduced predators be a significant factor in the management of Atlantic salmon. While the addition of another large predator suggests predation-risk will increase, it is important to understand that muskellunge numbers are low and their spatial and temporal habitats minimally overlap with Atlantic salmon. The greater challenge for managers and the river ecosystem will most probably be the muskellunge's longer-term, potential threats to native fishes and possible conflicts with the growing recreational fishery for muskellunge.

Acknowledgements We thank M. Gautreau, A. Halford, K. Roach, and C. Cronin for muskellunge collections and tracking and T. Goff and S. Ratelle, Fisheries and Oceans, Canada—Mactaquac Biodiversity Centre, for muskellunge and fish collections at the Mactaquac Dam. Reviews by R. Jones, J. Bagnall, C. Ayer, and two anonymous reviewers were most useful.

References

Anderson C, Cabana G (2005) δ^{15}N in riverine food webs: effects of N inputs from agricultural watersheds. Can J Fish Aquat Sci 62:333–340

Bakshtanskiy EL, Nesterov VD, Neklyudov MN (1980) The behavior of young Atlantic salmon, *Salmo salar*, during downstream migration. J Ichthyol 20:93–100

Bevelhimer MS, Stein RA, Carline RF (1985) Assessing significance of physiological differences among three Esocids with a bioenergetics model. Can J Fish Aquat Sci 42:57–69

Blackwell BF, Juanes F (1998) Predation on Atlantic salmon smolts by striped bass after dam passage. N Am J Fish Manage 18:936–939

Connolly RM, Hindell JS, Gorman D (2005) Seagrass and epiphytic algae support nutrition of a fisheries species, *Sillago schomburgkii*, in adjacent intertidal habitats. Mar Ecol Prog Ser 286:69–79

Curry RA, Munkittrick KR (2005) Fish community responses to multiple stressors along the Saint John River, New Brunswick, Canada. In: JN Rinne, R Calamusso, R Hughes (eds) Historical changes in large river fish assemblages of the Americas, vol 45. American Fisheries Society Symposium, pp 505–521

Curry RA, van de Sande J, Whoriskey FG (2006) Temporal and spatial habitats of anadromousbrook charra in the Laval River and its estuary. Env Biol Fish 76:301–370

Curry RA, Sparks D, VandeSande J (2002) Movement patterns of a riverine population of brook trout. Trans Am Fish Soc 131:551–560

Dombeck MP (1986) Muskellunge habitat with guidelines for habitat management. Am Fish Soc Spec Publ 15:208–215

Doucett RR, Power G, Barton DR, Drimmie RJ, Cunjak RA (1996) Stable isotope analysis of nutrient pathways leading to Atlantic salmon. Can J Fish Aquat Sci 53: 2058–2066

Faangstam H (1993) Individual downstream swimming speed during the natural smolting period among young of Baltic salmon (*Salmo salar*) Can J Zool 71:1782–1786

France R (1995a) Critical examination of stable isotope analysis as a means for tracing carbon pathways in stream ecosystems. Can J Fish Aquat Sci 52:651–656

France R (1995b) Stable nitrogen isotopes in fish: literature synthesis on the influence of ecotonal coupling. Estuar Coast Shelf Sci 41:737–742

He X, Kitchell JF (1990) Direct and indirect effects of predation on a fish community: a whole-lake experiment. Trans Am Fish Soc 119:825–835

Hesslein RH, Hallard KH, Ramlal P (1993) Replacement of sulfur, carbon, and nitrogen in tissue of growing broad whitefish (*Coregonus nasus*) in response to a change in diet traced by δ^{34}S, δ^{13}C, and δ^{15}N Can J Fish. Aquat Sci 50:2071–2076

Huntingford FA, Metcalfe NB, Thorpe JE (1988) Choice of feeding station in Atlantic salmon, *Salmo salar*, parr: Effects of predation risk, season and life history strategy. J Fish Biol 33:917–924

Jardine TD, Gray MA, McWilliam SM, Cunjak RA (2005) Stable isotope variability in tissues of temperate stream fishes. Trans Am Fish Soc 134:1103–1110

Jepsen N, Aarestrup K, Okland F, Rasmussen G (1998) Survival of radio-tagged Atlantic salmon (*Salmo salar* L.) and trout (*Salmo trutta* L.) smolts passing a reservoir during seaward migration. Hydrobiologia 371/372:347–353

Jones, RA, Anderson L, Goff T (2004) Assessment of Atlantic salmon stocks in southwest New Brunswick: an update to 2003. Canadian Science Advisory Secretariat Research Document 2004/019, pp 70

Kelly MH, Hagar WG, Jardine TJ, Cunjak RA (2006) Non-lethal sampling of sunfish and slimy sculpin for stable isotope analysis: how scale and fin tissue compare with muscle tissue. N Am J Fish Manage 26:921–925

Krueger CC, Jones ML, Taylor WW (1995) Restoration of lake trout in the Great Lakes: challenges and strategies for future management. J Great Lakes Res 21:1

LaPan SR, Schiavone A, Werner RG (1996) Spawning and post-spawning movements of St. Lawrence River muskellunge, pp 73–82. In: Kerr SJ, Olver CH (eds) Managing Muskies in the '90s. Workshop Proceedings. Ontario Ministry of Natural Resources, Southern Region Science and Technology Transfer Unit Workshop Proceedings. WP-007, 169pp

Livingstone DA (1951) The freshwater fishes of Nova Scotia. N S Inst Sci 23:1–89

Madenjian CP, Fahnenstiel GL, Johengen TH, Nalepa TF, Vanderploeg HA, Fleischer GW, Schneeberger PJ, Banjamin DM, Smith EB, Bence JR, Rutherford ES, Lavis DS, Robertson DM, Jude DJ, Ebener EP (2002) Dynamics of the Lake Michigan food web, 1970–2000. Can J Fish Aquat Sci 59:736–753

McCutchan JH Jr, Lewis WM Jr, Kendall C, McGrath CC (2003) Variation in trophic shift for stable isotope ratios of carbon, nitrogen, and sulfur. Oikos 102:378–390

McKinley RS, Power G (1992) Transmitter attachment/implant—laboratory manual. University of Waterloo, Environmental Research Department, Waterloo, ON

Miller ML, Menzel BW (1986) Movement, activity, and habitat use patterns of muskellunge in West Okoboji Lake, Iowa. Am Fish Soc Spec Publ 15:51–61

Mohler JW, Millard MJ, Fletcher JW (2002) Predation by captive wild brook trout on calcein-marked versus nonmarked Atlantic salmon fry. N Am J Fish Manage 22:223–228

Neal JW, Noble RL, Rice JA (1999) Fish community response to hybrid striped bass introduction in small warmwater impoundments. N Am J Fish Manage 19:1044–1053

Naughton GP, Bennett DH, Newman KB (2004) Predation on juvenile salmonids by smallmouth bass in the lower Granite Reservoir system, Snake River N Am. J Fish Manage 24:534–544

Phillips DL, Gregg JW (2003) Source partitioning using stable isotopes: coping with too many sources. Oecologia 136:261–269

Phillips DL, Newsome SD, Gregg JW (2005) Combining sources in stable isotope mixing models: alternative methods. Oecologia 144: 520–527

Post DM (2002) Using stable isotopes to estimate trophic position: models, methods, and assumptions. Ecology 83:703–718

Scott WB, Crossman EJ (1973) Freshwater fishes of Canada. Bulletin 184. Fisheries Research Board of Canada, Ottawa

Simonson TD, Hewett SW (1999) Trends in Wisconsin's muskellunge fishery. N Am J Fish Manage 19:291–299

Stocek RF, Cronin PJ, Seymour PD (1999) The muskellunge, *Esox masquinongy*, distribution and biology of a recent addition to the ichthyofauna of New Brunswick. Can Field Nat 113:230–234

Strand RF (1986) Identification of principle spawning areas and seasonal distribution and movement of muskellunge in Leech Lake Minnesota. Am Fish Soc Spec Publ 15:62–73

Tanner HA, Tody WH (2002) History of the Great Lakes salmon fishery: a Michigan perspective. In: Lynch KD, Jones ML, Taylor WW (Eds) Sustaining North American salmon: perspectives across regions and disciplines. American Fisheries Society, Bethesda, pp 139–154

Townsend CR (2003) Individual, population, community, and ecosystem consequences of a fish invader in New Zealand streams. Conserv Biol 17:38–47

Vanderklift MA, Ponsard S (2003) Sources of variation in consumer-diet δ^{15}N enrichment: a meta analysis. Oecologia 136:169–182

VanderZanden MJ, Casselman JM, Rasmussen JB (1999) Stable isotope evidence for the food web consequences of species invasions in lakes. Nature 401:464–467

Warner K, AuClair RP, DeRoche SE, Havey KA, Ritzi CF (1968) Fish predation on newly stocked landlocked salmon. J Wildl Manage 32:712–717

Wynne F (1995) Outlook for the culture of muskellunge, northern pike and chain pickerel in Kentucky, USA. World Aquac 26:41–44

Environ Biol Fish (2007) 79:61–69
DOI 10.1007/s10641-006-9120-6

CROSSMAN

Characteristics of Ontario muskellunge (*Esox masquinongy*) fisheries based on volunteer angler diary information

Steven J. Kerr

Received: 7 September 2005 / Accepted: 29 June 2006 / Published online: 29 August 2006

Abstract This paper consolidates and summarizes information on Ontario muskellunge, *Esox masquinongy*, sport fisheries derived from angler diary programs sponsored by the Ontario Ministry of Natural Resources (OMNR) and Muskies Canada Incorporated (MCI) from 1979 to 2004. Interest in muskellunge as a sport fish has increased substantially over the past 10–15 years. Muskellunge catches were found to be highly correlated with angling effort. Catch rates, expressed in terms of catch-per-unit-of-effort (CUE), have improved over the past decade to the point where, in 2001, the provincial CUE was 0.119. The long term catch rate is 0.069. This improvement in angling quality is attributed to new minimum size limit regulations and increased catch-and-release angling practices. Over the period from 1979–2004, release rates by muskellunge anglers have averaged 94%. Based on a sample size of 9,499 fish, the mean size of angled muskellunge over the past 26 years was 37.0 inches (94 cm). Numerous fish exceeding 50 inches (127 cm) are angled from Ontario waters each year. The incidence of lymphosarcoma, a highly contagious, malignant blood cancer, has averaged only 2% since 1979. Based on an analysis of this information, Ontario's muskellunge fisheries appear to be stable and sustainable. Volunteer angler diary programs provide an accurate and cost-effective means to monitor the status of muskellunge fisheries in Ontario.

S. J. Kerr (✉)
Ontario Ministry of Natural Resources, 300 Water Street, 7000, K9J 8M5 Peterborough, Ontario, Canada
e-mail: steve.kerr@mnr.gov.on.ca

Keywords Esox masquinongy · Angler diary · Recreational fishery · Catch-and-release

Introduction

Muskellunge, *Esox masquinongy*, are a highly valued species for many anglers. There are 302 lakes and 105 streams and rivers in Ontario which are known to support muskellunge populations (OMNR 1987, Kerr 2001). Muskellunge are found in the southern portion of the province, the Great Lakes, and northwestern Ontario (Fig. 1).

Muskellunge were stocked in over 275 southern Ontario waters between 1940 and 1990. Introductions served to establish muskellunge populations in at least 71 inland lakes. Stocking was discontinued in 1990 and muskellunge fisheries are currently based on naturally reproducing stocks.

Regulations for muskellunge include catch and possession limits, closed seasons and minimum size limits. Minimum size limits must conform to one of five benchmarks based on the biological characteristics of the population and management

Fig. 1 Distribution of muskellunge lakes in Ontario. Map reproduced from Mandrak and Crossman (1992) with the permission of N. E. Mandrak

objectives for the fishery (OMNR et al. 1999). There are several Ontario waters which have a 54 inch (137 cm) minimum size limit and have the potential to produce muskellunge of world record proportions (Casselman et al. 1999).

Assessment of muskellunge regulations or evaluation of muskellunge population status is difficult. The low density of many populations and the solitary nature of the fish make muskellunge difficult to capture using traditional sampling techniques such as index netting. Muskellunge are relatively difficult to angle and traditional creel surveys do not adequately estimate muskellunge angling effort or success. Finally, muskellunge anglers comprise a relatively small proportion of licenced anglers in the province.

Angler diaries have been found to be useful for specialized fisheries, such as muskellunge, where angling effort is sporadic and where catches are often low (Cooke et al. 2000). Angler diary information is usually based on completed trips so that accurate information can be obtained on angling effort and catches (Anderson and Thompson 1991).

Angler diaries have been used to monitor muskellunge fisheries in other jurisdictions including Illinois (Ferencak 2004), Minnesota (Younk and Cook 1992), North Carolina (Borawa 1990), and Wisconsin (Hacker 1973). In Ontario, angler diary programs have been used for muskellunge in several waters including Lake of the Woods (Duffy and Mosindy 2001), Lac Seul (Foster and Colby 1999), Lake St. Clair (MacLennan 1996), Rice Lake (Lewis and Potter 1980), the St. Lawrence River (Hart 1980, Schiavone 1987), and the Kawartha Lakes (OMNR 1999).

Muskies Canada Incorporated (MCI), an angling organization dedicated to protecting and enhancing muskellunge sport fisheries, initiated a volunteer angler diary program in 1979 and it's members have maintained records since that time. This paper has been prepared to consolidate and summarize angler diary information for muskellunge which has been collected from both Ontario Ministry of Natural Resources (OMNR) sponsored programs as well as the MCI angler diary program.

Materials and methods

Muskellunge angler diary information was obtained from two basic sources: OMNR sponsored projects and the MCI angler diary program.

OMNR projects involved the cooperation of anglers and tourist lodge operators and were directed at specific fisheries. In northwestern Ontario, guides and lodge owners maintained logs for their guests on lakes including Lake of the Woods and Lac Seul. In southern Ontario, volunteer anglers recorded information on their muskellunge fishing activities. Long-running angler diary programs have been conducted on several southern Ontario waters including Lake St. Clair and the St. Lawrence River. In both cases, information was submitted by participants at the end of the fishing season.

Traditionally, MCI members used two forms for recording information. The angler log was used to record information on fishing activity such as the number of active anglers, the length of time fishing, and their catch. The release form was used to record biological information, such as total length, girth, incidence of lymphosarcoma, and sex of fish, for a muskellunge which had been caught. After a joint review of diary design in

2004, a new format, which included a small, plasticized log book with waterproof record forms, was implemented. Modifications to the diary included a highlighted section illustrating mandatory information requirements, instructions for completing the form including the importance of recording unsuccessful fishing trips, and an example of a properly completed log. In recent years, MCI has also instituted incentive programs for those who have completed and submitted their angler diary by a specified date. Completed log forms are collected by Release Directors of each MCI chapter and forwarded to the National Release Director who then provides this information to a designated contact with the Ministry of Natural Resources.

In both programs, participants were specialist anglers who were targeting muskellunge. Participating anglers were asked to enter information at the end of their fishing trip and frequently reminded to record information for all fishing trips regardless of success. In recent years, an annual summary has been prepared and disseminated to participants in an effort to promote continued participation.

For each year, data were pooled for all waterbodies and used to calculate total reported angling effort and catches. Pooled data were also used to compute a provincial catch rate and determine the mean size of angled muskellunge.

Angling quality was described by both catch-per-unit-of-effort (CUE) and the number of hours required to catch a fish. CUE was calculated in terms of the number of muskellunge caught per rod hour of angling effort.

Results

Muskellunge angling information was collected for a total of 93 different waterbodies. Data was assembled for individual waterbodies on an annual basis. Some data were poorly reported and, in a few instances, some information was omitted. Entries missing key information were not used in some analysis.

The number of participants varied among programs and from year to year. The OMNR program on Lake of the Woods included up to 20 different fishing resorts. On Lac Seul, a total of 89 fishing parties participated in the three year program. Participants in the multi-year Lake St. Clair program ranged from 9–43 individuals annually. Participation rates by MCI members has improved in recent years to the point where, in 2004, completed logs were received from 169 different members.

Reported angling effort and catches

Between 1979 and 2004, volunteer anglers reportedly exerted 360,851 rod hours of angling effort seeking muskellunge on 93 different waterbodies (Table 1, 2). The total reported muskellunge catch during that time period was 24,799 fish. Angling effort and catches varied substantially among years.

Reported fishing effort was greatest on waterbodies which included the Ottawa River, Georgian Bay, Rideau River, Lac Seul, Lake of the Woods, Lake St. Clair and several of the Kawartha lakes.

There was a clear and distinct relationship between fishing effort by MCI anglers and their reported catch. Muskellunge catches increased as fishing effort increased and decreased in years of reduced angling effort. Similar observations have been reported by Foster and Colby (1999).

Muskellunge seasons in Ontario traditionally open in mid-late June and extend until the late fall. Angling effort usually peaked early in the season but was relatively consistent throughout the remainder of the fishing season. There are indications that fishing activity during the fall (e.g., October-November) is increasing.

Angling success

Angling success, expressed in terms of catch-per-unit-of-effort (CUE), varied among lakes and between years. Based on a consolidation of reported effort and catches over the past twenty-six years, the provincial CUE for muskellunge was 0.069 meaning it took an average of 14.6 h to catch a muskellunge. This is substantially higher than CUE values of 0.038 and 0.027, respectively, reported from muskellunge angler diary programs in North Carolina (Borawa 1990) and Minnesota (Younk and Cook 1992).

Table 1 Angling effort and catches reported by cooperative anglers during volunteer angler diary programs in Ontario, 1979–2004. CUE is expressed in terms of the number of muskellunge caught per rod hour of fishing effort. Angling effort values have been rounded

Year	# Waters fished	Reported angling effort (rod hours)	Reported muskellunge catch	Hours/fish	Catch-per-unit-effort (CUE)
1979	11	1,185	71	16.7	0.060
1980	14	1,631	66	24.7	0.040
1981	25	2,365	103	23.0	0.044
1982	25	2,356	131	18.0	0.056
1983	30	3,415	141	24.2	0.041
1984	28	3,390	139	24.4	0.041
1985	25	12,377	585	21.2	0.047
1986	6	6,664	220	30.3	0.033
1987	1	6,120	197	31.1	0.032
1988	24	15,606	712	21.9	0.046
1989	22	21,185	959	22.1	0.045
1990	30	15,755	1,138	13.8	0.072
1991	27	14,750	876	16.8	0.059
1992	36	19,768	1,207	16.4	0.061
1993	26	18,885	1,336	14.1	0.071
1994	31	25,320	1,899	13.3	0.075
1995	22	33,114	2,082	15.9	0.063
1996	24	18,557	1,806	10.3	0.097
1997	26	18,259	1,918	9.5	0.105
1998	25	17,614	1,503	11.7	0.085
1999	28	34 648	2,059	16.8	0.059
2000	27	13,785	1,305	10.6	0.095
2001	34	11,256	1,337	8.4	0.119
2002	35	17,675	1,129	15.7	0.064
2003	36	9,103	833	10.9	0.092
2004	49	16,068	1,047	15.3	0.065
1979–2004	93 different waterbodies	360,851 rod hours	24,799 fish	14.6	0.069

There is every indication that angling success for muskellunge has improved over the past twenty-five years (Fig. 2). This can probably be attributed to the institution of larger minimum length limits in 1988 (Kerr 1998) and again in 2001 (OMNR et al. 1999) as well as an increase in catch-and-release angling practices.

Table 2 Monthly angling effort and catches reported by volunteer anglers. Data represents a consolidation of information from Muskies Canada (1979–2004) as well as OMNR angler diary programs on various waters. CUE is expressed in terms of the number of muskellunge caught per rod hour of fishing effort. Some records were not included because insufficient data was recorded

Month	Reported angling effort (rod hours)	Reported muskellunge catch	Hours/fish	CUE
June	38,559	3,055	12.6	0.079
July	48,382	3,919	12.3	0.081
August	33,696	2,536	13.3	0.075
September	32,970	2,471	13.3	0.075
October	32,780	2,170	15.1	0.066
November	8,097	612	13.2	0.076
December	672	47	14.3	0.070

Release rates

Most muskellunge anglers are strong proponents of the catch-and-release fishing ethic (Gasbarino 1986). Release rates among dedicated muskellunge anglers have increased over the past twenty-six years to the point where almost all fish caught are subsequently released (Table 3). This observation is supported by data from the 2000 national recreational fishing survey which indicated that 97.2% of angled muskellunge were released (DFO 2003).

It is important to realize that these are values reported by dedicated muskellunge anglers. Release rates are known to vary according the

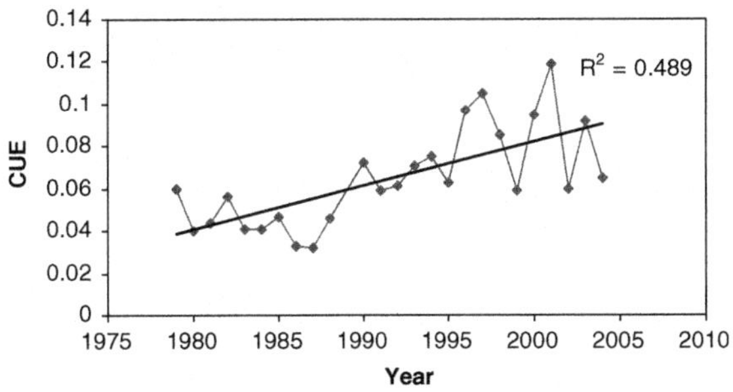

Fig. 2 Angling success, expressed in terms of catch-per-unit-of-effort (CUE), reported by muskellunge anglers from various volunteer angler diary programs in Ontario, 1979–2004. Catch rates are based on pooled data for each year

type of angler. Margenau and Petchenik (2004) reported that over 90% of Wisconsin anglers who targeted muskellunge released their catch compared with only 58% of more generalist anglers who perhaps caught a muskellunge incidentally while fishing for other species.

Table 3 Release rates reported by muskellunge anglers during voluntary angler diary programs, 1979–2004

Year	% of Released fish	Sample size (no. fish)
1979	94.2	104
1980	83.3	66
1981	82.7	104
1982	91.6	131
1983	93.6	141
1984	93.5	139
1985	94.9	585
1986	90.0	220
1987	95.4	197
1988	94.4	712
1989	94.9	959
1990	97.5	1,138
1991	98.7	867
1992	98.7	1,112
1993	98.5	1,298
1994	98.8	1,899
1995	99.2	1,902
1996	99.2	1,459
1997	99.1	1,601
1998	99.3	1,096
1999	99.7	1,590
2000	97.9	763
2001	99.7	899
2002	99.5	766
2003	97.7	511
2004	99.9	1,047
1979–2004	94.0	21,284

Size of angled fish

Based on a sample size of 9,499 fish angled over a twenty-six year period, the mean size of angled muskellunge in Ontario was 37.0 inches (94 cm). The smallest muskellunge reported was 10.3 inches (26.2 cm) in length. Overall, the mean size of the catch was relatively consistent over the sample period with no apparent trends except, perhaps, the last two years (Table 4). One might expect that, with decreased exploitation (i.e., increased catch-and-release), the mean size of fish would increase. The fact that it did not could be accounted for by either the pooled data set failing to detect these trends or that harvest rates by more generalist (i.e., non-muskellunge) anglers have not declined. Simonson and Hewett (1999) found that non-muskellunge anglers accounted for much of the muskellunge harvest in Wisconsin.

Many Ontario waters provide anglers with the opportunity to catch a 'trophy' muskellunge. For most anglers a trophy is deemed to be a fish measuring at least 40–45 inches (101–114 cm) in length (Margenau et al. 1992, Younk and Pereira 2003). Based on data collected during this program, there were at least twenty-five waters where muskellunge exceeding 40 inches (101 cm) in length have been reported and there were an additional twenty-three waters where muskellunge greater than 50 inches (127 cm) in length have been documented. The largest muskellunge reported during this program was a fish measuring 60 inches (152 cm) which was angled from the St. Lawrence River during the fall of 2004.

Incidence of lymphosarcoma

Lymphosarcoma is a malignant blood cancer which is highly contagious. The disease is associated with the skin and is manifested by skin lesions and tumours. Although some superficial lesions may regress, the tumour usually results in the death of the fish. Sonstegard and Hnath (1978) reported that lymphosarcoma may infect as high as 16% of feral populations of muskellunge. The incidence of lymphosarcoma in Ontario muskellunge is relatively low. Based on a sample size of 6,633 fish, the long term incidence was 2.0% (Table 5).

Table 4 Size of angled muskellunge reported by anglers during volunteer angler diary programs in Ontario, 1979–2004. Mean sizes are based on a pooled sample for each year. Waters where no fish were captured are not included

Year	Sample size			
	No. of waterbodies	No. of fish	Mean length of catch (Inches)	Largest fish (Inches)
1979	9	414	38.2	56.0
1980	10	471	38.0	52.9
1981	20	101	36.0	53.1
1982	17	210	37.0	53.0
1983	20	169	39.5	54.5
1984	12	117	36.5	51.5
1985	15	168	36.2	50.5
1986	19	274	35.2	53.5
1987	8	106	34.9	51.0
1988	8	240	38.3	53.0
1989	11	121	34.8	53.0
1990	17	267	36.4	55.5
1991	12	204	37.7	55.5
1992	16	218	37.8	57.0
1993	16	237	38.3	56.6
1994	21	232	41.1	53.0
1995	19	242	34.9	53.0
1996	24	602	35.9	54.0
1997	19	309	35.3	54.0
1998	18	359	38.5	55.0
1999	23	1,215	37.5	54.5
2000	21	534	37.9	53.5
2001	22	428	34.6	52.0
2002	27	812	34.6	54.0
2003	31	398	37.9	58.5
2004	49	1,051	37.4	60.0
1979–2004 Summary	93 different waterbodies	9,499 fish	37.0	60.0

Table 5 Incidence of lymphosarcoma reported from muskellunge angled by Muskies Canada Inc. members, 1979–2004

Year(s)	Number of fish examined	Number of infected fish (% of total)
1979–94	2,468	47 (1.9)
1995	168	5 (3.0)
1996	195	3 (1.5)
1997	280	10 (3.6)
1998	470	11 (2.3)
1999	511	9 (1.8)
2000	417	17 (4.1)
2001	315	8 (2.5)
2002	316	11 (3.5)
2003	442	5 (1.1)
2004	1,051	10 (1.0)
1979–2004	6,633	136 (2.0)

Discussion

Angler diaries have been used as a tool to monitor numerous recreational fisheries across Ontario and elsewhere in North America. There are several potential pitfalls with volunteer angler diary programs. These include recording errors (Thompson et al. 1988, Anderson and Thompson 1991, Younk and Cook 1992), angler bias (Sztramko et al. 1991, Duffy and Mosindy 2001, Walker and Schramm 2004), and low participation rates (Haas 1978, Borawa 1990, Bray and Schramm 2001). Results can also differ according to the type of angler who is participating (e.g., local versus non-resident, specialized versus generalist, guided or unguided). Success is usually greater for anglers who are specialists (Bijsterveld 1983, Younk and Cook 1992), those who use guides (Knapp and Goeman 2005) or those who are familiar with the waterbody they are fishing.

None of these factors are believed to have been a significant issue with Ontario muskellunge angler diary programs. The success of this long-running program may be attributed to the sound design and implementation of OMNR sponsored projects, and to the enthusiasm and commitment of MCI members. While low participation has been a problem in many other volunteer programs, participation in the MCI program is currently at its highest. This sustained level of participation is attributed to the use of incentives, the summary and distribution of data on an annual basis, log book design, and regular contacts with participants. Data confidentiality is another important factor. Angler names or the exact locations of fishing sites are not reported. These factors are all known to contribute to a successful program (Cooke et al. 2000). Having established a long term data series, it is recommended that angler diary programs continue to be used to monitor the status of Ontario's muskellunge resource.

Ontario provides numerous high quality recreational fisheries for muskellunge. This is evidenced by the diversity of opportunities, the quality of fishing experience, an increase in popularity among anglers, and the number of waters where a fish of 'trophy' size may be angled.

For the lakes examined, there is every indication that Ontario's muskellunge fisheries are being managed on a sustainable basis. Despite increasing interest and angling effort, muskellunge catches and overall angling quality have improved in recent years particularly in larger waterbodies such as Georgian Bay, Lake St. Clair, Lake of the Woods, the Ottawa River, and the St. Lawrence River. The incidence of lymphosarcoma is low and, recently, there appears to be a greater number of large (e.g., >50 inches (127 cm)) fish being reported. Finally, all muskellunge stocks in Ontario are sustained on the basis of natural reproduction.

It is believed that two factors are primarily responsible for the improvement in Ontario's muskellunge fisheries: the implementation of new minimum size limit regulations, and an increase in catch-and-release angling practices. Currently minimum size limits in Ontario must conform to one of the following: 36 inches (91 cm), 40 inches (102 cm), 44 inches (112 cm), 48 inches (122 cm) or 54 inches (137 cm). The institution of relatively high minimum size limits is known to increase muskellunge abundance and size of fish (Cornelius and Margenau 1999, Simonson and Hewitt 1999, Younk and Pereira 2003). MacLennan (1996) reported an immediate and dramatic increase in the numbers of Lake St. Clair muskellunge in the newly protected size group following an increase in the minimum size limit from 30 to 40 inches.

Catch-and-release angling plays an important role in muskellunge management and contributes to increased size structure and abundance (Margenau et al. 1992, Cornelius and Margenau 1999, Ferencak 2004). Dent (1986) reported that the voluntary release of muskellunge in Pomme de Terre Lake, Missouri, was responsible for maintaining a quality fishery even in the presence of increased fishing pressure. Ontario's muskellunge anglers are currently releasing approximately 99% of their catch.

Summary and conclusions

Angler diaries are a cost-effective means of monitoring the status of Ontario's muskellunge resource. This volunteer program also provides anglers and the tourism industry with the opportunity to be involved in the collection of valuable information. It is recommended that OMNR volunteer angler diary projects be expanded particularly with the tourist industry in northwestern Ontario. The Muskies Canada angler diary program should continue to be fully supported by OMNR.

With few exceptions, the status of Ontario's muskellunge resource seems to be healthy. It is recommended that the angler diary program be expanded in some areas to provide more complete coverage in order to continuously monitor the state of this valued resource.

Acknowledgements I am indebted to all anglers, too numerous to mention, who have participated in angler diary programs in Ontario over the years. Mike Sadiwynk, Victor Beacham and Chris Purdy, National Release Program Directors with Muskies Canada Inc., are gratefully acknowledged for providing MCI angler diary information on an annual basis. I thank Tom Mosindy and Don MacLennan, OMNR, for providing their angler diary information from Lake of the Woods and Lake St. Clair, respectively. Valuable suggestions and comments on this manuscript were received from three anonymous reviewers.

References

Anderson LE, Thompson PC (1991) Development and implementation of the angler diary monitoring program for Great Bear Lake, Northwest Territories. Am Fish Soc Symp 12:457–475

Bijesterveld L (1983) The 1981 British Columbia tidal water sportfishing diary program. Canadian Manuscript Report of Fisheries and Aquatic Sciences 1717. Vancouver, British Columbia, p 25

Borawa JC (1990) Muskellunge fishery angler diary program. Federal Aid in Fish Restoration Project F-24-15. North Carolina Wildlife Resources Commission. Raleigh, North Carolina, p 7

Bray GS, Schramm HL (2001) Evaluation of a statewide volunteer angler diary program for use as a fishery assessment tool. North Am J Fish Manag 21:606–615

Casselman JM, Robinson CJ, Crossman EJ (1999) Growth and ultimate length of muskellunge from Ontario waterbodies. North Am J Fish Manag 19:271–290

Cooke SJ, Dunlop WI, MacLennan D, Power G (2000) Applications and characteristics of angler diary programmes in Ontario, Canada. Fish Manag Ecol 7:473–487

Cornelius RR, Margenau TL (1999) Effects of length limits on muskellunge in Bone Lake, Wisconsin. North Am J Fish Manag 19:300–308

Dent RJ (1986) A case history of voluntary catch-and-release of muskellunge with management implications for fishery programs. Am Fish Soc Special Public 15:316–322

Department of Fisheries and Oceans (2003) 2000 survey of recreational fishing in Canada. Economic and Commercial Analysis Report No. 165. Ottawa, Ontario

Duffy M, Mosindy T (2001) 1988–1999 Lake of the Woods musky angler diary surveys. Northwest Science and Technology Aquatics Update 2001–01. Ontario Ministry of Natural Resources. Kenora, Ontario, p 6

Ferencak J (2004) Musky creel project, 1987–2003. Illinois Department of Natural Resources. Springfield, Illinois, 26 p + appendices

Foster RF, Colby PJ (1999) Status of the Lac Seul muskellunge fishery. Northern Bioscience Ecological Consulting. Thunder Bay, Ontario, p 58

Gasbarino PL (1986) Catch and release of muskellunge—philosophy and methods. Am Fish Soc Special Publ 15:300–308

Haas RC (1978) The muskellunge in Lake St. Clair. Am Fish Soc Special Public 11:334–339

Hacker VA (1973) The results of a ten year voluntary muskellunge creel census at Little Green Lake, Green Lake County, Wisconsin, 1963–1972. Fisheries Management Report No. 58. Wisconsin Department of Natural Resources. Oshkosh, Wisconsin, p 14

Hart ML (1980) 1979 musky angler diary volunteer program, St. Lawrence River. Ontario Ministry of Natural Resources. Brockville, Ontario, p 5

Kerr SJ (1998) Muskellunge regulations in Ontario during the past fifty years (1949–1998). Fish and Wildlife Branch. Ontario Ministry of Natural Resources. Peterborough, Ontario, p 34

Kerr SJ (2001) Atlas of muskellunge streams and rivers in Ontario. Fish and Wildlife Branch. Ontario Ministry of Natural Resources. Peterborough, Ontario, p 16

Knapp ML, Goeman TJ (2005) Catch and harvest by fishing guides compared with other anglers in five central Minnesota lakes. N Am J Fish Manag 25:885–889

Lewis C, Potter B (1980) A summary of maskinonge harvest from Rice Lake, 1979. Ontario Ministry of Natural Resources. Lindsay, Ontario, p 5

MacLennan D (1996) Changes in the muskellunge fishery and population of Lake St. Clair after an increase in the minimum size limit, pp 19–27. In: Kerr SJ, Olver CH (eds) "Managing Muskies in the 90s" Workshop Proceedings. Ontario Ministry of Natural Resources. Kemptville, Ontario, 169 pp

Mandrak NE, Crossman EJ (1992) A checklist of Ontario freshwater fishes annotated with distribution maps. Royal Ontario Museum. Toronto, Ontario, p 176

Margenau TL, Petchenik JB (2004) Social aspects of muskellunge management in Wisconsin. North Am J Fish Manag 24:82–93

Margenau TL, Meiller L, Nelson E, Stedman RC, Johnson D (1992) Opinions of anglers who fish muskellunge in Wisconsin. Wisconsin Department of Natural Resources. Spooner, Wisconsin, p 43

Ontario Ministry of Natural Resources (1987) Atlas of muskellunge lakes in Ontario. Fisheries Branch. Toronto, Ontario, p 34

Ontario Ministry of Natural Resources (1999) 1998 Kawartha lakes angler diary program. Kawartha Lakes Fisheries Assessment Unit Update 1999–5. Lindsay, Ontario, p 2

Ontario Ministry of Natural Resources, Muskies Canada Inc., Northern Ontario Tourist Outfitters, Ontario Federation of Anglers and Hunters, and the Royal Ontario Museum (1999) Size limit regulations for Ontario muskellunge: a new approach. Peterborough, Ontario, 12 p + appendices

Schiavone A (1987) 1986 muskellunge cooperator report. New York State Department of Environmental Conservation. Watertown, New York, p 3

Simonson TD, Hewett SW (1999) Trends in Wisconsin's muskellunge fishery. N Am J Fish Manag 19:291–299

Sonstegard RA, Hnath JG (1978) Lymphosarcoma in muskellunge and northern pike: Guidelines for disease control. Am Fish Soc Special Public 11:235–237

Sztramko LK, Dunlop WI, Powell SW, Sutherland RG (1991) Applications and benefits of an angler diary program on Lake Erie. Am Fish Soc Symp 12:520–528

Thompson PC, Anderson LE, Ellison HR (1988) Angler catch and fishing effort diary: 1983 and 1984 pilot study at Plummer's Great Slave Lake lodge, Northwest Territories. Canadian Technical Report of Fisheries and Aquatic Sciences 1631. Winnipeg, Manitoba, p 18

Walker CA, Schramm HL (2004) Evaluation of catch card reporting at Mississippi state lakes. 2004 Proceedings of the Annual Conference of the Southeastern Association of Fish and Wildlife Agencies 58:123–129

Younk JA, Cook MF (1992) Applications of an angler diary for muskellunge, *Esox masquinongy*. Fisheries Investigational Report 420. Minnesota Department of Natural Resources. St. Paul, Minnesota, p 21

Younk JA, Pereira DL (2003) An examination of Minnesota's muskellunge waters. Fisheries Investigational Report 498. Minnesota Department of Natural Resources. St. Paul, Minnesota, p 30

Environ Biol Fish (2007) 79:71–83
DOI 10.1007/s10641-006-9167-4

ORIGINAL PAPER

The use of angler diary surveys to evaluate long-term changes in muskellunge populations on Lake of the Woods, Ontario

Thomas E. Mosindy · Mark J. Duffy

Received: 16 September 2005 / Accepted: 19 October 2006 / Published online: 15 November 2006

Abstract Reported improvements in the muskellunge angling fishery on Lake of the Woods, Ontario over the last two decades have paralleled an increase in the practice of catch and release angling, and an increase in minimum size regulations for this species. The overall status of muskellunge populations in this large, complex lake has proven difficult to monitor using standard assessment methods. A volunteer muskellunge angler diary program, established in 1988, has provided a cost effective method of gathering a large amount of information with which to track this fishery and associated populations. Results from angling diaries indicated increased angling effort, catch and success rates for muskellunge on Lake of the Woods since the early 1990s. Although the month of July accounted for the majority of angling effort and catch, angling success rates and sizes of fish reported in diaries improved monthly into the late fall. Angling success rates were consistently higher in angling diaries than from creel surveys, but both survey types showed similar long-term trends in the fishery. Angler diary data, incorporating both the numbers of fish caught and/or seen by anglers, were used to calculate catch equality indices which proved to be sensitive to changes in population abundance. Increased minimum length regulations for muskellunge during 1987–2001 have been largely responsible for a decline in harvest rates from an estimated 36% in 1986 to 0% since 1999. Although higher size limits have yet to produce more quality-sized fish in angler catches, diary survey data, supported by recent improvements in catch rates from assessment gill nets, would indicate that muskellunge recruitment has increased. This article concludes with a brief review of how muskellunge angler diary data has been used in the past, including recommendations to minimize biases associated with this survey method.

Keywords *Esox masquinongy* · Volunteer angler diaries · Catch equality indices · Catch and release rates · Minimum size limits

T. E. Mosindy (✉)
Ontario Ministry of Natural Resources, Lake of the Wood Fisheries Assessment Unit, 808 Robertson Street, P9N 3X9 Kenora, ON, Canada
e-mail: tom.mosindy@mnr.gov.on.ca

M. J. Duffy
Saskatchewan Environment, Box 5000, S0J 1L0 La Ronge, SK, Canada

Introduction

Angler diaries are used by fisheries management agencies to provide information on sport fisheries for various species (Anderson and Thompson 1991; Sztramko et al. 1991; Cooke et al. 2000).

Volunteer diary surveys rely on anglers themselves to report on their fishing activity. The fact that diary data are self-reported lends both advantages and disadvantages to this data collection method. Angler diary surveys have proven to be cost effective programs that encourage volunteer participation and resource stewardship through angler involvement in monitoring fisheries. Information on catches and angling success may be more accurate than other creel survey methods since it is based on completed trips. Anderson and Thompson (1991) found that the accuracy of data from an angler diary program run on Great Bear Lake was comparable to that from roving creel surveys but was produced at 20% of the cost. However, Pollock et al. (1994) recommended against the use of angler diary programs, unless all other sampling techniques were ruled impractical, owing to large biases that can be introduced into the data by anglers themselves. For example, anglers contribute to these biases by not responding to surveys, by exaggerating the number and size of the fish reported, by not reporting unsuccessful trips or by misinterpreting survey questions and methods. In a review of angler diary programmes conducted in Ontario, Cooke et al. (2000) found that most diary surveys failed when agencies were not firmly committed to the program from the start or did not incorporate steps to identify and address potential biases or problems associated with angler diaries. Given current declines in government funding and staffing, the authors concluded that angler diary programs have the potential to become the primary source of information to manage specialized and remote fisheries (Cooke et al. 2000). Kerr (2004) recommended that volunteer angler diary programmes continue to be used to monitor status of muskellunge, *Esox masquinongy*, fisheries in Ontario in the future.

Anglers for muskellunge have a long history of keeping careful records of their catches and other observations made during fishing trips (Kerr 2004). Both Muskies Inc. and Muskies Canada publish annual summaries of catch information which has been forwarded by their members. This usually includes fish weights, lengths, fishing location and date, and methods used. Fisheries management agencies have developed muskellunge angler diary programs to collect similar information but for management purposes (Younk and Cook 1992; MacLennan 1996; Kerr 2004). Owing to low catch rates, the variable nature of muskellunge angling and low population densities, information to manage muskellunge fisheries is often missed, or underestimated during regular creel survey and standard assessment netting programmes (Hanson et al. 1986; Kerr 2004). Although muskellunge anglers frequently handle more fish than biologists, both the accuracy and usefulness of information collected by these anglers has been questioned in the past, owing to potential biases associated with recording data. Oehmcke et al. (1986) observed that information from angler catch reports provided a reliable index of catch but could not be related to long-term trends in muskellunge abundance in Chautauqua Lake, New York.

This study evaluates the utility of angler diary surveys and the information that they generate, in determining current population status and long-term effects of regulation changes, specifically increased minimum size limits, on size structure and relative abundance of muskellunge populations in Lake of the Woods, Ontario. The Lake of the Woods Muskie Angler Diary Survey has been run in conjunction with resort owners and fishing guides on Lake of the Woods since 1988. Changes to muskellunge size regulations and possession limits have occurred during this period. Minimum size regulations were increased from 711 mm (28 in.) to 1,020 mm (40 in.) in 1987, along with a reduction in the daily bag limit from two fish per day to one fish per day and two in possession. The minimum size regulation was again raised to 1,220 mm (48 in.) in 1992 and 1,370 mm (54 in.) in 2001.

Materials and methods

Study area

Lake of the Woods is a 385,000-ha lake, extending 106 km north–south and 105 km east–west. It borders Ontario, Minnesota and Manitoba with about 65% of its surface area within Ontario. This complex lake is made up of distinct basins which range from oligotrophic lake trout, *Salvelinus namaycush*, waters on the deeply cut Precambrian shield in the north and east to the shallow mesotrophic basin in the south. Here, rugged landscape is replaced with the gentle relief of a glacial outwash plain and overlain by lacustrine sediments left by glacial Lake Agassiz. The Ontario portion of Lake of the Woods has been divided into seven sectors for fisheries assessment and management purposes, based on distinct physical and ecological characteristics. Muskellunge which are part of the native fish community along with northern pike, *Esox lucius*, are found in all the Ontario waters of the lake. They are managed as self-sustaining populations to produce record-sized fish.

Sampling methods and analysis

Muskellunge angler diaries were distributed to resorts and guides on Lake of the Woods before the season opening in 1988, 1995, 1996, 1999, 2002 and 2005. The open season for muskellunge on the Ontario portion of Lake of the Woods runs from the third Saturday in June until 30 November. Resort managers were encouraged to have their guests who were muskellunge anglers complete a diary after each fishing trip. Survey information included: the number and origin of anglers; date and duration of the trip; whether a trip was guided or not; general location fished; number, length and weight of muskellunge and northern pike caught; and number and estimated length of muskellunge and northern pike seen but not landed. A cover letter was also sent out with diaries, outlining their purpose and importance of the information to be collected. Diaries were returned to the Lake of the Woods Fisheries Assessment Unit at the end of each muskellunge angling season for data summary and analysis. Participating resort owners and guides were sent a summary of survey results along with a small gift as a token of appreciation.

Muskellunge angler diary data were compared with results from open water creel surveys of the Ontario waters of Lake of the Woods that were conducted from the opening of walleye angling season (third Saturday in May) until 30 September during 1983–2005. These creel surveys employed a roving design with non-uniform probability sampling (Malvestuto et al. 1978). Creel crews interviewed a subsample of anglers, proportional to the time available to cover a sampling unit, during each survey day. The total number of angler-hours fished per boat, residency of anglers, base of operations (e.g. guest of a Canadian or American resort, cottager) and targeted species were established. In addition, anglers were asked a number of optional questions, outside of the standard interview format, that could be used to characterize current angling practices and activities related to the fishing trip. All fish kept by anglers were tallied and weighed by species. Total length (TL) and appropriate tissues for age interpretation (Mann 1993) were taken from a random subsample of these.

Mean catches of muskellunge in fish community index gill nets were also used to evaluate changes in relative abundance during 1980–2004 from all Ontario sectors of Lake of the Woods. Standard mesh multifilament gill nets, measuring 121.9 m long and 1.8 m deep and consisting of eight 15.2 m spaced panels in a random mesh order (i.e. 64, 114, 38, 89, 51, 102, 76 and 127 mm stretch) were employed. Nets were set perpendicular to shore and left overnight for 20–24 h at 7–10 pre-selected index sites in each sector. They were fished monthly from May through September for two consecutive years out of every 6 years from 1980 to 2004. Index netting coverage of all Ontario sectors of Lake of the Woods, except Shoal Lake, was completed every 6 years. All fish caught in each mesh size were identified to species and counted. TL, fork length, weight, sex and state of maturity were recorded. Primary species including muskellunge were sampled for ageing tissues that included scales and at least one other calcified tissue, such as cleithra.

Statistical significance was tested using standard analyses techniques (McClave et al. 1997) and statistical analysis software (Statistix 7 2000). Type-1 error rates in all statistical analyses were set at $\alpha = 0.05$. Diary entries missing key information were excluded. The statistical significance of differences between calculated means were established using the analysis of variance (ANOVA) and a comparison of means test (Bonferroni).

Differences in size structure of reported catches between various diary years were compared using the two-sample Kolmogorov–Smirnov test. Differences among specific size classes were tested using the two-sample proportion procedure. Trends in angling success rates or catch-per-unit effort (CPUE), calculated from diary survey data, were analysed using the Wilcoxon rank-sum test, since catch rates displayed a Poisson distribution and parametric tests were inappropriate. Changes in relative abundance of muskellunge in fish community index gill net catches from 1980 to 2004 were determined by comparing the mean number of muskellunge caught per net lift during successive 6-year intervals that corresponded to a full rotation in coverage of all Ontario sectors. CPUE data were normalized by applying a $\log_{10}$ (catch + 1) transformation to the mean number of muskellunge per net lift and analysed using ANOVA and Bonferroni comparison of means tests. Both Lorenz curves and a Gini coefficient (Smith 1990) were used to test for changes in the distribution of catch rates, in this case if the reported muskellunge catch became more evenly distributed amongst anglers over time. The Gini coefficient was calculated by taking the ratio of the total area between the actual distribution (the cumulative percent of fishing trips versus the cumulative percent of muskellunge caught) and a 45° line, divided by the total area under the 45° line. As the Gini coefficient increased, the catch departed more from perfect equality. Lorenz curves and Gini coefficients were calculated for the five periods when the muskellunge angler diary program was run.

Results

Data from 1995 to 1996 were combined for the purpose of analysis, due to a small number of participants in 1995. Data were available from 8 resorts/guides in 1988, 2002 and 2005, 16 resorts/guides in 1995/1996 and 20 resorts/guides in 1999.

Effort estimates from open water creel surveys showed a steady increase in muskellunge angling activity from 40,929 angler-h ± 9,921 in 1992 to 129,615 angler-h ± 32,206 (estimate and 95% confidence limits) in 2005 (Table 1). Anglers targeting muskellunge represented over 10% of total effort during the open water survey in 2005 (Spendlow 2006). Effort reported by angler diary participants also increased from about 4,667 angler-h in 1988 to 12,847 and 7,362 angler-h in 2002 and 2005. Effort calculated from angler diaries accounted for 6–27% of all muskellunge angling effort estimated during open water creel surveys in 1999, 2002 and 2005 when both surveys were run concurrently (Table 1). The number of muskellunge reportedly caught by diary participants increased from 273 in 1988 to 442 in 1995/1996, 932 in 1999, 567 in 2002 and 469 in 2005.

Over 95% of all muskellunge anglers who either participated in the muskellunge angler diary surveys during 1988–2005 or were contacted during open water surveys in 1983–2005 were non-residents of Canada from the United States, with the remainder from Ontario and other provinces. The number of guided trips reported in angler diaries has varied, from 17% in 1988 to 34% in 2002.

Angler diaries indicated that the month of July had the highest angling pressure (Fig. 1) and the most muskellunge caught in each survey year (Fig. 2). July accounted for between 40 and 56% of all muskellunge angling and 34–64% of all muskellunge reported in diaries. Although angling effort declined, angling success rates (CPUE) appeared to improve monthly through the season in most years (Fig. 1). With the exception of 1988, the highest CPUEs were observed during late fall (Wilcoxon rank-sum, $P < 0.001$). Angling success rates during the October–November period ranged from 0.055 fish angler-h^{-1} in 2002 to 0.104 fish angler-h^{-1} in 2005. Mean TLs of muskellunge at capture varied significantly by month in 1988 and 2002 (ANOVA, $F = 2.90$, $P < 0.001$), but not in 1995/1996, 1999 or 2005. When monthly catches were combined across all survey years, mean lengths of muskellunge in the reported catch (Fig. 2) increased monthly through the angling season (ANOVA, $F = 7.07$, $P < 0.001$, comparison of means, $t = 2.81$), with highest mean lengths from August (964.7 mm) and October–November (971.4 mm).

Table 1 Comparisons of estimated muskellunge angling effort (angler-h), catches, harvest rates and CPUE (number of fish per angler-hour) during 1983–2005 from Lake of the Woods, Ontario

Year	No. diaries/interviews	Effort (angler-h)	Catch (no. fish)	Harvest (no. fish)	% Kept	CPUE (no. fish h^{-1})
1983 (SE)	130	56,805 (8,945)	1,222 (346)	133	10.9	0.025
1986 (SE)	184	96,950 (11,369)	1,309 (342)	474	36.2	0.015
1987 (SE)	223	85,017 (8,608)	849 (237)	123	14.5	0.021
1988*	107	4,667	266	21	7.9	0.057
1992 (SE)	116	40,929 (5,062)	1,016 (444)	15	1.5	0.022
1993 (SE)	158	74,441 (10,265)	1,202 (263)	25	2.1	0.016
1995/1996*	210	11,639	442	2	0.5	0.038
1999*	427	21,708	932	0	0	0.043
1999 (SE)	225	80,991 (12,548)	1,800 (532)	0	0	0.024
2002*	193	12,847	567	0	0	0.044
2002 (SE)	315	130,192 (17,866)	3,275 (673)	0	0	0.029
2005*	165	7,362	469	0	0	0.064
2005 (SE)	309	129,615 (16,432)	3,282 (743)	0	0	0.032

Angler diary surveys are noted by an asterisk

SE standard error of estimate

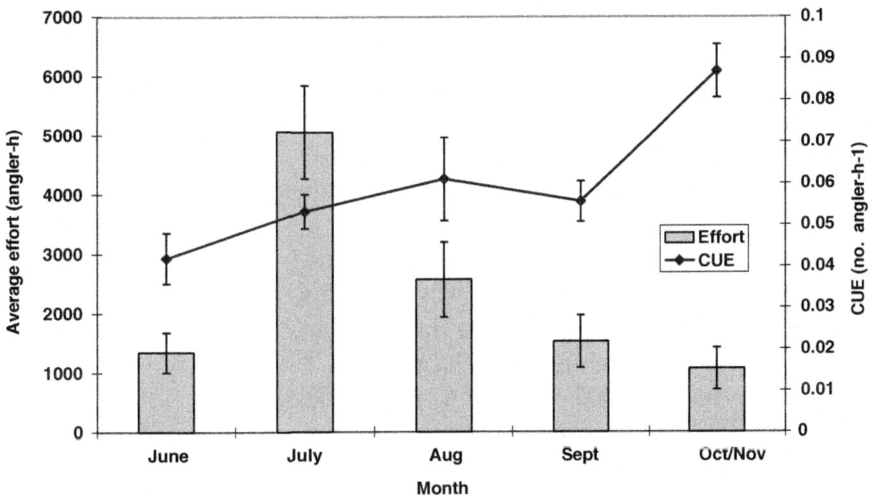

Fig. 1 A comparison of average monthly effort (angler-h ± standard deviation) and angling success rates (CPUE, number of fish caught per angler-hour ± standard deviation) during muskellunge angler diary surveys, 1988–2005, Lake of the Woods, Ontario

The percent of muskellunge harvested by anglers decreased dramatically during the last two decades (Table 1). Harvest rates observed during open water creel surveys declined from 36% in 1986 to about 2% in 1993. Harvest rates reported by muskellunge anglers during diary surveys dropped from about 8% in 1988 to <1% in 1995/1996. No harvested muskellunge have been observed in both the open water creel and angler diary surveys since 1999.

Angling success rates for muskellunge, expressed as the number of fish caught per angler-hour, have varied over the long term. CPUEs from creel surveys were consistently lower than those calculated from angler diaries (Table 1). Although angling success rates declined from about 0.057 to 0.038 fish angler-h^{-1} between 1988 and 1995/1996 in diary surveys, CPUEs from both open water creel surveys and diary surveys subsequently increased through the 1990s. Angling success rates from diary surveys increased significantly from 0.038 fish angler-h^{-1} in 1995/1996 to 0.64 fish angler-h^{-1} in 2005 (Wilcoxon rank-sum test, $z = 5.573$, $P < 0.001$). Angling CPUEs from open water creel surveys also increased steadily from 0.016 to 0.032 fish angler-h^{-1} during 1993–2005.

Lorenz curves and Gini coefficients of variation (CV) were calculated from both the numbers of muskellunge landed, as well as fish reportedly

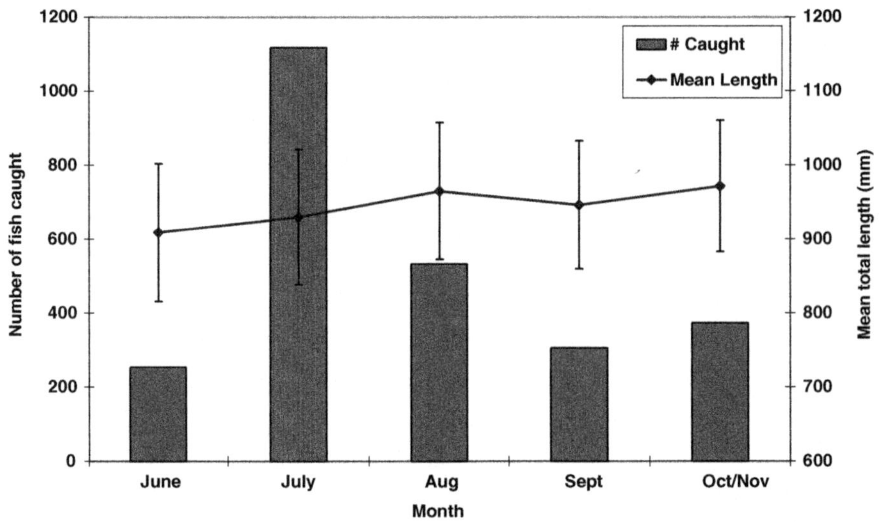

Fig. 2 The number and mean TLs (mm ± standard deviation) of muskellunge reportedly caught by month during muskellunge angler diary surveys, 1988–2005, Lake of the Woods, Ontario

seen and/or lost by angler diary participants. The area between the 45° line and Lorenz curves (Fig. 3), calculated from diary survey data, declined during 1988–2005. Similarly, Gini CVs decreased from 0.75 in 1988 to 0.051 in 1999 and 0.46 in 2005. A decline in both measures indicated that the reported catch had become more equally distributed among anglers during the study period.

A significant decrease in average TL of muskellunge reported in angler diaries was observed during 1988–2005, from 1,007 mm in 1988 to 970 mm in 1995/1996, 969 mm in 1999, 872 mm in 2002 and 933 mm in 2005 (ANOVA, $F = 39.67$, $P < 0.001$; Bonferroni comparison of means, $t = 2.809$). A comparison of length at capture throughout the diary program (Table 2) indicated that an increasing percentage of smaller fish were being caught. Size structure decreased between all sampling periods (Kolmogorov–Smirnov test, $P < 0.001$), except 2002–2005, when size structure increased. The representation of muskies smaller than 1,000 mm (39 in.) increased significantly from about 45% of the reported catch in 1988 to 56% in 1999 (two-sample proportion test, $z = 3.23$, $P < 0.001$) and 74% in 2002 ($z = 6.8$, $P < 0.001$). This proportion dropped to 66% of the reported catch in 2005, although this decline was not significant (two-sample proportion test, $z = 2.55$, $P = 0.005$). The largest increase in-

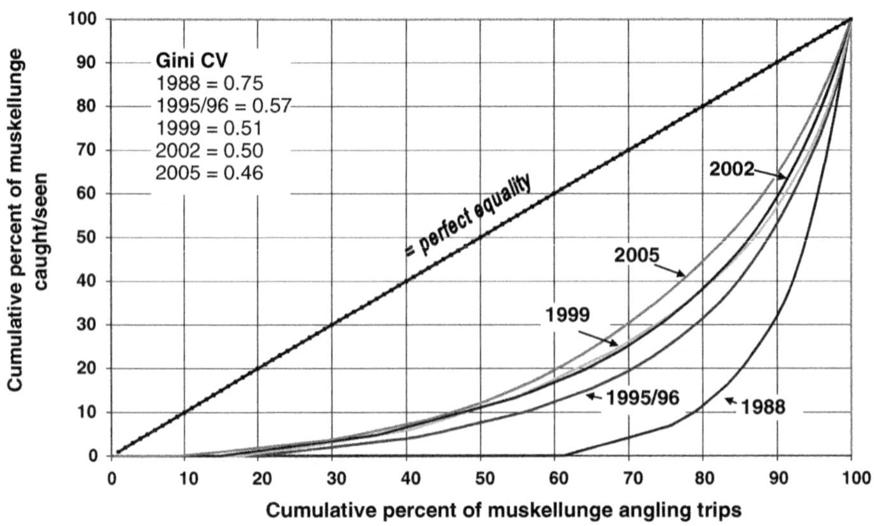

Fig. 3 A comparison of Lorenz curves and Gini CV calculated from muskellunge angler diary data from 1988 to 2005, Lake of the Woods, Ontario

Table 2 Percent representation of muskellunge by various TL classes (mm) reportedly caught during muskellunge angler diary surveys, 1988–2005, Lake of the Woods, Ontario

TL (mm)	Percent frequency				
	1988	1995/1996	1999	2002	2005
≤799	11.7	17	13.2	36.5	16.4
800	11.4	13.6	15.6	21.1	23.4
900	21.6	23.1	27.3	16	26.1
1,000	26.4	23.1	22.3	15.4	20.5
1,100	16.1	12.2	12.6	6.5	8
1,200	12.1	9.5	7.6	3.9	5.2
≥1,300	0.7	1.6	1.5	0.7	0.5
n	273	442	932	570	440
Mean	100.7	97	96.9	87.2	93.3
SD	16.69	18.28	17.99	18.11	16.16

volved length classes smaller than 800 mm TL which increased from about 13% in 1999 to almost 37% of the reported catch in 2002. These same trends were evident in a comparison of angling success rates (CPUEs) for various size classes during 1988–2005 diary surveys (Fig. 4). Increased angling success for fish smaller than 1,000 mm was observed during the two most recent surveys in 2002 and 2005. Following a decline between 1988 and 1995/1996 surveys, CPUEs for fish larger than 1,000 mm have varied only slightly between years, with the exception of an apparent increase in angling success for fish in the 1,000–1,100 mm interval in 2005. Angling success rates for the largest size classes

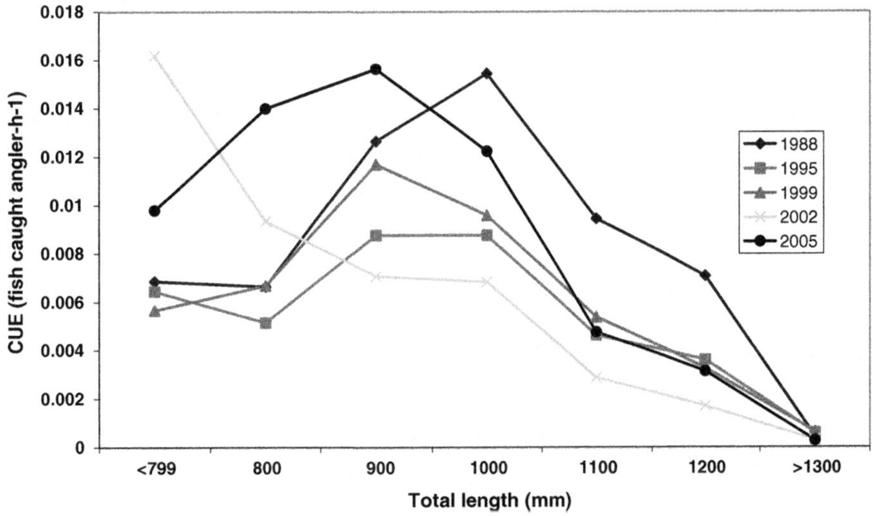

Fig. 4 A comparison of angling success rates (CPUE) expressed as fish caught per angler-hour by various TL classes from muskellunge angler diary surveys during 1988–2005, Lake of the Woods, Ontario

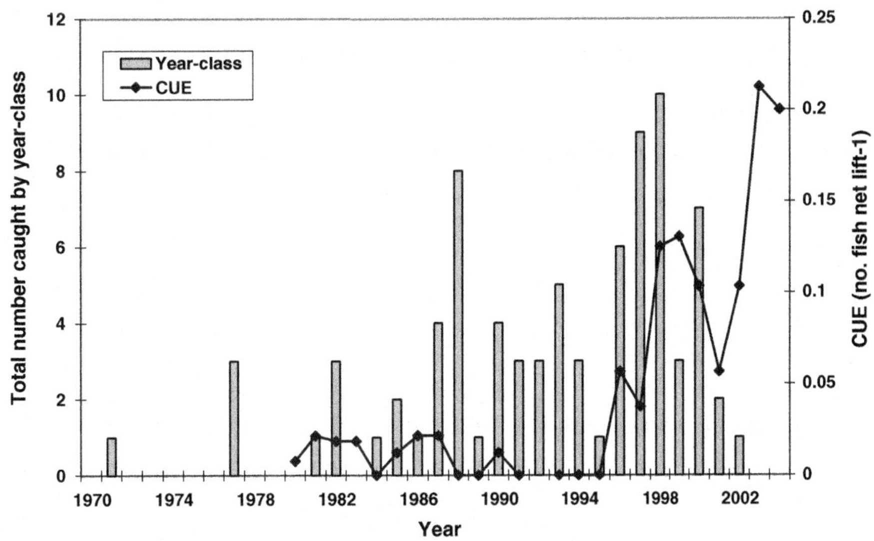

Fig. 5 The cumulative muskellunge catch by year-class and the mean number of muskellunge caught per net lift in fish community index gill nets during 1980–2004, Lake of the Woods, Ontario

(>1,300 mm) have remained unchanged throughout the study period at <0.001 fish angler-h^{-1}.

Mean catches of muskellunge in fish community index gill nets (Fig. 5) increased significantly in Ontario sectors of the lake during 1980–2004 (ANOVA, $df = 3$, $F = 23.49$, $P < 0.001$; Bonferroni comparison of means, $t = 2.642$). Prior to 1992, index net catches of muskellunge averaged 0.0042 fish net lift^{-1} in 1980–1985 and 0.0025 fish net lift^{-1} in 1986–1991 with no muskellunge caught in several years. Mean catches of 0.0133 and 0.0431 fish net lift^{-1} were observed in 1993–1998 and 1999–2004. Since the mid–1990s, muskellunge have been consistently caught from each sector during the annual FCIN program. A cumulative age distribution of muskellunge from index gill nets by year-class (Fig. 5) indicated an increased frequency of successful year-class production since 1984. Year-classes exceeding an average representation of 4.124 individuals (SD = 2.865, $n = 74$) during 1984–2002 were produced in 1988, 1993, 1996, 1997, 1998 and 2000. The occurrence of stronger than average year-classes would appear to have increased in recent years.

Discussion

The success of the muskellunge angler diary survey in Lake of the Woods reflects many of the same benefits as angler diary programs for muskellunge conducted elsewhere. These programs can be effective tools for gathering large amounts of information with which to characterize current muskellunge angling activity and to evaluate long-term trends in a fishery.

Collection of useful information with which to manage the muskellunge sport fishery on Lake of the Woods has been limited in the past, owing to a number of factors. Muskellunge anglers tend to be loners and their overall density on a large lake, such as Lake of the Woods, is low. Even though muskellunge angling was estimated to represent almost 10% of all open water angling activity on Lake of the Woods, Ontario in 2005 (Spendlow 2006), angling effort for muskellunge averaged 0.38 angler-h ha^{-1} and varied from 0.13 to 0.95 angler-h ha^{-1} by sector. In addition, muskellunge are not easy to catch and angling success rates are extremely low when compared to other species. Muskellunge catch rates averaged 0.03 fish angler-h^{-1} while catch rates for walleye, *Sander vitreum*, were estimated at 0.9 fish angler-h^{-1} during the 2005 open water creel survey on Lake of the Woods (Spendlow 2006). As a result, angling effort and catch data are often missed or underestimated during regular creel surveys. The status of muskellunge populations on Lake of the Woods was also a challenge to assess using standard index netting methods. Muskellunge are a solitary predatory (Scott and Crossman 1973) and their overall densities are low in most waters. In the past, too few fish were caught in index gill nets or other assessment gear from Lake of the Woods to provide enough fish to estimate growth, mortality or maturity rates, let alone to track relative abundance throughout the lake.

The muskellunge angler diary survey in Lake of the Woods has proven to be an important source of data with which to index long-term trends in relative abundance and to evaluate impacts of changes in angling regulations on this fishery. Although angling success rates (CPUEs) are often used to assess population status and the quality of an angling fishery (Ricker 1975), their reliability as a measure of relative abundance has recently been questioned (Korver et al. 1996), especially for species such as walleye which display schooling behaviour. The increased ability of anglers to find concentrations of fish, aided by improvements in angling technologies (i.e. sonar locators, underwater cameras, etc.) can produce catch rates that are independent of the overall population density in a water body (Rietveld et al. 2000). However, this does not appear to be the case for a species such as muskellunge which are low density predators and tend not to congregate, except during spawning (Scott and Crossman 1973). Angling success rates for muskellunge may not be as dependent on locating concentrations of fish as other factors. Long-term improvements in angling success rates (CPUE) from both muskellunge angler diaries and open water creel surveys on Lake of the Woods have paralleled the increased mean catch of muskellunge from fish community index gill nets in all lake sectors since the early to mid–1990s. Although angling success rates for Lake of the Woods muskellunge were below the provincial average of 0.068 fish angler-h^{-1} (Kerr 2004), they likely reflect the low abundance of muskellunge compared to other species in this fish community, as observed in index gill net catches. An increase in the number of participants between 1988 and 1995/1996 diary surveys could account for an observed decline in CPUEs from 0.057 to 0.038 fish angler-h^{-1}, rather than indicating an actual decline in the fishery over this period. Both Lorenz curves and Gini indices decreased during 1988–2005, suggesting that the reported catch had become more equally distributed among anglers. Along with increased CPUEs, these indices provide evidence that the muskellunge fishery has improved. More anglers were seeing and/or catching muskellunge on Lake of the Woods than ever before. Both Lorenz curves and Gini CV were calculated not only from the numbers of muskellunge successfully landed but also included fish which were seen and/or lost by angler diary participants. Unsuccessful hook-ups are considered to be part of the overall muskellunge angling experience and are usually noted since so few fish are normally encountered.

A decrease in mean size and a steady increase in the percent representation of fish <1,000 mm in the reported diary catch from Lake of the Woods suggest that improvements in the fishery have resulted from increased recruitment, especially in recent years. A cumulative age distribution of muskellunge from the fish community index gill net catch also indicated that successful year-classes have been produced annually since 1984. Strong year-classes in recent years, especially 1996, 1997 and 1998, would account for increased numbers of smaller fish (<800 mm TL) in the reported catch during 2002 and 2005 diary surveys. MacLennan (1996) and Casselman et al. (1999) reported similar increases in muskellunge recruitment and population densities, along with higher angler catch rates, following implementation of the 1,020-mm minimum size limit on Lake St Clair and the Niagara River in 1987.

The original intent of the minimum size increase in 1987 from 711 mm (28 in.) to 1,020 mm (40 in.), along with a reduction in the daily creel limit from two fish to one, was to enhance protection for female muskellunge throughout Ontario for at least two repeat spawnings (Lewis 1996). Subsequent minimum size increases on Lake of the Woods and other Ontario waters in 1992 and 1999 were to afford muskellunge even greater protection and to manage for world class trophy fisheries. Although angling success rates for smaller fish have steadily increased, largely as the result of stronger recruitment, improvements in the Lake of the Woods fishery have yet to result in more and/or larger-sized fish in the angling creel. Following an initial decline after 1988, CPUEs for larger size classes (≥1,000 mm) have remained largely unchanged. Both MacLennan (1996) and Casselman et al. (1999) reported improvements in angling success rates and an increase in the proportion of legal-sized muskies (>102 cm) in angling catches from the Lake St Clair and the Niagara River as early as the mid–1990s, following establishment of a 102-cm minimum size limit in 1987. Cornelius and Margenau (1999) attributed an increase in mean length and abundance of adult muskies, along with a corresponding decline in relative weight (W_r), to an increase in minimum length limits from 30 to 40 in. during 1964–1995 in Bone Lake, Wisconsin. In the case of Lake of the Woods, a recruitment response to the 102-cm minimum size limit has lagged behind more southerly waters and increases in overall abundance have only occurred within the last decade. Barring a significant decline in survival, these improvements should result in more quality-sized fish in angling catches in the future. Although growth potential for Lake of the Woods muskellunge ranks among the highest in Ontario, Casselman et al. (1996, 1999) observed that exceptionally large year-classes and at least 20–25 years are still required to produce record-sized fish from this lake.

Reliable data form the basis for sound management decisions. Muskellunge angler diary surveys can generate data that is often more accurate than standard creel surveys. Unlike roving creel survey techniques, diaries are based on completed trips which can provide better estimation of angling effort, catch and angling success rates (CPUEs). Muskellunge angler diary surveys also have the advantage of targeting specific anglers in the overall population. Schraeder (1989) found that angling success rates reported by specialist anglers or those completing diaries tended to be higher than average. Muskellunge angling success rates from angler diaries were also consistently higher than those from open water creel surveys on Lake of the Woods which involved interviews with all angler groups. Foster et al. (1999) observed that angling success rates for muskellunge along with the number of fish successfully landed were directly related to an angler's muskellunge fishing experience.

Although the fact that muskellunge anglers have a long tradition of participation in angling diary programs lends itself to more accurate record keeping, it does not necessarily mean that associated biases will be completely eliminated. Oehmcke et al. (1986) identified several potential sources of angler bias in muskellunge diary programs that can be reduced by adjustments to survey design and questionnaire. Anglers are often reluctant to accurately report information, especially angling locations or lure types and techniques used, owing to concerns over confidentiality of data. In the case of the Lake of the Woods diary program, anglers were clearly informed about the survey's objectives and data requirements. Data were only used by the Ministry of Natural Resources for management purposes and remained confidential. Participants were not asked for exact angling locations but the general area or sector. On a large lake such as Lake of the Woods, this is sufficient for the level of analysis. The reliability of information on landed fish versus fish lost is sometimes questioned since the latter are often recorded as fish caught but released. For the purposes of calculating Lorenz curves and Gini CV from Lake of the Woods data, all fish successfully landed along with those reportedly lost and/or seen were included. By treating this information equally, a potential reporting bias was minimized but information that could be used as an index of relative abundance was still retained. In order to prevent recall biases, anglers were encouraged to complete diaries at the end of each fishing trip. Anglers who report only successful fishing trips and ignore unsuccessful outings introduce a significant bias that contributes to overestimates of angling success rates and catch. This bias, along with those associated with accurately recording angling effort or fish lengths and weights, can be addressed by providing clear instructions on how and why information should be recorded in a particular way. This can be included as part of the diary booklet or through training workshops held prior to the program's commencement (Oehmcke et al. 1986). Diary participants need to be educated on the importance of recording all fishing trips, whether successful or not, and be assured that information will remain confidential. Non-response remains a common problem in voluntary surveys (Pollock et al. 1994). Both the use of mandatory reporting or relying upon so-called "index" anglers to collect a smaller but more reliable number of samples have been suggested as possible solutions. Oehmcke et al. (1986) found that mandatory reporting by muskellunge anglers in Chautauqua and Cattaraugus counties in New York proved largely unsuccessful, since reporting rates fell dramatically over time in the absence of enforcement. The muskellunge angler diary program in Lake of the Woods has had good response from participants since it has relied on fishing guides and resort operators to distribute and return completed diaries from their guests at the end of each season. Many resort operators on the lake have kept camp diaries in the past that record muskellunge angling catches of their guests. The Lake of the Woods muskellunge angler diary program has also benefited from continuous communication between the MNR and diary cooperators, from the initial stages of setting up the survey through to its conclusion with the release of survey results to participants.

Muskellunge angler diary surveys can provide more detailed data on a muskie fishery than is available through other survey methods. Angler diaries are often the only source of catch data, including numbers and sizes of both harvested and released fish, which is essential to assessing the current status of a fishery and interpreting the impacts of factors such as regulation changes on muskellunge populations. Agencies and clubs can ask anglers to record a wide variety of other information and observations from their fishing trips. Richards and Ramsell (1986) evaluated the survival of muskellunge caught and released by anglers based on recapture data reported in diaries on over 1,600 fish that had been originally tagged by anglers from US muskellunge fishing clubs during six separate tagging studies. Observations of lymphosarcoma on angled muskellunge recorded by participants in the Muskies Canada diary program showed the incidence of this disease to be about 2% in Ontario (Kerr 2004). Angler diary programs have been primary contributors to the Cleithrum Project (Casselman and Crossman 1986), which focuses on collection of biological data on angled, trophy-sized muskellunge and northern pike–muskellunge hybrids in North America. Data from this project have been instrumental in validating management practices such as catch and release amongst muskellunge anglers and the use of minimum size regulations based on population growth potential (Casselman et al. 1999). Angler diary programs sponsored by Muskies Inc. and Muskies Canada have provided a variety of other information which has proven useful to management agencies, including incidence of natural hybrid tiger muskellunge (Kerr 2004), popularity of specific lure types and angling methods such as "quick strike" presentations with live bait (Margenau and Petchenik 2004). Participants in the muskellunge angler diary program in Lake of the Woods have been asked to record catch and size data on any northern pike that are caught incidentally during muskellunge angling. This information has been used to evaluate the long-term response of pike populations to changes in size and possession limit regulations on the lake (Duffy and Mosindy 2001).

Angler diary programs indicated that early season (June–July) accounted for both the highest angling effort and reported catch on Lake of the Woods and other Ontario waters (Kerr 2004). However, highest angling success rates and the largest mean size of fish were reported from October–November on Lake of the Woods. Younk and Cook (1992) also observed that muskellunge catch rates from Minnesota angler diary surveys were highest during fall. Kerr (2004) noted comparable results for muskellunge in other jurisdictions, including Ontario; they are becoming an increasingly popular species, CPUEs have improved over the last 25 years, and release rates have increased. An increase in release rates can be attributed to a combination of increasing minimum size limits and greater adoption of catch and release angling for muskellunge, popularized by groups such as Muskies Inc. and Muskies Canada. Although catch and release practices became strongly entrenched in the muskellunge angling community in the early 1980s (Richards and Ramsell 1986), incidental harvest of muskellunge by other anglers remained high until increasing minimum size limits reduced this option (Mosindy 1996). Margenau and Petchenik (2004) observed that over 90% of muskellunge anglers released their catch versus 58% of so-called "generalist" anglers who may have taken a muskellunge incidentally while fishing for other species.

Angler diary programs for muskellunge, such as the one conducted on Lake of the Woods, can provide a large amount of information on a fishery that is often difficult to monitor using standard assessment methods. However, care should be taken when interpreting the results from diary surveys. Due to the nature of the angling fishery, information is recorded primarily on larger fish in the population, which may bias population structure estimates. Biological information, especially calcified tissues to allow age interpretation, is not usually collected from fish during angler diary programs. This information is vital in identifying changes within population dynamics (e.g. growth, mortality, recruitment) that ultimately shape population age and size class structures. Ideally, angler diary data should be combined with information from other assessment programs (index netting, creel surveys, spawning ground and nursery habitat assessments, etc.) to manage muskellunge populations.

References

Anderson LE, Thompson PC (1991) Development and implementation of the angler diary monitoring program for Great Bear Lake, Northwest Territories. Am Fish Soc Symp 12:457–475

Casselman JM, Crossman EJ (1986) Size, age, and growth of trophy muskellunge and muskellunge–northern pike hybrids—the Cleithrum Project, 1979–1983, vol 15. American Fisheries Society, Special Publication, Bethesda, pp 93–110

Casselman JM, Robinson CJ, Crossman EJ (1996) Assessing sustainability of trophy muskellunge fisheries. In: Kerr SJ, Olver CH (eds) Managing Muskies in the '90s workshop proceedings. Ontario Ministry of Natural Resources, Southern Region Science and Technology Transfer Unit Workshop Proceedings WP–007, 169 pp

Casselman JM, Robinson CJ, Crossman EJ (1999) Growth and ultimate length of muskellunge from Ontario water bodies. North Am J Fish Manag 19:271–290

Cooke SJ, Dunlop WI, MacLennan D, Power G (2000) Applications and characteristics of angler diary programmes in Ontario, Canada. Fish Manag Ecol 7:473–487

Cornelius RR, Margenau TL (1999) Effects of length limits on muskellunge in Bone Lake, Wisconsin. North Am J Fish Manag 19:300–308

Duffy MJ, Mosindy T (2001) 1988–1999 Lake of the Woods Muskie angler diary surveys. Ontario Ministry of Natural Resources, Northwest Science and Technology, Aquatics Update 2001-1, 7 pp

Foster RF, Colby PJ, Grabowski R (1999) Status of the Lac Seul muskellunge fishery. Report prepared for the Ministry of Natural Resources by Northern Bioscience Ecological Consulting, Thunder Bay, Ontario, 58 pp

Hanson DA, Axon JA, Casselman JM, Haas RC, Schiavone A, Smith MR (1986) Improving musky management: a review of management and research needs, vol 15. American Fisheries Society, Special Publication, Bethesda, pp 333–341

Kerr SJ (2004) Characteristics of Ontario muskellunge fisheries based on volunteer angler diary information. Fish and Wildlife Branch. Ontario Ministry of Natural Resources, Peterborough, Ontario, 19 pp., appendices

Korver RM, Lester NP, Shuter BJ, Jones ML (1996) Density-dependent catchability in angling fisheries (or Do you suffer from q-suction?). Poster presentation at the American Fisheries Society, 126th annual meeting, Dearborn

Lewis CA (1996) An overview of muskellunge management in Ontario. In: Kerr SJ, Olver CH (eds) Managing Muskies in the '90s workshop proceedings. Ontario Ministry of Natural Resources, Southern Region Science and Technology Transfer Unit Workshop Proceedings WP-007, 169 pp

MacLennan D (1996) Changes in the muskellunge fishery and population of Lake St. Clair after an increase in the minimum size limit. In: Kerr SJ, Olver CH (eds) Managing Muskies in the '90s workshop proceedings. Ontario Ministry of Natural Resources, Southern Region Science and Technology Transfer Unit Workshop Proceedings WP-007, 169 pp

Malvestuto SP, Davies WD, Shelton WL (1978) An evaluation of the roving creel survey with non-uniform probability sampling. Trans Am Fish Soc 107(2):255–262

Mann S (1993) Collection techniques for fish ageing structures. Ontario Ministry of Natural Resources. Northwest Science and Technology. NWST Technical Report TR–73, 20 pp

Margenau TL, Petchenik JB (2004) Social aspects of muskellunge management in Wisconsin. North Am J Fish Manag 24:82–93

McClave JT, Dietrich FH, Sincich T (1997) Statistics, 7th edn. Prentice-Hall, Upper Saddle River, 823 pp

Mosindy T (1996) Recent trends in the Lake of the Woods muskellunge fishery. In: Kerr SJ, Olver CH (eds) Managing Muskies in the '90s workshop proceedings. Ontario Ministry of Natural Resources, Southern Region Science and Technology Transfer Unit Workshop Proceedings WP-007, 169 pp

Oehmcke AA, Addis JT, Mooradian SR, Ogden K, Stange D (1986) The role of anglers and private organizations in muskellunge management. In: Managing Muskies, vol 15. American Fisheries Society, Special Publications, Bethesda, pp 323–334, 372

Pollock KH, Jones CM, Brown TL (1994) Logbooks, diaries, and catch cards. In: Angler survey methods and their applications in fisheries management, vol 25. American Fisheries Society, Special Publication, Bethesda, pp 127–131, 371

Richards K, Ramsell R (1986) Quantifying the success of muskellunge catch-and-release programs: a summary of cooperative angler tagging studies. In: Managing Muskies, vol 15. American Fisheries Society, Special Publication, Bethesda, pp 309–315, 372

Ricker WE (1975) Computation and interpretation of biological statistics of fish populations, vol 191. Bulletin Fisheries Research Board of Canada, Ottawa, 382 pp

Rietveld HJ, Armstrong EW, Baccante DA, Korver RM, Leith R, Mathers RA, Mosindy T, Reid DM, Ritchie BJ, Seyler J, Wepruk RL (2000) Regulatory control of walleye (*Stizostedion vitreum*) sport fisheries in Ontario. Percid Community Synthesis, Harvest Control Working Group. Ontario Ministry of Natural Resources, Peterborough, Ontario, 39 pp

Schraeder HA (1989) Analysis of volunteer angler diary data for the Thames River watershed. File Report. Ontario Ministry of Natural Resources, Aylmer, Ontario

Scott WB, Crossman EJ (1973) Freshwater fishes of Canada, vol 184. Bulletin Fisheries Research Board of Canada, Ottawa, 966 pp

Smith CL (1990) Resource scarcity and inequality in the distribution of catch. North Am J Fish Manag 10:269–278

Spendlow I (2006) Lake of the Woods open water creel survey: 2005. Ontario Ministry of Natural Resources,

Northwest Science and Information, NWSI Technical Report, TR–138, 16 pp
Statistix 7 (2000) Statistix for Windows. Analytical Software, Tallahassee
Sztramko LK, Dunlop WI, Powell SW, Sutherland RG (1991) Applications and benefits of an angler diary program on Lake Erie, vol 15. American Fisheries Society, Special Publication, Bethesda, pp 520–528
Younk JA, Cook MF (1992) Applications of an angler diary for muskellunge (*Esox masquinongy*). Investigational Report 420. Minnesota Department of Natural Resources, St Paul, 21 pp

Environ Biol Fish (2007) 79:85–98
DOI 10.1007/s10641-006-9090-8

CROSSMAN

Evaluation of temperature-selection differences among juvenile muskellunge originating from different latitudes

Curtis P. Wagner · David H. Wahl

Received: 7 September 2005 / Accepted: 8 June 2006 / Published online: 26 September 2006

Abstract Genetic differences among muskellunge *Esox masquinongy* populations are related to residence in major river drainages, suggesting the existence of divergent stocks. By use of radio-telemetry we compared different seasonal and diel temperature selection in a southern Illinois reservoir for three geographically and genetically distinct stocks of age-2 muskellunge from throughout the latitudinal range of the species. Muskellunge from the Upper Mississippi River drainage were represented by the Leech Lake, Minnesota, population and the Ohio River drainage was represented by the Cave Run Lake, Kentucky, population. Progeny from North Spring Lake, Illinois, an interstock, or mixed-origin stock, were also evaluated. No differences in temperature selection were observed among stocks of juvenile muskellunge across seasons or diel periods. The seasonal mean temperatures of the water varied significantly—spring 21.7°C, summer 28.4°C, and fall 14.8°C, with an overall average temperature of 24.1°C, in agreement with previously published values obtained in laboratory trials. This lack of different temperature-selection patterns among stocks is in contrast with other life-history characteristics, for example growth, metabolism, and conversion efficiency, which have been shown to differ among populations and stocks.

Keywords *Esox masquinongy* · Temperature · Stock differentiation · Adaptation · Physiology

C. P. Wagner · D. H. Wahl
Kaskaskia Biological Station, Center for Aquatic Ecology and Conservation, Illinois Natural History Survey, Rural Route 1, Box 157, Sullivan, IL 61951, USA

C. P. Wagner (✉) · D. H. Wahl
Department of Natural Resources and Environmental Sciences, University of Illinois, 1816 South Oak Street, Champaign, IL 61820, USA
e-mail: cwagner@uiuc.edu

Introduction

Temperature is one of the most important ecophysiological variables affecting the distribution, ecology, and physiology of ectotherms, including fish (Coutant 1987). Temperatures experienced by poikilothermic organisms affect all aspects of behavior and physiology, including locomotion, immune function, sensory input, foraging rates and ability, courtship, reproduction, metabolism, and growth (Fry 1971; Angilletta et al. 2002). Because all physiological processes are directly or indirectly affected by temperature, it follows that temperature is an important component of habitat for an organism (Magnuson et al. 1979). Most macrohabitats are thermally heterogeneous and behavior, morphology, and physiology interact to determine the realized body temperature (Tracy 1982; Huey 1991). Behavioral thermoregulation places an organism in temperatures that optimize physiological processes and buffer any negative effects of temperature (Hernandez et al. 2002; Freidenburg and Skelly 2004).

Phylogenetic theory predicts that, across a species range, intraspecific populations are locally adapted, because of natural selection (Mayr 1975). Many fish species have broad latitudinal distribution and different populations, or stocks, have evolved under different environmental conditions, for example temperature conditions (Fry 1971). As a result, local populations have different physiological and behavioral responses to their environment; these gradually become adaptive genetic differences among stocks (Ihssen et al. 1981; MacLean and Evans 1981; Philipp et al. 1993). Local adaptation suggests the possibility of different temperature optima, and thus preferred temperatures (Beitinger and Fitzpatrick 1979; McCauley and Casselman 1981), among populations (Hall et al. 1978; Koppelman et al. 1988; Angilletta et al. 2002; Freidenburg and Skelly 2004).

The taxonomy of the muskellunge *Esox masquinongy* has undergone substantial revision over the last century (Crossman 1978, 1986). During the late 1800s and early 1900s, apparent correlations between markings and location led, for a short time, to classification of the muskellunge as three separate species (Crossman 1978). By the late 1970s, the idea had been established that all variations were, indeed, a single species and there was insufficient evidence to warrant separate species or subspecies classification (Crossman 1978). More recent genetic analysis of various populations has revealed three distinct clusters related to separation by major river drainages, suggesting the existence of divergent stocks (Koppelman and Philipp 1986). Existing information indicates muskellunge persisted through the Wisconsin glacial period in the Mississippi refugium and, on glacial recession, moved north up the Mississippi valley and established its current range via the Mississippi and Ohio River systems and through the precursors to tributaries of the Great Lakes (Crossman 1978, 1986). Muskellunge were isolated by major river drainages and experienced different environmental conditions and thermal histories. As these isolated groups diverged through post-glacial recolonization, genetic processes such as natural selection resulted in stocks of muskellunge that are genetically dissimilar, and probably physiologically and behaviorally different from one another (Altukhov 1981; MacLean and Evans 1981; Ihssen et al. 1981; Clapp and Wahl 1996; Begg et al. 1999). The currently identified genetically distinct muskellunge stocks are the Upper Mississippi River drainage stock, the Great Lakes/St Lawrence River drainage stock, and the Ohio River drainage stock (Koppelman and Philipp 1986; Clapp and Wahl 1996; Fig. 1).

Previous research has examined physiological performance characteristics among populations of

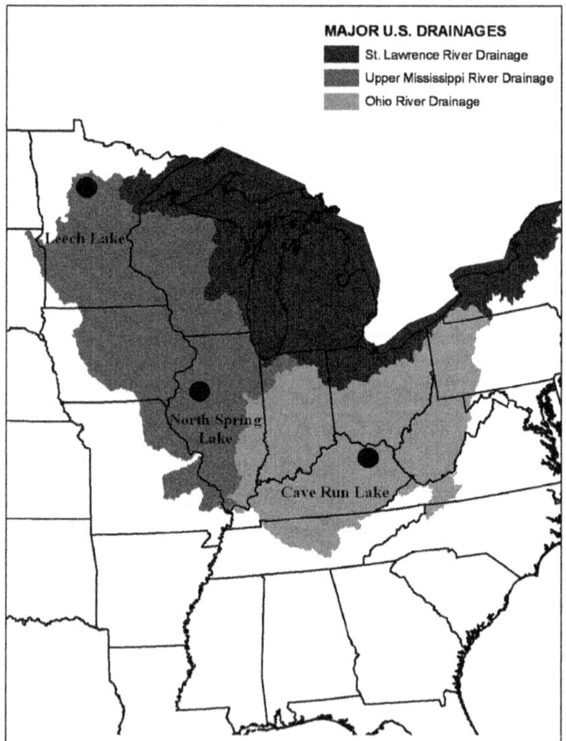

Fig. 1 Major US river drainages, including the Upper Mississippi River, the Ohio River, and the Great Lakes/St Lawrence River drainages, which form the native muskellunge range. Study populations (*large black dots*) include Leech Lake, Minnesota; Cave Run Lake, Kentucky; and North Spring Lake, Illinois. Although geographically situated in the Upper Mississippi River drainage, the Illinois population is a mixed-origin brood stock containing genotypes from several major river drainage sources

muskellunge in natural systems, but only for a limited number of populations from a small part of the native range (Younk and Strand 1992; Margenau and Hanson 1996, 1997). Growth and survival differences were observed among populations, revealing population differentiation among relatively local demes. Because the temperature conditions encountered by the populations in these studies varied only slightly, the effect of latitude on performance characteristics could not be adequately addressed. Although differences between food consumption, growth, and metabolism among muskellunge populations and stocks have been observed in the laboratory, bioenergetic differences among populations cannot be explained solely in terms of temperature adaptation or previously defined genetic groupings (Clapp and Wahl 1996). Additional research is needed to further understand behavioral and physiological differences among populations and stocks of muskellunge in relation to latitudinal variation.

Although laboratory determinations of temperature preference and avoidance by fish are well developed (Richards et al. 1977), there are comparatively few quantitative field approaches for such determinations (Coutant and Carroll 1980; Headrick and Carline 1993; Barwick et al. 2004; Bettoli 2005). The location of a fish in the water column can be affected by factors other than temperature (e.g. light, prey density, social behavior) (Coutant and Carroll 1980; Jobling 1981; Koppelman et al. 1988; Barwick et al. 2004). Because an organism may be affected by any number of these factors, temperature selection will be used in our study as an approximation of preference. The objective of our study was to determine if genetically distinct stocks of muskellunge from different latitudes select different temperatures in a common field environment. To evaluate potential differences between temperature selection among muskellunge stocks, we used temperature-sensing radio telemetry in an Illinois reservoir. If local adaptation has occurred, the preferred temperature and the temperature for physiological optima should differ among populations and reflect adaptive responses to the environment in which they evolved.

Study site

Forbes Lake (226 ha and 36 km of shoreline) is a warm-water, shallow (9.5 m maximum depth, 4.6 m average depth) reservoir in Marion County, Illinois (W 88.7, N 38.7) (Fig. 2). Located within the Wabash River drainage basin, Forbes Lake has a watershed catchment size of 5,585 ha and a retention time of 219 days (Austen et al. 1993). The eutrophic reservoir has anoxic conditions in the hypolimnia during summer and secchi disk visibility was typically less than or equal to 0.6 m during our study. Emergent vegetation is dominated by *Justica americana* and submergent vegetation includes *Ceratophyllum demersum* and *Potamogeton* spp. The sport fishery is dominated by largemouth bass *Micropterus salmoides*, bluegill *Lepomis macrochirus*, crappies *Pomoxis* spp., and channel catfish *Ictalurus punctatus*. Other fish fauna include hybrid striped bass *Morone saxatilis*× *M. chrysops*, saugeye *Sander vitreum*×*S. canadense*,

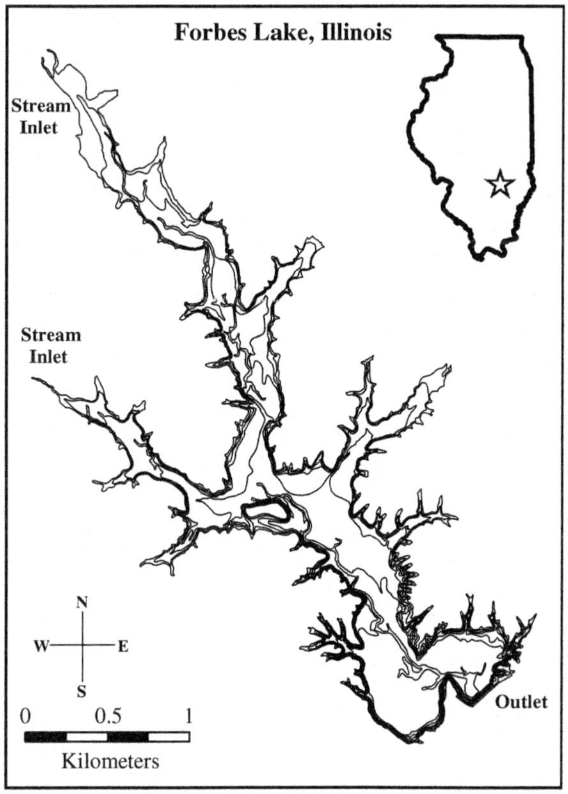

Fig. 2 Map of Forbes Lake, located in Marion County, Illinois, used for radio telemetry of juvenile muskellunge from three populations

muskellunge, and the forage fish gizzard shad *Dorosoma cepedianum*.

Methods

Muskellunge stocks and transmitter implants

Muskellunge used for this study included representatives from three source stocks—Leech Lake, Minnesota, population from the Upper Mississippi River drainage stock (MISS), Cave Run Lake, Kentucky, population from the Ohio River drainage stock (OH), and the North Spring Lake, Illinois, progeny, a mixed-origin stock (IL) (Fig. 1, Table 1). Genetic analysis using electrophoresis techniques has confirmed substantial divergence of the Upper Mississippi River drainage and Ohio River drainage stocks (Koppelman and Philipp 1986). The North Spring Lake population was established throughout the 1980s and early 1990s by multiple stocking of muskellunge from throughout the Ohio and Upper Mississippi River drainages (W. Herndon, Jr, IL DNR, personal communication). All muskellunge used in this study were reared in hatchery environments in close proximity to source population locations. Early life stages were reared in tanks before transfer to hatchery ponds with minnow forage. Muskellunge were acquired as young-of-the-year (YOY) in fall 2002 and held in ponds at the Sam Parr Biological Station, Kinmundy, Illinois. Radio transmitters were implanted in 20 age-2 muskellunge (5 MISS, 7 OH, 8 IL) on 23–24 March 2004, with implantation in two additional muskellunge (IL) on 14 May 2004 and in one muskellunge (OH) on 21 September 2004 (Table 2).

Temperature selection was measured using temperature-sensing radio transmitters (Model RI-2CT, Holohil Systems, Ontario, Canada) implanted intraperitoneally. Transmitters were cylindrical, 32 mm long and 10 mm diameter, and were equipped with a 22 cm long external whip antenna. The weight of the transmitters was 5.8 g in air, and constituted up to 2.1% (Brown et al. 1999) of fish body weight; expected battery lifetime was 12 months. Radio frequency transmitters were chosen because of the relatively shallow depths and moderate conductivity (200 μSiemens cm^{-1}) of Forbes Lake (Winter 2000). Each transmitter broadcasted a unique frequency within an operating band of 148–150 MHz. Before field use the pulse rates of these temperature-sensing transmitters were calibrated in the laboratory at known water temperatures (0.05°C accuracy).

Muskellunge were collected the day of implantation and transported to Forbes Lake (5 km). Surgical procedures were similar to those described previously (Ross and Kleiner 1982; Summerfelt and Smith 1990; Cooke et al. 2003). Fish were anaesthetized in a 60-ppm induction bath of clove oil and ethanol (Anderson et al. 1997) in lake water until the opercular rate became slow and irregular and the fish were unresponsive. After loss of equilibrium (4 min average time), fish were measured (total length, mm, and weight, g) and placed ventral side up on a foam-padded surgery table. A recirculating maintenance dose of anesthetic (30 ppm) in oxygenated water was continuously pumped over the gills. A 15–25 mm incision was made slightly dorsal to the ventral midline, anterior to the pelvic girdle and the transmitter was gently

Table 1 Sources of muskellunge stocks used for evaluation of temperature selection

Population (abbreviation)	Source water	Drainage (stock)	Latitude (north)	Cooling degree days (CDD)	Heating degree days (HDD)	Mean annual air temp. (°C)
Kentucky (OH)	Cave Run Lake	Ohio	37°35′	1,154	4,713	12.9
Illinois (IL)	North Spring Lake	Mixed	40°40′	998	6,097	10.4
Minnesota (MISS)	Leech Lake	Mississippi	46°35′	347	9,495	4.4

The Kentucky population was from the Ohio River drainage (Ohio stock) and the Minnesota population was from the Upper Mississippi River drainage (Mississippi stock). The Illinois population is progeny of a mixed-stock brood source. Cooling (CDD) and heating (HDD) degree days for each source population were calculated using a base temperature of 65°F, with 1961–1990 data from the National Oceanic and Atmospheric Administration and the Midwest Climate Center

Table 2 Contact history for age-2 muskellunge implanted with radio transmitters in Forbes Lake, IL

Fish id	Stock	Length (mm)	Days at large	Contacts	
				Final	Total
1	IL	522	257	12/06/04	46
2	IL	481	162	09/02/04	36
3	IL	519	70	06/02/04	16
4	IL	500	257	12/06/04	63
5	IL	520	212	10/22/04	53
6	IL	488	161	09/01/04	40
7	IL	503	9	04/02/04	1
8	IL	473	124	07/26/04	28
9	IL	476	19	06/02/04	9
10	IL	498	73	07/26/04	22
11	MISS	610	213	10/22/04	54
12	MISS	582	244	11/22/04	60
13	MISS	597	258	12/06/04	63
14	MISS	536	244	11/22/04	65
15	MISS	524	258	12/06/04	60
16	OH	399	90	06/22/04	18
17	OH	430	163	09/03/04	46
18	OH	438	135	08/06/04	42
19	OH	407	97	06/29/04	20
20	OH	420	257	12/06/04	56
21	OH	409	257	12/06/04	56
22	OH	430	257	12/06/04	67
23	OH	421	76	12/06/04	4

Stocks included the Illinois population (IL), the Upper Mississippi River drainage stock (MISS), and the Ohio River drainage stock (OH). All fish were implanted on 23–24 March 2004 with the exception of fishes 9 and 10 (14 May 2004) and fish 23 (21 September 2004)

inserted until completely inside the peritoneal cavity. After insertion, the viscera was shielded with a scalpel handle, and a 16-gauge hypodermic needle was inserted through the body-cavity wall to enable the antenna wire to pass to the outside of the body. The incision was closed with three simple, interrupted 3/0 monofilament sutures (Ethicon). The average time for the procedure was 7 min per fish (range 5–10 min). Before release the fish were left to recover overnight for 12–18 h post-surgery in a holding pen (1.5 m×2.5 m×1.5 m) submerged in Forbes Lake (Weller and Winter 2001). Water temperature during stocking was 11°C.

Fish locations

Muskellunge were located by boat using an SRX400 radio receiver (Lotek Engineering, Ontario, Canada), a hand-held three-element Yagi directional antenna, and aviation headphones. Initial tracking was conducted by traveling parallel to the shoreline while scanning for all the frequencies at large in the lake. Transmitter frequency was used to identify individual fish. After establishing the general location of the fish, final positions were determined from signal strength by use of a 27 cm whip antenna (Hanson and Margenau 1992; Zigler et al. 2003). Preliminary trials indicated that transmitters could be located within a 2 m radius by use of this method. To minimize disturbance, fish were approached at low speed with the receiver over the front of the boat (Jepsen et al. 2001). Tracking was continued until all fish were located or the entire lake had been covered. Boat launch location and direction of travel was randomly chosen each tracking session. When multiple trackings for an individual muskellunge indicated no movement or a suspicious location, an attempt was made to startle the fish in an effort to evoke movement, followed by attempts to recover the transmitter or the presumably deceased fish. If efforts to find the transmitter or elicit movement failed, the fish was assumed to be dead or the transmitter expelled, and was removed from the search list. Data were examined to determine the last valid location and subsequent data were eliminated from analyses.

Occupied water temperature was determined by recording the time interval between transmitter pulses (pulse width, ms) as measured automatically by the radio receiver software. Water temperature and dissolved oxygen profiles were recorded at 0.5-m intervals after each successful location (Yellow Springs Instrument Model 550). To further describe ambient environmental conditions, temperature and dissolved oxygen profiles were measured at 1-m intervals at two fixed stations, near the dam and mid-lake, during each tracking event and routinely every two weeks from April to October. A semi-continuous temperature logger (Onset Computer Corporation, Hobo Water Temp Pro) placed 1 m below the surface was also used to record temperature at 2-h intervals for the duration of the study.

All fish were located at least once every two weeks during the daytime, typically starting at approximately 0900 h, from 2 April 2004 until 6 December 2004. No data were used until 8 days after surgery, to enable the fish to become accustomed to the transmitters (Clapp et al. 1990; Markham et al. 1991). Similarly to other studies (Hanson and Margenau 1992), much movement was observed during the first 3–4 days after release; this subsequently declined to levels typical of the remainder of the study period. Seasonal variation of temperature selection among muskellunge stocks was assessed by defining seasons as spring (15 March–14 June), summer (15 June–14 September), and fall (15 September–14 December) (Burrell et al. 2000). The delineation of seasons by date corresponded well with increasing, stationary, and decreasing trends in ambient air and water temperatures throughout the study. To assess diel temperature selection, all radio-tagged muskellunge were monitored during ten diel tracking events (three in spring, four in summer, three in fall). Diel periods were defined as crepuscular (2 h before and after sunrise and sunset), day (2 h after sunrise until 2 h before sunset), and night (2 h after sunset until 2 h before sunrise) (Zigler et al. 1999). During each diel tracking event, individual fish were located approximately every 3 h during a period of 14–24 h spanning all three diel periods.

Statistical analysis

The effects of stock, diel period, and season on temperature selection by the muskellunge were tested by use of combined day and diel tracking data and three-factor analysis of variance (ANOVA) with repeated measures (Maceina et al. 1994; Littell et al. 1998) using the Proc Mixed procedure in SAS (SAS Institute 1999). Individual fish were treated as subjects with sample date specified as the repeated factor. Stock, diel period, and season were fixed factors and individual fish were a random factor. Although analysis of model residuals suggested deviation from normality (Shapiro–Wilk: $P<0.01$), graphical inspection of the residuals (e.g. normal probability plot, spread location plot) indicated a substantial level of normality. Multiple transformations were imposed on the temperature data with no improvements in model residuals. Because analysis of variance procedures are usually robust to violations of the normality assumption (Box and Anderson 1955; Srivastava 1959; Tiku 1971; Sokal and Rohlf 1981), untransformed data were used for all analyses. The Satterthwaite's approximation for the degrees of freedom was implemented in the model statement to account for unbalanced treatment-sample size (Steel and Torrie 1980; Freund et al. 1986; Yandell 1997). When ANOVA procedures detected significant differences ($\alpha<0.05$), a Tukey's multiple comparison test was used to examine how factors differed.

Results

A total of 925 occupied temperature measurements were recorded across 35 sampling dates for muskellunge from the three stocks (Table 2). The mean temperatures selected were very similar among stocks (24.0–24.2°C, Fig. 3) and did not differ statistically (ANOVA: P=0.89, Table 3). Muskellunge (stocks pooled) selected a mean temperature of 24.1°C over the duration of the study. Initial stocking sizes (length, TL mm and

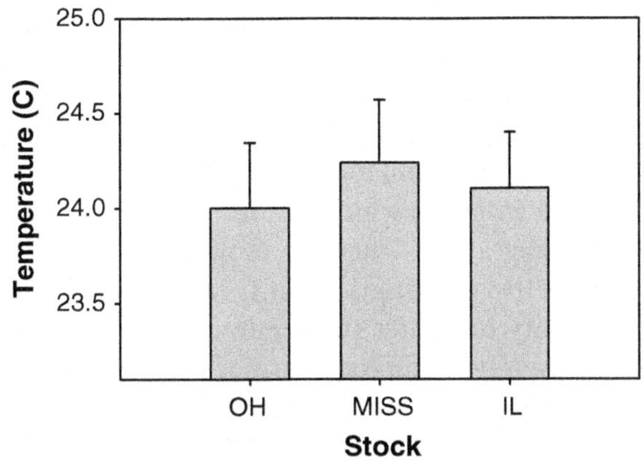

Fig. 3 Overall mean selected temperatures of muskellunge from the Ohio River drainage stock (*OH*), the Upper Mississippi River drainage stock (*MISS*), and the North Spring Lake population (*IL*) in Forbes Lake, Illinois, from 2 April through 6 December 2004. *Vertical lines* represent ±1 standard error

weight, g) were significantly different among three stocks (ANOVA: $P<0.01$, Table 2), with muskellunge from the Upper Mississippi drainage the largest, Illinois progeny intermediate, and the Ohio River drainage muskellunge the smallest. Some studies with other species have suggested that smaller juvenile fishes select and prefer significantly warmer temperatures than larger adult individuals (McCauley and Huggins 1979; Edsall 1999). Although all individuals in our study were of similar age (2 years), we investigated the potential effect of size on temperature selection by using initial length and weight as covariates. Neither initial length (ANCOVA: P=0.81) nor weight (ANCOVA: P=0.56) significantly contributed to observed variation in temperature selection among muskellunge stocks.

None of the two-way and three-way interaction terms combining stock, season, and diel period accounted for significant amounts of variation in temperature selection among muskellunge stocks (Table 3). As expected, however, the main effect of season did account for a significant amount of variation (ANOVA: $P<0.01$, Table 3). During the spring, muskellunge (stocks pooled) selected an average temperature of 21.7°C (SE=0.2, Fig. 4). Mean summer temperature selected was 28.4°C (SE=0.1); that in the fall was 14.8°C (SE=0.2, Fig. 4).

Seasonal frequency distributions of selected temperatures are usually approximately normally distributed about the mean (Fig. 5). To avoid frequency biases, multiple temperature measurements for individual fish were averaged from the diel tracking events. Distributions were slightly skewed toward cooler temperatures during the spring and fall. This is partially explained simply by the seasonal warming and cooling of the water but might also be a function of a wider range of cooler oxygenated water temperatures available when the lake is less thermally stratified. The mode of selected temperatures has been suggested as corresponding well to preferred temperatures (McCauley 1977). During the spring, all stocks selected a bimodal distribution of temperatures, with peaks of 17°C and 23–24°C (Fig. 5). Summer modes for the fish were between 27°C and 29°C. In the fall, bimodal distribution centered around 12°C and 15°C and around 13°C and 16°C were observed for the Ohio River drainage stock and the Illinois progeny muskellunge, respectively. A unimodal distribution, with a selected temperature mode of 15°C, was observed

Table 3 Analysis-of-variance type III tests of the fixed effects of stock, diel period, and season on temperature selected by juvenile muskellunge

Source of variation	Num *DF*	Den *DF*	*F*	*P*
Stock	2	18.1	0.11	0.89
Diel period	2	884	1.07	0.34
Season	2	892	1188.27	<0.01*
Stock×Diel period	4	884	0.34	0.85
Stock×Season	4	883	2.04	0.09
Season×Diel period	4	882	2.24	0.06
Stock×Season×Diel period	8	883	0.22	0.99

Satterthwaite's approximation was used for the denominator degrees of freedom. Asterisks denote significance of $P≤0.05$

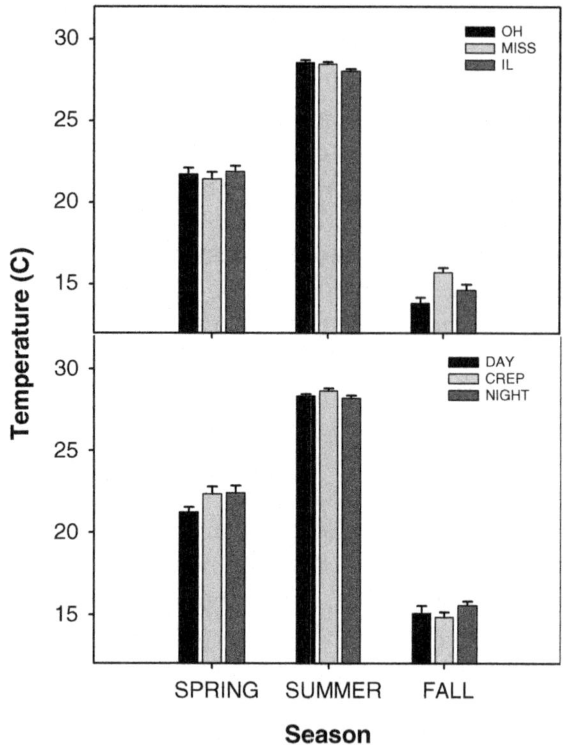

Fig. 4 Seasonal mean temperatures selected by muskellunge from three stocks (*top panel*, *OH*: Ohio River drainage stock, *MISS*: Upper Mississippi River drainage stock, *IL*: Illinois population) and of stocks pooled by diel period (*bottom panel*, *DAY*: day, *CREP*: crepuscular, *NIGHT*: night). *Vertical lines* represent ±1 standard error

for the Upper Mississippi River drainage stock (Fig. 5).

Across seasons, no variation in temperature selection among stocks was explained by the effect of diel period (ANOVA; P=0.34, Table 3). Also, both within and across seasons, no distinctive variation in the magnitude of the range of occupied temperatures (stocks pooled) was observed throughout the 24-h diel range. Not only did the mean temperature selected not vary by diel period, but the variation around the mean selected temperature did not vary across the diel range.

Water temperatures at 1 m depth in Forbes Lake ranged from a low in spring and fall of approximately 8°C to a summer high of approximately 30°C (Fig. 6). Surface temperatures were usually very similar to the 1-m temperatures; bottom temperatures ranged from 8°C to 19°C. Thermal stratification, and subsequent formation of a dissolved oxygen gradient, created an oxygen-limiting effect on available temperatures for approximately 5 months from mid-May to October. Thermal stratification and the presence of adequately oxygenated (>3 mg L^{-1}) water was available from the water surface to approximately 3°C cooler than surface temperatures during much of the growing season (Fig. 6). Despite the availability of cooler oxygenated waters, muskellunge usually occupied warmer waters near the surface (Fig. 6). Surface temperatures occasionally warmed to 35°C for brief periods during the summer; muskellunge avoided temperatures warmer than 32°C. During early spring and most of the fall the entire water column contained adequate dissolved oxygen and muskellunge typically chose a broader range of temperatures (Figs. 5 and 6).

Discussion

Variation in latitude, and consequently ambient temperatures and length of the growing season, is probably one of the most important environmental conditions leading to population differentiation (Fry 1971; Philipp 1992). Different latitudes and temperature conditions have been linked with intraspecific variations in growth (Clapp and Wahl 1996; Power and McKinley 1997; Galarowicz and Wahl 2003), survival (MacLean et al. 1981; Philipp and Whitt 1991; Philipp and Claussen 1995), and other life history characteristics (Ihssen et al. 1981). It has recently been suggested that temperature preference and selection are under the same selective pressure as other life-history characteristics and can be expected to vary with latitude of origin among a species group (Angiletta et al. 2002).

Two competing models explain the manner in which conditions, including the range of temperatures in which an organism will spend most of its time in a gradient or free-choice environment (Reynolds 1977), vary across a latitudinal gradient (i.e. among stocks). Local temperature adaptation predicts that the temperatures for optimum growth and energetic processes are adapted to the local temperature conditions and are directly

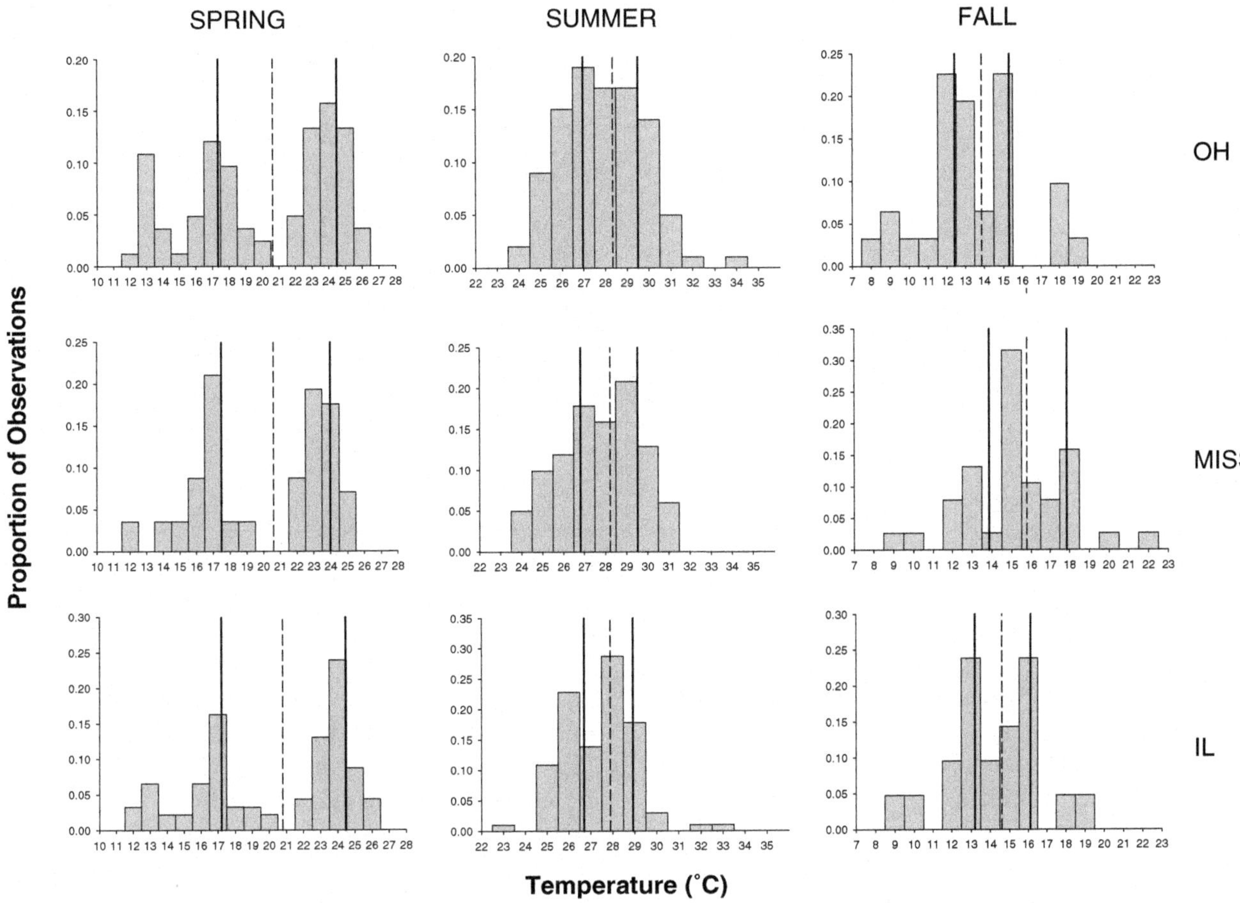

Fig. 5 Frequency histograms representing temperatures of water occupied by muskellunge of different stocks (*OH*: Ohio River drainage stocks, *MISS*: Upper Mississippi River drainage stocks, *IL*: Illinois population) in different seasons. The *dashed vertical lines* represent the unadjusted means and the *solid vertical lines* represent the 25–75% interquartile range

positively related and, consequently, individuals from cooler environments should select cooler temperatures (Levinton 1983; Yamahira and Conover 2002). Alternatively, countergradient variation focuses on differences in the length of the growing season across latitudes and predicts that individuals from cooler environments should select similar or warmer temperatures than conspecifics from warmer environments when presented with a temperature gradient (Conover and Schultz 1995; Freidenburg and Skelly 2004; Belk et al. 2005). Selection of warmer temperatures would increase growth rates and, ultimately, result in a large body at the beginning of a long winter (Conover and Present 1990).

Previous studies have drawn different conclusions about the existence and nature of temperature selection and preference differences among intraspecific stocks of freshwater fishes. A laboratory investigation of the white perch *Marone americana* final preferenda concluded that southern populations choose higher temperatures than their northern counterparts, with a middle latitude population preferring an intermediate temperature, suggesting conformation with the theory of local adaptation (Hall et al. 1978). Studies with largemouth bass, however, revealed no measurable difference in the final thermal preferenda of geographically distinct stocks (Reynolds and Casterlin 1976; Koppelman et al. 1988). We did not detect differences in temperature selection among genetically distinct stocks of muskellunge in a free-choice field environment. We hypothesized, on the basis of the theory of local adaptation, that

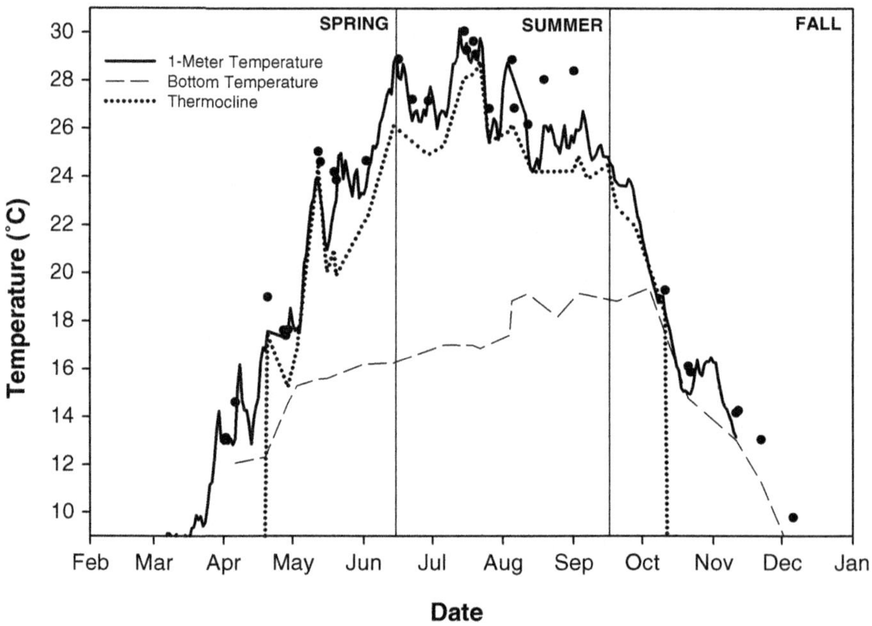

Fig. 6 Ambient temperature and oxygen conditions in Forbes Lake, Illinois, during the study. Temperature at 1 m is represented by a *solid line*, mean bottom temperature with a *dashed line*, and the mean temperature at which dissolved oxygen was ≤3 mg L^{-1} (thermocline) is shown with a *dotted line*. *Vertical lines* represent delineation of seasons based on date (Burrell et al. 2000). Mean temperatures selected by sampling date for the combined stocks are indicated by *black dots*

muskellunge from the Leech Lake, Minnesota, population would select cooler temperatures than the Cave Run Lake, Kentucky, population given the same range of available temperatures. Conversely, countergradient variation suggests that muskellunge from the Leech Lake population would occupy similar or warmer temperatures than those from the Cave Run Lake population. We observed very little variation in temperature selection, potentially lending support to the countergradient variation model for temperature selection by juvenile muskellunge. It is also possible that temperatures across the natural muskellunge range are not sufficiently heterogeneous to provide the necessary environmental grain for adaptation and differentiation of temperature selection to have occurred (MacLean and Evans 1981; Clapp and Wahl 1996).

Our study is the first to evaluate the temperature selection behavior of sub-adult muskellunge in a field environment. Across stocks, muskellunge selected a mean water temperature of 24.1°C, a value that corresponds well with previously determined final preference temperatures ranging from 24.0°C to 25.1°C in laboratory settings for juvenile muskellunge (Coutant 1977; Reynolds and Casterlin 1979; Jobling 1981). Fish do not remain stationary at a single temperature, but instead make exploratory movements into both cooler and warmer waters and it may be more appropriate to consider temperature preference as a range, rather than a fixed temperature (Jobling 1981). To this end, McCauley (1977) noted that modes (and we propose interquartile ranges) of frequency distributions correspond to the range of preference temperatures, or those temperatures occupied most frequently. Modal temperature selection across seasons and stocks, ranged from 12°C to 29°C in our study, with interquartile ranges generally encompassing 7°C in the spring, 3°C in the summer, and 3°C during the fall. Our results are also in agreement with laboratory-derived avoidance temperatures for sub-adult muskellunge. Upper avoidance and upper lethal temperatures of 32–34°C and critical thermal maximum temperatures of 33.2–36.1°C have previously been proposed (Scott and Crossman 1973; Hassan and Spotila 1976; Reynolds and Casterlin 1979; Jobling 1981). In our study, muskellunge usually avoided temperatures ranging from 32°C to 35°C, despite the availability of these temperatures near the surface for a limited time in the summer.

The direct correlation of laboratory-derived preferred temperatures with the optimum temperatures for physiological processes, typically measured as growth potential, is well established (Brett 1971; Coutant 1975; Beitinger and Fitzpatrick 1979; Beitinger and Magnuson 1979; Magnuson et al.

1979; Jobling 1981; McCauley and Casselman 1981). Clapp and Wahl (1996) observed the highest rates of consumption and growth were at 25°C for three stocks of juvenile muskellunge in the laboratory; this temperature corresponded well with published preference temperatures (24.0–25.1°C, Jobling 1981). Both growth and consumption at 25°C differed substantially between the Leech Lake, Minnesota, and Cave Run Lake, Kentucky, populations (Clapp and Wahl 1996). On the basis of these observations we expected that the temperature selected and the optimum physiological temperature would also differ. Because thermoregulatory behavior in a free-choice environment did not differ among stocks, any observed differences in growth, survival, or other performance characteristics would be a direct function of different physiological energetic factors (Koppelman et al. 1988).

In our study the methods used to determine the temperature selected give a measure of realized temperature selection as opposed to fundamental temperature selection (Coutant and Carroll 1980). Numerous factors can affect the observed, or realized, temperature selected, for example prey availability, competition, structural habitat selection, intraspecific and interspecific interactions, and the availability of adequately oxygenated water (Reynolds 1977). Only by examining stock differences in a common environment, however, can the effects of genetic and latitude of origin on phenotypic expression be differentiated (Galarowicz and Wahl 2003). Throughout the summer, only a narrow range (2–4-degree range) of oxygenated water was usually available. A limited availability of suitable water temperatures has the potential to mask temperature-selection differences among stocks; given the small variation around seasonally selected temperatures, however, we feel that the range of temperatures available was adequate to enable observation of temperature selection differences. It is possible that the method used was not sufficiently precise to detect differences among stocks. The equipment we used could measure temperatures to ±0.05°C, and should have been sufficient to detect differences, had they existed. Last, it is possible, although not presumed, that different survival or developmental acclimation during early ontogeny in the rearing ponds before the study resulted in individuals adapted to the local thermal regime. Such an effect would, however, require that individual phenotypic plasticity is more important than innate genotypic differences among stocks.

Our goal was to examine temperature selection differences in a natural field setting. Muskellunge are an important sportfish commonly stocked throughout much of the lower Midwestern United States. Awareness of the populations and stocks of fishes most appropriate for use in hatchery operations has recently increased(Clapp and Wahl 1996; Pearsons and Hopley 1999). Native stocks should always be considered first, when such populations exit (Philipp et al. 1993). Only when a proposed introduction outside the current range of the muskellunge or in a region where muskellunge have been exterminated, should transplanted stocks be considered. Although we found no differences in temperature selection among genetically distinct stocks of muskellunge, knowledge of behavioral temperature regulation can aid the identification of waters in which the temperatures preferred by muskellunge occur regularly (Beitinger and Fitzpatrick 1979; Hernandez et al. 2002). Additional evaluation of potential differences in growth, survival, and other life history characteristics are needed before recommendations regarding appropriate stocks can be made.

Future research efforts should be directed toward further understanding the relative importance of temperature, dissolved oxygen, physical structure, and prey availability in determining the distribution of juvenile muskellunge. A large body of literature addresses these potential tradeoffs for striped bass *Morone saxatilis* (Coutant and Carroll 1980; Cheek et al. 1985; Coutant 1985, 1990; Bettoli 2005). Northern pike *Esox lucius*, a species closely related to the muskellunge, have been reported to occupy the coolest water available with dissolved oxygen concentrations ≥3 mg L^{-1} during the summer months in a southern Ohio reservoir (Headrick and Carline 1993). Individuals lost weight during this period in an attempt to behaviorally thermoregulate for the coolest available temperatures. Smallmouth bass have been shown to occupy water at temperatures greater than those preferred, for extended periods, when forage and cover were disproportionately high at the warm end of a temperature gradient (Bevelhimer 1996). It is possible that muskellunge in our study periodically abandoned preferred temperatures in exchange for remaining in locations of suitable physical structure or where prey was abundant. Fish fed to satiation prefer a warmer temperature than individuals feeding with limited rations (Magee et al. 1999, Despatie et al. 2001). Muskellunge with high prey resources, for example gizzard shad in Forbes Lake, may have the ability to cope metabolically with the demands associated with occupation of warm water. Trade-offs between preferred temperature and oxygen, physical habitat, and prey may have masked among-stock differences in temperature selection. Also, juvenile fish typically prefer water warmer than that preferred by mature conspecifics (McCauley and Huggings 1979; Edsall 1999; Magee et al. 1999). Future assessments of possible temperature-selection differences among adult stocks of muskellunge in similar field settings are needed.

Acknowledgements We thank M. Anderson, B. Bauer, J. Godbout, S. Ostrand, P. Port, and M. Wegener for field assistance. S. Butler made useful comments that improved the manuscript. We thank M. Hearn and the staff of the Minor E. Clark Fish Hatchery, Kentucky Department of Fish and Wildlife, for providing Cave Run Lake, KY, muskellunge. The Central Illinois Muskie Hunters (CIMH), Muskies, Inc. supported the study and purchased the Leech Lake, MN, fish. We thank J. Sacco and S. Rusteberg for help initiating the study and CIMH past-presidents J. Gillis, L. Nevling, D. Serck, and J. Sacco for ongoing support. Support was provided in part through funds administered by the Research Committee of Muskies, Inc. International and were coordinated by S. Budnik. Additional support was provided by Federal Aid in Sport Fish Restoration project F-151-R administered by the Illinois Department of Natural Resources. In particular, S. Pallo, S. Krueger, L. Dunham, and S. Stuewe helped coordinate activities with the Division of Fisheries, Illinois Department of Natural Resources. The Illinois Natural History Survey and the Department of Natural Resources and Environmental Sciences, University of Illinois, also provided support.

References

Altukhov YP (1981) The stock concept from the viewpoint of population genetics. Can J Fish Aquat Sci 38:1523–1538

Anderson WG, McKinley RS, Colavecchia M (1997) The use of clove oil as an anesthetic for rainbow trout and its effects on swimming performance. North Am J Fish Manage 17:301–307

Angilletta MJ, Niewiarowski PH, Navas CA (2002) The evolution of thermal physiology in ectotherms. J Thermal Biol 27:249–268

Austen DJ, Petersen JT, Newman B, Sobaski ST, Bayley PB (1993) Compendium of 143 Illinois Lakes: Bathymetry, physico-chemical features, and habitats, Vol 2—Lakes of Regions 4 and 5. Aquatic Ecology Technical Report 93/9, Illinois Department of Natural Resources, Champaign, Illinois, USA

Barwick DH, Foltz JW, Rankin DM (2004) Summer habitat use by rainbow trout and brown trout in Jocassee Reservoir. North Am J Fish Manage 24:735–740

Begg GA, Friedland KD, Pearce JB (1999) Stock identification and its role in stock assessment and fisheries management: an overview. Fish Res 43:1–8

Beitinger TL, Fitzpatrick LC (1979) Physiological and ecological correlates of preferred temperature: preferenda versus optima. Am Zool 19:319–329

Beitinger TL, Magnuson JJ (1979) Growth rates and temperature selection of bluegill, Lepomis macrochirus. Trans Am Fish Soc 108:378–382

Belk MC, Johnson JB, Wilson KW, Smith ME, Houston DD (2005) Variation in intrinsic individual growth rate among populations of leatherside chub (Snyderichthys copei Jordan & Gilbert): adaptation to temperature or length of growing season? Ecol Freshwater Fish 14:177–184

Bettoli PW (2005) The fundamental thermal niche of adult landlocked striped bass. Trans Am Fish Soc 134:305–314

Bevelhimer MS (1996) Relative importance of temperature, food, and physical structure to habitat choice by smallmouth bass in laboratory experiments. Trans Am Fish Soc 125:274–283

Box GEP, Anderson SL (1955) Permutation theory in the derivation of robust criteria and the study of departures from assumptions. J Roy Statist Soc B17:1–34

Brett JR (1971) Energetic responses of salmon to temperature: a study of some thermal relations in the physiology and freshwater ecology of sockeye salmon Oncorhynchus nerka. Am Zool 11:99–113

Brown RS, Cooke SJ, Anderson WG, McKinley RS (1999) Evidence to challenge the '2% rule' for biotelemetry. North Am J Fish Manage 19:867–871

Burrell KH, Isely JJ, Bunnell DB, Van Lear DH, Dolloff CA (2000) Seasonal movement of brown trout in a southern Appalachian river. Trans Am Fish Soc 129:1373–1379

Cheek TE, Van Den Avyle MJ, Coutant CC (1985) Influences of water quality on distribution of striped bass in a Tennessee River impoundment. Trans Am Fish Soc 114:67–76

Clapp DF, Clark RD Jr, Diana JS (1990) Range, activity, and habitat of large, free-ranging brown trout in a Michigan stream. Trans Am Fish Soc 119:1022–1034

Clapp DF, Wahl DH (1996) Comparison of food consumption, growth, and metabolism among muskellunge—an investigation of population differentiation. Trans Am Fish Soc 125:402–410

Conover DO, Present TMC (1990) Countergradient variation in growth rate: compensation for length of growing season among Atlantic silversides from different latitudes. Oecologia 83:316–324

Conover DO, Schultz ET (1995) Phenotypic similarity and the evolutionary significance of countergradient variation. Trend Ecol Evol 10:248–252

Cooke SJ, Graeb BDS, Suski CD, Ostrand KG (2003) Effects of suture material on incision healing, growth and survival of juvenile largemouth bass implanted with miniature radio transmitters: case study of a novice and experienced fish surgeon. J Fish Biol 62:1366–1380

Coutant CC (1975) Responses of bass to natural and artificial temperature regimes. In: Clepper H (ed) Black bass biology and managements. Sport Fishing Institute, Washington DC, USA, pp 272–285

Coutant CC (1977) Compilation of temperature preference data. J Fish Res Board Can 34:739–745

Coutant CC (1985) Striped bass, temperature, and dissolved oxygen: a speculative hypothesis for environmental risk. Trans Am Fish Soc 114:31–61

Coutant CC (1987) Thermal preference: when does an asset become a liability? Environ Biol Fishes 18:161–172

Coutant CC (1990) Temperature-oxygen habitat for freshwater and coastal striped bass in a changing climate. Trans Am Fish Soc 119:240–253

Coutant CC, Carroll DS (1980) Temperatures occupied by ten ultrasonic-tagged striped bass in freshwater lakes. Trans Am Fish Soc 109:195–202

Crossman EJ (1978) Taxonomy and distribution of North American esocids. Am Fish Soc Special Publication 11:13–26

Crossman EJ (1986) The noble muskellunge: a review. Am Fish Soc Special Publication 15:1–13

Despatie SP, Castonguay M, Chabot D, Audet C (2001) Final thermal preferendum of Atlantic cod: effect of food ration. Trans Am Fish Soc 130:263–275

Edsall TA (1999) Preferred temperatures of juvenile lake whitefish. J Great Lakes Res 25:583–588

Freidenburg LK, Skelly DK (2004) Microgeographical variation in thermal preference by an amphibian. Ecol Lett 7:369–373

Freund R, Little R, Spector P (1986) SAS system for linear models. SAS Institute, Cary, North Caroline, USA

Fry FEJ (1971) The effect of environmental factors on the physiology of fish. In: Hoar WS, Randall DJ (eds) Fish physiology, vol 6. Academic Press, New York, New York, USA, pp 1–98

Galarowicz TL, Wahl DH (2003) Differences in growth, consumption, and metabolism among walleyes from different latitudes. Trans Am Fish Soc 132:425–437

Hall LW Jr, Hocutt CH, Stauffer JR Jr (1978) Implication of geographic location on temperature preference of white perch, Morone americana. J Fish Res Board Can 35:1464–1468

Hanson DA, Margenau TL (1992) Movement, habitat selection, behavior, and survival of stocked muskellunge. North Am J Fish Manage 12:474–483

Hassan KC, Spotila JR (1976) The effect of acclimation on the temperature tolerance of young muskellunge fry. In: Thermal Ecology II U.S. National Technological Information Services, CONF-750425, Springfield, Virginia, USA, pp 136–140

Headrick MR, Carline RF (1993) Restricted summer habitat and growth of northern pike in two southern Ohio impoundments. Trans Am Fish Soc 122: 228–236

Hernandez M, Buckle LF, Espina S (2002) Temperature preference and acclimation in *Poecilia sphenops* (Pisces, Poeciliidae). Aquat Res 33:933–940

Huey RB (1991) Physiological consequences of habitat selection. Am Nat 137: S91–S115

Ihssen PE, Booke HE, Casselman JM, McGlade JM, Payne NR, Utter FM (1981) Stock identification: materials and methods. Can J Fish Aquatic Sci 38:1838–1855

Jepsen N, Beck S, Skov C, Koed A (2001) Behavior of pike (*Esox lucius* L.) >50 cm in a turbid reservoir and in a clearwater lake. Ecol Freshwater Fish 10:26–34

Jobling M (1981) Temperature tolerance and the final preferendum—rapid methods for the assessment of optimum growth temperatures. J Fish Biol 19:439–455

Koppelman JB, Philipp DP (1986) Genetic application in muskellunge management. In: Hall GE (ed) Managing Muskies. American Fisheries Society Special Publication, 15 Bethesda, Maryland, USA, pp 111–121

Koppelman JB, Whitt GS, Philipp DP (1988) Thermal preferenda of northern, Florida, and reciprocal F1 hybrid largemouth bass. Trans Am Fish Soc 117:238–244

Levinton JS (1983) The latitudinal compensation hypothesis: growth data and a model of latitudinal growth differentiation based upon energy budgets. I. interspecific comparison of Ophryotrocha (Polychaeta: Dorvilleidae). Biol Bull (Woods Hole) 165:686–698

Littell RC, Henry PR, Ammerman CB (1998) Statistical analysis of repeated measures data using SAS procedures. J Anim Sci 76:1216–1231

Maceina MJ, Bettoli PW, Devries DR (1994) Use of a split-plot analysis of variance design for repeated-measures fishery data. Fisheries 19:14–20

MacLean JA, Evans DO (1981) The stock concept, discreteness of fish stocks, and fisheries management. Can J Fish Aquatic Sci 38: 1889–1898

MacLean JA, Evans DO, Martin NV, DesJardine RL (1981) Survival, growth, spawning distribution, and movements of introduced and native lake trout Salvelinus namaycush in two inland Ontario Lakes. Can J Fish Aquatic Sci 38:1685–1700

Magee A, Myrick CA, Cech JJ (1999) Thermal preference of female threespine sticklebacks under fed and food-deprived conditions. Calif Fish Game 85:102–112

Magnuson JJ, Crowder LB, Medvick PA (1979) Temperature as an ecological resource. Am Zool 19: 331–343

Margenau TL, Hanson DA (1996) Survival and growth of stocked muskellunge: effects of genetic and environmental factors. Wisconsin Department of Natural Resources Research Report 172, pp 1–11

Margenau TL, Hanson DA (1997) Performance of Leech Lake, Minnesota, muskellunge in a Wisconsin lake. Wisconsin Department of Natural Resources Research Report 175, pp. 1–11
Markham JL, Johnson DL, Petering RW (1991) White crappie summer movements and habitat use in Delaware Reservoir, Ohio. North Am J Fish Manage 11:504–512
Mayr E (1975) Population, species, and evolution. Belknap Press of Harvard University Press, Cambridge, Massachusetts and London, England 410 pp
McCauley RW (1977) Laboratory methods for determining temperature preferences. J Fish Res Board Can 34:749–752
McCauley RW, Casselman JM (1981) The final preferendum as an index of the temperature for optimum growth in fish. In: Proceedings of the world symposium on aquaculture in heated effluents and recirculation systems, vol 2. Student Aquaculture Association of the United Nations, Berlin, pp 81–93
McCauley RW, Huggins NW (1979) Ontogenetic and nonthermal seasonal effects on thermal preferenda of fish. Am Zool 19:267–272
Pearsons TN, Hopley CW (1999) A practical approach for assessing ecological risks associated with fish stocking programs. Fisheries 24:16–23
Philipp DP, Whitt GS (1991) Survival and growth of northern, Florida, and reciprocal F1 hybrid largemouth bass in Central Illinois. Trans Am Fish Soc 120:58–64
Philipp DP (1992) Stocking Florida largemouth bass outside of its native range. Trans Am Fish Soc 121:686–691
Philipp DP, Epifanio JM, Jennings MJ (1993) Point counterpoint: conservation genetics and current stocking practices—are they compatible? Fisheries 18:14–17
Philipp DP, Claussen JE (1995) Fitness and performance differences between two Stocks of largemouth bass from different river drainages within Illinois. Am Fish Soc Symposium 15:236–243
Power M, McKinley RS (1997) Latitudinal variation in lake sturgeon size as related to the thermal opportunity for growth. Trans Am Fish Soc 126:549–558
Reynolds WW, Casterlin ME (1976) Thermal preferenda and behavioral thermoregulation in three centrarchid fishes. In: Esch GW, McFarlane RW (eds) Thermal ecology II. U.S. National Technological Information Services, CONF-750425, Springfield, Virginia, USA, pp 185–190
Reynolds WW, Casterlin ME (1979) Thermoregulatory rhythm in juvenile muskellunge (Esox masquinongy): evidence of a diel shift in the lower set-point. Comp Biochem Physiol 63A:523–525
Reynolds WW (1977) Temperature as a proximate factor in orientation behavior. J Fish Res Board Can 34:734–739
Richards FP, Reynolds WW, McCauley RW, Crawshaw LI, Coutant CC, Gift JJ (1977) Temperature preference studies in environmental impact assessments: an overview with procedural recommendations. J Fish Res Board Can 34:728–761
Ross MJ, Kleiner CF (1982) Shielded-needle technique for surgically implanting radio-frequency transmitters in fish. Progr Fish Culturist 44:41–43
SAS Institute Inc (1999) SAS OnlineDoc®, Version 8. Cary, North Carolina, USA
Scott WB, Crossman EJ (1973) Freshwater fishes of Canada. Fisheries Research Board of Canada, Ottawa, Canada
Sokal RR, Rohlf FJ (1981) Biotelemetry. Freeman, New York, New York, USA
Srivastava ABL (1959) Effects of non-normality of the power of the analysis of variance test. Biometrika 46:114–122
Steel RGD, Torrie JH (1980) Analysis of covariance. In: Principles and procedures of statistics: a biometrical approach. McGraw–Hill, New York, pp 401–437
Summerfelt RC, Smith LS (1990) Anesthesia, surgery, and related techniques. In: Schreck CB, Moyle PB (eds) Methods for fish biology. American Fisheries Society, Bethesda, Maryland, USA, pp 213–272
Tiku ML (1971) Power function of *F*-test under non-normal situations. J Am Statist Assoc 66:913–916
Tracy CR (1982) Biophysical modeling in reptilian physiology and ecology. In: Gans C, Pough FH (eds) Biology of the reptilia, volume 12, physiology C. Academic Press, London, pp 275 321
Weller RR, Winter JD (2001) Seasonal variation in home range size and habitat use of flathead catfish in Buffalo Springs Lake, Texas. North Am J Fish Manage 21:792–800
Winter JD (2000) Designing telemetry studies and other technical and analytical considerations. In: Eiler JH, Alcorn D, Neuman M (eds) Biotelemetry 15: Proceedings of the 15th International Society of Biotelemetry. Juneau, Alaska, U.S.A. International Society on Biotelemetry. Wageningen, the Netherlands, pp 229–247
Yamahira K, Conover DO (2002) Intra- vs. interspecific latitudinal variation in growth: adaptation to temperature or seasonality? Ecology 83:1252–1262
Yandell BS (1997) Practical data analysis for designed experiments. Chapman & Hall, London UK 437 pp
Younk JA, Strand RF (1992) Performance evaluation of four muskellunge Esox masquinongy strains in two Minnesota lakes. Minnesota Department of Natural Resources Section of Fisheries Investigational Report No. 418. 22 pp
Zigler SJ, Dewey MR, Knights BC (1999) Movement and habitat use by paddlefish in Navigation Pool 8 of the upper Mississippi River. North Am J Fish Manage 19:180–187
Zigler SJ, Dewey MR, Knights BC, Runstrom AL, Steingraeber MT (2003) Movement and habitat use by radio-tagged paddlefish in the upper Mississippi River and tributaries. North Am J Fish Manage 23:189–205

Environ Biol Fish (2007) 79:99–109
DOI 10.1007/s10641-006-9063-y

ORIGINAL PAPER

Expulsion of miniature radio transmitters along with eggs of muskellunge and northern pike—a new method for locating critical spawning habitat

Rodney B. Pierce · Jerry A. Younk · Cynthia M. Tomcko

Received: 7 September 2005 / Accepted: 11 April 2006 / Published online: 8 July 2006

Abstract Identification and protection of critical spawning habitat for muskellunge *Esox masquinongy* and northern pike *Esox lucius* is important for preserving the reproductive potential of both species. In this study, we implanted miniature radio transmitters through the oviduct into the egg masses of female muskellunge and northern pike just prior to spawning. This non-surgical procedure was a novel approach for identifying spawning sites when transmitters were expelled with the eggs during egg deposition. Preliminary studies in three lakes showed that muskellunge and northern pike deposited many of the transmitters in likely spawning habitat. An inability to find eggs limited our validation of this method, but nevertheless, a relatively high proportion (70%) of northern pike larger than 690 mm (27.2 inches) expelled transmitters in a previously known spawning area in Willow Lake, Minnesota. Shoreline vegetation in that area consisted primarily of sedges *Carex* spp., and the adjacent water was shallow with substrate consisting of large mats of water bulrush *Scirpus subterminalis*. A lower proportion (50%) of muskellunge expelled transmitters in Elk Lake, Minnesota. Water depth at likely spawning sites averaged 1.1 m (3.6 feet) and vegetative cover was variable, but *Chara* spp. was common to most sites. In Moose Lake, Minnesota, containing sympatric populations of muskellunge and northern pike, 60% of muskellunge and 90% of pike expelled transmitters. *Chara* spp. beds were the predominant substrate where transmitters were expelled in Moose Lake, but the two species deposited transmitters on deepwater bars (3.7–5.2 m) in addition to shallow near-shore habitat. These results suggest more flexibility in depths used for spawning than typically reported for muskellunge and northern pike.

Keywords Egg deposition · Oviduct · Radio telemetry · Spawning sites

R. B. Pierce (✉) · C. M. Tomcko
Minnesota Department of Natural Resources, Division of Fisheries and Wildlife, 1201 East Highway 2, Grand Rapids, MN 55744, USA
e-mail: rod.pierce@dnr.state.mn.us

J. A. Younk
Minnesota Department of Natural Resources, Division of Fisheries and Wildlife, 2114 Bemidji Avenue, Bemidji, MN 56601, USA

Introduction

Natural shorelines and their associated watersheds are under increasing pressure from human development, and development has caused degradation of fish habitat (Burns 1991; Cross and McInerny 1995; Engel and Pederson 1998; Radomski and Goeman 2001) that has included eutrophication and changes in aquatic vegetative cover. Populations of muskellunge *Esox masquinongy* and northern pike *Esox lucius* are sensitive to shoreline and related land development that removes vegetation, reduces water quality, and adds biological oxygen demand to sediments (Brynildson 1958; Dombeck et al. 1984). Eutrophication affects muskellunge eggs and larvae that are intolerant of low dissolved oxygen concentrations in a microzone along the bottom where they are dispersed (Dombeck et al. 1984). Northern pike are closely associated with wetlands and aquatic vegetation, especially for spawning (Becker 1983; Bry 1996). Northern pike can exhibit natal-site fidelity, and both muskellunge and northern pike exhibit spawning-site fidelity (Crossman 1990; Miller et al. 2001), which suggests that protection and management of traditional spawning sites is warranted. Identification of these critical habitats is an important step toward protecting the reproductive potential of both species.

A unique approach for identifying spawning habitat was first reported by Pierce (2004) who implanted miniature radio transmitters through the oviduct into the egg masses of mature female northern pike just prior to spawning. This nonsurgical procedure offered potential for identifying egg deposition sites when transmitters were expelled with eggs during spawning. Spawning sites were then found by locating the shed transmitters. The preliminary study by Pierce (2004) indicated that advances in miniaturization have led to transmitters that are small enough to fit up through the oviducts of northern pike. The study further suggested that oviduct insertion of miniature transmitters offered cost- and time-saving advantages over other approaches for finding spawning habitat, offered potential for avoiding stresses on the fish associated with other transmitter attachment methods such as surgery, and that small fish (< 650 mm total length) were unable to expel the transmitters.

For this study, we extended the work by Pierce (2004) with additional trials implanting miniature radio transmitters through the oviducts of both muskellunge and northern pike females just prior to spawning. Three additional trials were designed to (1) test the effectiveness of an alternate transmitter shape, (2) compare transmitter expulsion rates for different fish sizes, (3) compare three sizes of transmitters for use in muskellunge, and (4) apply the technique in a lake where muskellunge and northern pike coexist, but where spawning sites had not previously been identified. Using this novel approach, we described likely esocid spawning habitat in three Minnesota lakes.

Material and methods

Effects of fish size and alternate transmitter shape

During spring 2003, we tested the usefulness of altering transmitter shape by reforming the potting resin to obtain a more tapered or torpedo-shaped posterior (antenna) end. The tapered end was intended to allow fish to more readily expel transmitters compared with standard transmitters having a flat posterior surface. Northern pike were captured using 25–mm bar mesh trap nets. The nets had rectangular frames (0.9 m × 1.8 m), two throats, and 12-m leads that extended into shore. Radio transmitters were inserted in 20 northern pike trapped during 19–21 April while they were ripe (ready to expel eggs) and staging for spawning in Willow Lake. Willow Lake (47°9′ N latitude and 93°46′ W longitude) is a 96-ha lake with a maximum depth of 14 m and total alkalinity of 103 mg $CaCO_3$/l. The fish were trapped in a shallow bay in the southwest end of the lake where the ice recedes from the shoreline earliest in the spring. The bay also provides the principal known spawning area for northern pike.

Each fish was implanted with an Advanced Telemetry Systems (ATS) model F1410 miniature radio transmitter[1]. Each transmitter was 6 mm in diameter and 18 mm long, weighed 1.0 g, and had a thin 25-cm-long dipole antenna. The tapered portion extended 3–4 mm beyond the end of the transmitter. Transmitters were inserted approximately 3.5 cm (1.4 inches) up the oviduct and into the egg mass with the antenna trailing out through the oviduct (Fig. 1). Transmitted frequencies were between 48.470 and 48.971 mHz

[1] Use of trade names does not imply endorsement of product

with at least 10 kHz difference between frequencies.

We attempted to locate fish at least once per day with a receiver and hand-held loop antenna. At close range, a short (15 cm) whip antenna or underwater antenna (construction described in detail by Kenow et al. 1992) was used to more accurately locate transmitters. Each fish was tracked for up to 19 days or until the transmitter was expelled. Coordinates for daily transmitter locations were recorded using a geographic positioning system (GPS) receiver, and if a transmitter became stationary for several days, we assumed it was expelled. Daily locations for individual transmitters were transferred to a Landview (Minnesota Department of Natural Resources geographic information system) map of Willow Lake. The effect of transmitter shape on expulsion by small northern pike was tested by tagging 10 small females (501–639 mm; Table 1), and comparing their transmitter expulsion rates with rates from spring 2002 (data from Pierce 2004; Table 1). The effect of fish size on expulsion was tested by comparing expulsion rates for 10 large fish (691–831 mm; Table 1) with rates of the smaller fish and with rates from 2002 (data also from Pierce 2004). Statistical comparisons of transmitter expulsion rates were performed with traditional chi-square goodness-of-fit tests and comparisons of fish lengths were with two-sample T-tests. We accepted a significance level of $P < 0.10$ for statistical tests.

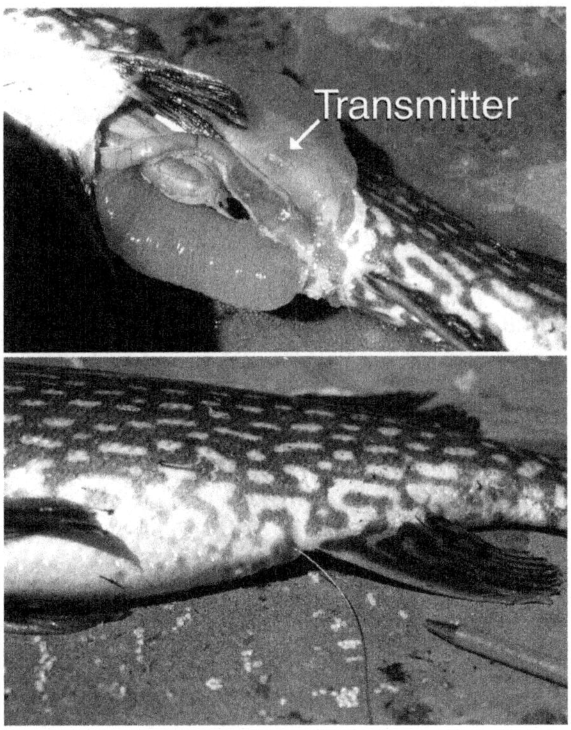

Fig. 1 Photographs illustrating the location of a transmitter within an ovary, and the antenna trailing out from the oviduct of a female northern pike

Comparison of transmitter sizes for muskellunge

We compared expulsion rates for three sizes of radio transmitters inserted through oviducts of adult female muskellunge. A total of 15 female muskellunge were trapped and implanted with radio transmitters while they were staging for spawning during 26 April–2 May 2003 in Elk Lake, Minnesota. Elk Lake (47°11′ N latitude and 95°13′ W longitude) is a 110-ha lake with a maximum depth of 28 m and total alkalinity of 155 mg $CaCO_3$/l. This population has been managed as brood stock for Minnesota's muskellunge hatchery program. Trap nets had 19-mm-bar mesh, 1.5 m × 1.8 m frames, two throats, and 30-m leads extending into shore. Gonad maturation stage for fish implanted with transmitters ranged from green to ripe. Fish were 1048–1211 mm total length when captured and released in Elk Lake (Table 2).

Transmitter size was randomly assigned, with five fish implanted with each transmitter size (Table 2). The smallest transmitter was ATS model F1410. Intermediate in size was ATS model F1420, which weighed 1.5 g and was 7.6 mm in diameter. Largest was ATS model F1630 weighing 2.5 g and having a diameter of 9 mm. Transmitter frequencies were 48.014–48.407 mHz with a minimum of 10 kHz between frequencies. Transmitter locations were monitored during 28 April–4 June 2003. Other procedures were similar to those used for northern pike.

Location of northern pike and muskellunge spawning habitat in Moose Lake

Use of radio transmitters was extended to the practical application of locating esocid spawning habitat in a lake where northern pike and muskellunge coexist. Ten female northern pike and 10

Table 1 Lengths and radio frequencies for implanted northern pike, along with the fate of the transmitters and dates they were deposited in Willow Lake, 2002–2003. Transmitters were implanted during 18–19 April 2002 and 19–21 April 2003

Total length (mm)	Frequency (mHz)	Deposition location or fate of transmitter	Date deposited
2002 Data from Pierce (2004), used standard (flat-shape) transmitters			
567	48.461	Still at large	–
577	48.491	Still at large	–
602	48.511	Recaptured with gill net	–
652	48.551	Recaptured with gill net	–
658	48.541	Located in spawning area	23 April
687	48.481	Recaptured with gill net	–
732	48.571	Still at large	–
734	48.591	Located in southwest end of lake	22 April
752	48.530	Located in spawning area	4 May
812	48.561	Located in spawning area	2 May
2003 Small northern pike, used tapered transmitters			
501	48.751	Located in northeast end of lake	4 May
541	48.891	Located in spawning area	26 April
580	48.931	Located in southwest end of lake	1 May
598	48.951	Located in spawning area	2 May
601	48.561	Still at large	–
622	48.731	Located in middle of south bay	3 May
625	48.771	Still at large	–
630	48.611	Located in spawning area	29 April
633	48.671	Located in middle of south bay	1 May
639	48.911	Still at large	–
2003 Large northern pike, used tapered transmitters			
692	48.711	Still at large	–
692	48.831	Located at north end of lake	2 May
700	48.791	Located in spawning area	2 May
700	48.851	Still at large	–
702	48.501	Located in spawning area	26 April
750	48.691	Located in spawning area	26 April
782	48.811	Located in spawning area	26 April
798	48.971	Located in spawning area	26 April
822	48.871	Located in spawning area	26 April
831	48.470	Located in spawning area	30 April

female muskellunge were trapped and implanted with transmitters in spring 2004. Fish were captured and released in Moose Lake (47°23′ N latitude and 93°41′ W longitude), a 512-ha lake with a maximum depth of 19 m and total alkalinity of 129 mg $CaCO_3$/l. Trap nets were the same nets used at Elk Lake, and were fished at locations uniformly located around the entire shoreline of the lake. Northern pike were implanted with transmitter model F1410 and muskellunge with model F1420, and all transmitters had a tapered posterior end. Northern pike were captured and transmitters inserted during 24–26 April in fish beginning to expel eggs and ranging in size from 675–879 mm total length (Table 3). Muskellunge that were still a bit green were implanted during 25–30 April and were 835–1160 mm total length (Table 3). Northern pike transmitters operated at frequencies of 48.010–48.191 mHz and muskellunge transmitters operated at 48.794–48.970 mHz, allowing at least 10 kHz between frequencies. Transmitter locations were monitored almost daily (27/33 days) during 26 April–28 May 2004. Location coordinates, depth and substrate type were recorded for transmitter deposition sites, then coordinates were transferred to a Landview map of Moose Lake. Substrates and aquatic vegetation at deposition sites were described from grab samples taken with an Ekman dredge. Scap nets and visual observations with an underwater viewing scope were used to try to find eggs in areas where transmitters were shed.

Table 2 Lengths and radio frequencies for three sizes of transmitters implanted in muskellunge, along with the fate of the transmitters and dates they were deposited in Elk Lake, 2003

Total length (mm)	Frequency (mHz)	Deposition location or fate of transmitter	Date implanted	Date deposited
Small transmitters				
1165	48.014	Located in southwest bay	27 April	16 May
1112	48.034	Still at large	27 April	–
1163	48.055	Expelled during recapture[a]	2 May	–
1048	48.075	Still at large	28 April	–
1081	48.095	Still at large	29 April	–
Intermediate transmitters				
1052	48.114	Still at large	2 May	–
1211	48.134	Located at northwest shore	27 April	19 May
1059	48.154	Located in southwest bay	28 April	19 May
1173	48.174	Located at northwest shore	30 April	16 May
1105	48.194	Still at large	1 May	–
Large transmitters				
1170	48.337	Located at west shore	29 April	19 May
1174	48.358	Found dead[a]	1 May	–
1194	48.371	Still at large	30 April	–
1181	48.397	Found dead[a]	26 April	–
1206	48.407	Located in mouth of creek	28 April	14 May

[a]Excluded from analysis

Table 3 Lengths and radio frequencies for implanted northern pike and muskellunge, along with the transmitter deposition location and depth, and dates they were implanted and found deposited in Moose Lake, 2004

Total length (mm)	Frequency (mHz)	Deposition location or fate of transmitter	Date implanted	Date deposited
Northern pike				
675	48.071	Deepwater bar, 5.8 m	26 April	7 May
682	48.010	Deepwater bar, 5.5 m	24 April	10 May
682	48.151	Lost signal	26 April	–
711	48.131	Deepwater bar, 5.2 m	25 April	5 May
715	48.051	Moose Creek	25 April	27 April
730	48.111	Near south shore, 1.8 m	26 April	9 May
752	48.191	Deepwater bar, 5.2 m	26 April	5 May
810	48.091	Deepwater bar, 4.9 m	25 April	27 April
821	48.031	Moose Creek	24 April	27 April
879	48.171	Expulsion due to handling?	26 April	27 April
Muskellunge				
835	48.911	Northeast shore, 0.6 m	28 April	9 May
900	48.870	Still at large	29 April	–
1000	48.930	Northeast shore, 1.5 m	25 April	10 May
1000	48.951	Still at large	25 April	–
1020	48.814	Deepwater bar, 3.7 m	25 April	7 May
1030	48.970	Deepwater bar, 3.7 m	30 April	20 May
1075	48.834	South shore, 2.1 m	27 April	27 May
1090	48.794	Still at large	28 April	–
1125	48.891	Northeast shore, 1.8 m	28 April	11 May
1160	48.854	Caught by angler	29 April	–

Results

Effects of fish size and transmitter shape

A relatively high proportion of large northern pike expelled their transmitters in the previously known spawning area in Willow Lake during spring 2003. Mean total length of the large northern pike was 747 mm (SE = 18 mm; Table 1). Eight out of 10 large fish expelled transmitters, and seven of them were shed in the relatively small known spawning area at the west side of the south bay of Willow Lake (Fig. 2). Shoreline vegetation along that portion of the lake consists primarily of sedges *Carex* spp., and the adjacent water is shallow and has large mats of water bulrush *Scirpus subterminalis* (where all seven transmitters were located). One transmitter was deposited in an unlikely spawning location, over sand substrate in water less than 1 m deep at the north end of the lake. Two transmitters were retained by large fish. Large northern pike expelled transmitters during the one-week period between 26 April and 2 May 2003, although most of the spawning appeared to occur on 26 April, a warm, sunny spring day. Five transmitters were shed in the spawning area on 26 April, one on 30 April, and one on 2 May. Although radio signals were located to within about a 2–5 m^2 area, we were unable to see or recover any of the transmitters.

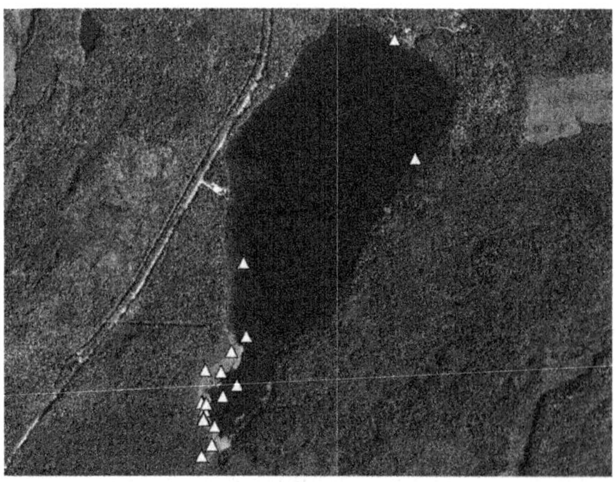

Fig. 2 Locations of transmitter deposition sites for all northern pike in Willow Lake, 2002–2003 (n = 19 locations). The cluster of points at the very southwest end of the lake contains 9 triangles, but because of the very close proximity of transmitters deposited in this important spawning location, several of the triangles are obscured by others

Fewer small northern pike shed transmitters in likely spawning habitat in Willow Lake. Mean total length of the small fish implanted with transmitters during spring 2003 was 597 mm (SE = 14 mm; Table 1). Although seven out of 10 fish expelled transmitters, only three were shed in the previously known spawning area. The three transmitters were shed on 26 April, 29 April, and 2 May. Three other transmitters were expelled in deeper (1.5–1.8 m) water near the previously known spawning area over sediments consisting of *Chara* spp. or muck with various plant fragments. A final transmitter was shed in a different bay at the northeast end of the lake over a bottom consisting of *Potamageton amplifolius, Chara* spp., and silt. None of the transmitters were recovered. In spite of low sample sizes, the proportion of large fish depositing transmitters in known spawning grounds in 2003 seemed to be greater than for small fish (Chi-square = 3.20, $df = 1$, $P = 0.07$). The most consistent deposition of transmitters at the known spawning area was for northern pike >700 mm total length (Table 1).

Altering the shape of the transmitters for spring 2003 appeared to make it easier for small northern pike to expel transmitters. Females expelled transmitters (regardless of location) at a greater rate in 2003 than in 2002 (Table 1, Chi-square = 3.52, $df = 1$, $P = 0.06$). Although some selection for fish size was imposed in 2003, there was no significant difference in the average length of fish implanted between the two years ($t = 0.16$, $df = 28$, $P = 0.88$). The difference in expulsion rates between the two years was primarily due to small fish because there was no apparent difference in expulsion rate for large (>691 mm) fish between the two years (Chi-square = 0.04, $df = 1$, $P = 0.84$). A difference in expulsion rates was evident for small fish (< 650 mm) between 2002 and 2003 (Chi-square = 4.55, $df = 1$, $P = 0.03$).

Effects of transmitter size for muskellunge

Three muskellunge were excluded from analyses in Elk Lake because of stress or mortality due to multiple recaptures and handling during sampling in spring 2003. The muskellunge with transmitter

48.055 expelled her transmitter in a holding tank when she was recaptured. Two muskellunge implanted with large transmitters (48.358 and 48.397) died before depositing their tags (Table 2). As a result of the three fish excluded from analyses and low sample sizes, no effect of transmitter size on expulsion rate was detected. Transmitter expulsion rates were 1/4 for small transmitters, 3/5 for intermediate-size transmitters, and 2/3 for large transmitters (Table 2). Significant differences in expulsion rates were not apparent for the different sized transmitters (Chi-square = 1.53, $df = 2$, $P = 0.46$). Combining all fish yielded a ratio of 6/12 transmitters expelled, or an equal probability ($P = 0.50$, 95% CI = 22.3–77.7%) of individuals depositing or retaining transmitters.

For muskellunge included in our analyses, expulsion rate seemed primarily related to fish size and the state of gonad maturation when the implants were made. Although sample sizes were small, muskellunge that expelled transmitters appeared to be larger ($n = 6$, mean length = 1164 mm, SE = 22 mm) than fish that retained transmitters ($n = 6$, mean length = 1099 mm, SE = 22) ($t = 2.08$, $df = 10$, $P = 0.06$). Females that were ripe and running eggs when implanted had a better chance of expelling radio transmitters. A significant difference was found in expulsion rates of fish that were green versus fish that were ripe when implanted (Chi-square = 4.00, $df = 1$, $P = 0.05$). All fish (3/3) implanted when gonad maturation stage was ripe expelled their transmitters. For fish implanted when gonad maturation stage was green, only 3/9 transmitters were expelled.

Elk Lake muskellunge spawning areas were not well defined prior to the current study. Prespawning study fish were extremely active, circling the lake numerous times with brief stops at the various bulrush beds and bays that are characteristic of Elk Lake. The period between implanting and transmitter deposition ranged from 15 to 25 days (mean = 18 days, SE = 1.6), providing evidence of prolonged and possibly fractional spawning. Based on transmitter deposition, likely spawning habitat appeared to be concentrated along the western and southern shores of Elk Lake (Fig. 3). Water depth at transmitter deposition sites averaged 1.1 m (SE = 0.2 m), and forms of aquatic vegetation varied among the sites, although *Chara* spp. was common to most of the sites. Other forms of vegetation represented included *Elodea canadensis*, *Ceratophyllum demersum*, *Potamogeton natans*, *P. zosteriformis*, *P. foliosus*, *Scirpus validus*, *Stuckenia pectinata*, *Typha angustifolia*, and *Zizania aquatica*.

Location of northern pike and muskellunge spawning habitat in Moose Lake

An unusually long cold period with windy, cloudy and rainy weather kept spring 2004 water temperatures 10°C or colder through 15 May, and probably delayed and protracted spawning in Moose Lake. Ice was moving off the lake on 23 April and was gone after 24 April. Nine of the 10 northern pike transmitters were deposited between 27 April and 10 May (Table 3). The tenth northern pike presumably migrated out into the Deer River as we were unable to locate it after 1 May. Four northern pike expelled tags on 27 April, two on 5 May, and the others between 7 and 10 May.

Northern pike deposited transmitters along an outlet stream or in deeper water within the lake basin. Two transmitters were expelled along Moose Creek, a narrow channel that runs out through a cattail and sedge meadow to the Deer River (Fig. 4). One transmitter (48.171) may have come out when the fish was released, as the transmitter was located the next day in an unlikely spot in deep water where the fish was

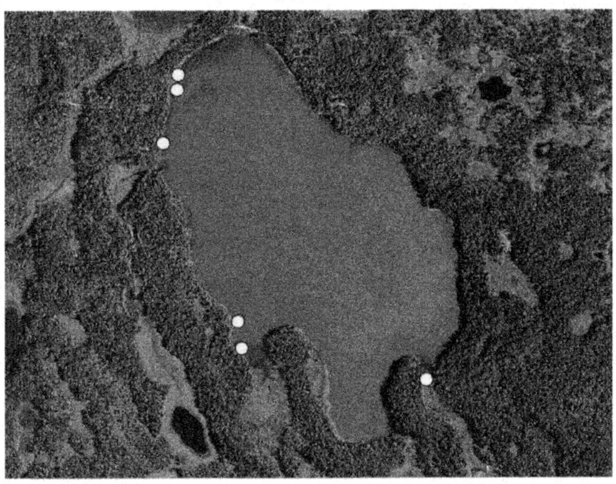

Fig. 3 Locations of transmitter deposition sites for muskellunge in Elk Lake, 2003

Fig. 4 Locations of transmitter deposition sites for muskellunge (circles) and northern pike (triangles) in Moose Lake, 2004

released. Other transmitters were deposited in the lake, usually in deeper water over bars extending out from shore, or a bar rising up from deeper water farther out in the lake (Fig. 4). Water depths for six of the transmitters were 4.9–5.8 m, and substrates for all but one were beds of *Chara* sp. The exception was found on a bar extending south from the north shore of the lake, where the substrate at 5.2 m depth was a dense bed of *Nitella* spp. Two transmitters were deposited within about 5 m of each other at depths of 4.9–5.2 m on a bar protruding from a point of land on the south side of the lake. These two northern pike were originally trapped in different nets, and deposited their transmitters in nearly the same spot on different dates. A final transmitter was shed near the south shore in 1.8-m water depth, also in a bed of *Chara*. None of the transmitters were recovered.

Muskellunge spawning apparently began later and lasted longer than northern pike spawning, and the muskellunge were less successful at expelling radio transmitters. The first muskellunge transmitter was not dropped until 7 May. The last expelled transmitter we were able to find was dropped 27 May. The peak of activity was likely during 7–11 May, as the other transmitters were dropped during 9, 10, 11, and 20 May (Table 3). We found no evidence that four of the 10 females dropped their transmitters, and in fact, transmitter 48.854 was recovered 24 October 2004 when the fish was caught by angling. The recovered transmitter did not come back out through the oviduct very easily.

Muskellunge appeared to use both near-shore and offshore habitat for spawning. Three muskellunge shed transmitters near shore along the northeast end of Moose Lake in proximity to two inlet streams (Fig. 4). The two streams provide inflow from Fawn and Little Moose lakes during the spring. These transmitters were in 0.6–1.8-m depth where the predominant vegetation was *Chara* spp. Two more muskellunge shed transmitters in deeper water over offshore bars. Water depth where both tags were located was 3.7 m. One was in a bed of *Chara* spp. whereas vegetation on the other bar was *Elodea canadensis*. The final transmitter was shed in *Chara* spp. in 2.1-m deep water on a drop off near the south shore of Moose Lake. None of the transmitters were recovered, even though one of them was in water only 0.6 m deep in a relatively sparse bed of *Chara* spp. An intensive search was made especially for that transmitter, but even with use of an underwater viewing scope and underwater radio antenna, we were unable to see or recover the shed transmitter.

We continued to track muskellunge that retained transmitters, with some locations obtained through 10 June. One fish was regularly located, and even observed, in shallow water among bulrushes along the southeast bay of Moose Lake. However, three individuals were only sporadically located after mid-May, presumably because they made movements off shore into deepwater habitat that we were unable to monitor with our radio receivers.

Discussion

Oviduct insertion of radio transmitters offers several advantages over other methods for locating spawning sites: (1) because the transmitters are implanted during the spawning run, they do not need to have a long battery life. This reduces the size of transmitter. (2) Oviduct insertion requires minimal effort for implanting. Small transmitters slide easily up the oviduct, requiring less than a minute to actually implant the transmitter, reducing handling time and eliminating use of anesthetics. (3) Transmitters do not need to be monitored continuously, but rather need only be located several times at the end of the spawning period to ensure that the transmitter is stationary. (4) Because implantation is non-surgical, the technique should be less intrusive. Previous research on telemetry has compared the effects of transmitter attachment methods on fish health and behavior, and shown that attachment methods can have adverse effects on some fish (Winter 1983; Mellas and Haynes 1985; Adams et al. 1998; Jepsen et al. 2001).

Although our studies were the first attempts to have radio transmitters expelled with eggs, Peake et al. (1997) used oviduct insertion to track adult female Atlantic salmon, *Salmo salar*, and rainbow trout, *Onchorynchus mykiss* (1.7–3.6 kg each). Because the transmitters used by Peake et al. for salmon and trout were intended to remain in the fish, they were much larger (9.9 g) than the transmitters we used. They found that 69% of transmitters were retained in Atlantic salmon for 60 days, and 83% of transmitters were retained in rainbow trout for 45 days. The larger transmitters affected egg expulsion and viability in the rainbow trout. Both the Peake et al. (1997) study and our work show that even if transmitters are not expelled with eggs, oviduct insertion can be a viable method of attachment for short-term biotelemetry studies. Daily monitoring of muskellunge and northern pike illustrated their early spring movements and use of habitat in all three lakes that we studied.

Locations of expelled tags in our study were considered only to be potential spawning sites because we were unable to recover transmitters or esocid eggs in the vicinity (about 2–5 m^2 areas) with the strongest radio signal. This remains the most serious problem in validating transmitter locations as spawning sites. Miniature transmitters were not found in the sediments and plant debris, even with intensive effort. Northern pike eggs, which adhere to vegetation, were not found at four of the shallowest transmitter locations in Willow Lake. Muskellunge eggs were also not found at the most readily examined shallow location in Moose Lake. Most helpful would be radio equipment that could better pinpoint shed transmitters, making it easier to sample eggs near the transmitters.

Size of the transmitter in relation to fish size was important. Our most consistent success with the smallest transmitters was for northern pike >700 mm total length. A more tapered transmitter design also appeared to facilitate expulsion. Although we were unable to discriminate between expulsion rates for different size transmitters in muskellunge, large transmitters were more difficult to insert (more resistance encountered from the oviduct) and so their size may have impeded expulsion through the oviduct. We had less success with muskellunge expelling their transmitters than with northern pike. Only 50% of muskellunge deposited transmitters in Elk Lake and 60% in Moose Lake. Transmitter sizes were larger for muskellunge, but maybe just as important for interpreting these results is their more extended spawning period along with their more fractional egg-laying behavior (Lebeau et al. 1986). For both esocids, transmitters seemed easiest to insert in ripe females, though ripe fish need to be handled carefully when released so that eggs and transmitters are not inadvertently extruded. We do not know to what extent transmitters were shed incidentally although we noted at least one northern pike that probably expelled its transmitter when the fish was released. It is also conceivable that contact between the trailing antenna and bottom substrates might be enough to pull out the transmitter. Because the transmitters were so easily implanted, we found no mortality while transmitters were active, but we do not know if there could be longer-term consequences for fish retaining transmitters. In view of all these considerations, we recommend (1) using the smallest available transmitter, (2) using transmitters that are tapered at the posterior end, (3) using the largest fish available, and (4) inserting transmitters in ripe (as opposed to green) females.

The utility of oviduct insertion of radio transmitters as a resource management technique was demonstrated in Moose Lake. Information from this telemetry study identified how esocids used various habitats in Moose Lake during spring spawning. Moose Lake does not have marshes with sedges and grasses that might be considered classic northern pike spawning habitat. While two northern pike dropped transmitters in the outlet channel, most of the other fish appeared to adopt a different strategy of spawning over bars with *Chara* spp. and *Nitella* spp. in deeper water. Sympatric northern pike and muskellunge appeared to use slightly different spawning habitat in Moose Lake, especially in deep water. Both muskellunge transmitters deposited over offshore bars were at 3.7 m depth, compared to an average depth of 5.3 m for northern pike transmitters deposited on deepwater bars.

Locations of expelled transmitters in our study, especially in Moose Lake, indicate that *Chara* spp. can be an important substrate, and suggest that esocids have greater flexibility in their use of deep water for spawning than has been previously reported in the literature. *Chara* spp. was a very important substrate where both muskellunge and northern pike transmitters were found in Moose Lake. Previous studies have shown that *Chara* spp. is used by spawning muskellunge (Strand 1986; Farrell et al. 1996) and northern pike (McCarraher and Thomas 1972; Farrell et al. 1996). In Leech Lake, Minnesota, *Chara* spp. was the dominant vegetation in muskellunge spawning sites that were distinctly off shore in depths of 1–2 m (Strand 1986). Leech Lake spawning sites were characterized by nearly pure stands of *Chara* spp. (attaining heights of up to 25 cm) interspersed with bare patches of marl that resulted in 67% vegetative cover. Werner et al. (1996), however, suggested that dense beds of *Chara* spp. in the St. Lawrence River system might hinder survival of egg and larval stages of muskellunge. Haas (1978) reported muskellunge spawning in open water with a minimum depth of 3 m in Lake St. Clair, but in contrast to Leech Lake and our study, spawning areas in Lake St. Clair had little or no vegetation until mid summer. Offshore spawning in Lake St. Clair and Leech Lake was different than the shoreline spawning commonly described by Crossman (1990) and others. Dombeck et al. (1984) found muskellunge spawning areas in seven Wisconsin and Michigan lakes that were located in shallows near shore (mean depth of the spawning sites was 74 cm). Zorn et al. (1998) described muskellunge spawning habitat in four Wisconsin lakes as occurring in shallow (< 1 m) bays over mucky bottoms covered with dead vegetation. Spawning areas for that study were determined visually (at night with a spotlight) which would tend to limit observations in deeper water compared with using telemetry to identify spawning sites.

Northern pike have traditionally been characterized as using shallow waters with live or decaying vegetation for spawning (Bry 1996). However, northern pike have also been found to spawn in deeper water in Lake Windermere, England, and in the St. Lawrence River. Pike in Windermere apparently used beds of *Elodea*, *Myriophyllum*, and *Nitella* in 2.0–3.5-m water depths (Frost and Kipling 1967). Farrell et al. (1996) and Farrell (2001) found northern pike eggs in water up to 3.7 m deep in Point Marguerite Marsh and Rose Bay in the upper St. Lawrence River. Farrell (2001) further suggested that water level regulation in the St. Lawrence River system has shifted reproductive habitat for northern pike toward deeper water, which may have led to poorer survival for swim-up fry.

Identification and protection of critical habitat is a high priority for proactive conservation of esocids. In this study, we showed how oviduct insertion of radio transmitters can be used to identify important habitat prior to shoreland alteration by humans. For example, development such as wetland destruction over a relatively small area in Willow Lake would have a very large impact on northern pike production. In Moose Lake, a combination of existing and future development along the northeast shore and near the two inlets has the potential to degrade muskellunge spawning habitat in the lake. Our telemetry work provides impetus for more careful use and monitoring of shoreland surrounding the inlets.

Summary

This paper, in combination with a management brief by Pierce (2004), described the first attempts to use oviduct insertion of radio transmitters to find spawning grounds of fish. Our preliminary studies with esocids showed that they deposited transmitters in likely spawning habitat in three lakes. An inability to find eggs limited the usefulness of this work and indicated that future validation efforts should couple this method with a well-designed egg collection study. Nevertheless, oviduct implantation of miniature radio transmitters was a quick and easy method of attaching radio tags to fish and locating likely spawning habitat. With sufficient validation, efforts to identify spawning grounds could be narrowed to finding transmitters at egg deposition sites immediately after the spawning season. If these sites are critical habitat, they could be afforded extra protection from human development. Our preliminary telemetry studies pointed to differences in spawning habitat used among different Minnesota lakes, and suggested some flexibility in depths used during spawning periods by both muskellunge and northern pike.

Acknowledgments We thank D. Arola, D. Dustin, B. Herwig, P. Jacobson, D. Pereira, G. Phillips, R. Thompson, J. Tillma, and P. Wingate for help with the manuscript or fieldwork. This study was funded in part by Federal Aid in Sport Fish Restoration, project F-26-R, Minnesota.

References

Adams NS, Rondorf DW, Evans SD, Kelly JE, Perry RW (1998) Effects of surgically and gastrically implanted radio transmitters on swimming performance and predator avoidance of juvenile chinook salmon (*Oncorhynchus tshawytscha*). Can J Fish Aquat Sci 55:781–787

Becker GC (1983) Fishes of Wisconsin. University of Wisconsin Press, Madison

Bry C (1996) Role of vegetation in the life cycle of pike. In: Craig JF (ed) Pike biology and exploitation. Chapman and Hall, London, pp 45–67

Brynildson C (1958) What's happening to northern pike spawning grounds?. Wisconsin Conserv Bull 23:1–3

Burns DC (1991) Cumulative impacts of small modifications to habitat. Fisheries 16:12–14

Cross, TK, McInerny MC (1995) Influences of watershed parameters on fish populations in selected Minnesota lakes of the central hardwood forest ecoregion. Minnesota Department of Natural Resources, Section of Fisheries Investigational Report 441, St. Paul

Crossman EJ (1990) Reproductive homing in muskellunge, *Esox masquinongy*. Can J Fish Aquat Sci 47:1803–1812

Dombeck MP, Menzel BW, Hinz PN (1984) Muskellunge spawning habitat and reproductive success. Trans Am Fish Soc 113:205–216

Engel S, Pederson JL Jr (1998) The construction, aesthetics, and effects of lakeshore development: a literature review. Wisconsin Department of Natural Resources, Technical Bulletin 170, Madison

Farrell JM (2001) Reproductive success of sympatric northern pike and muskellunge in an Upper St. Lawrence River Bay. Trans Am Fish Soc 130:796–808

Farrell JM, Werner RG, LaPan SR, Claypoole KA (1996) Egg distribution and spawning habitat of northern pike and muskellunge in a St. Lawrence River marsh, New York. Trans Am Fish Soc 125:127–131

Frost WE, Kipling C (1967) A study of reproduction, early life, weight-length relationship and growth of pike, *Esox lucius* L., in Windermere. J Animal Ecol 361:651–693

Haas RC, (1978) The muskellunge in Lake St. Clair. American Fisheries Society Special Publication 11:334–339

Jepsen N., Davis LE, Schreck CB, Siddens B (2001) The physiological response of chinook salmon smolts to two methods of radio-tagging. Trans Am Fish Soc 130:495–500

Kenow KP, Green WL, Boysen AF (1992) Probe developed for underwater retrieval of radio transmitters. U. S. Fish and Wildlife Service Research Information Bulletin No. 39. LaCrosse, Wisconsin

Lebeau B, Pageau G, Crossman EJ (1986) The muskellunge as a multiple spawner: an adaptive strategy for production of a large number of eggs. American Fisheries Society Special Publication 15: 342. Bethesda, Maryland

McCarraher DB, Thomas RE (1972) Ecological significance of vegetation to northern pike, *Esox lucius*, spawning. Trans Am Fish Soc 101:560–563

Mellas EJ, Haynes JM (1985) Swimming performance and behavior of rainbow trout (*Salmo gairdneri*) and white perch (*Morone americana*): effects of attaching telemetry transmitters. Can J Fish Aquat Sci 42:488–493

Miller LM, Kallemeyn L, Senanan W (2001) Spawning-site and natal-site fidelity by northern pike in a large lake: mark-recapture and genetic evidence. Trans Am Fish Soc 130:307–316

Peake S, McKinley RS, Beddow TA, Marmulla G (1997) New procedure for radio transmitter attachment: oviduct insertion. North Am J Fish Manage 17:757–762

Pierce RB (2004) Oviduct insertion of radio transmitters as a means of locating northern pike spawning habitat. North Am J Fish Manage 24:244–248

Radomski P, Goeman TJ (2001) Consequences of human lakeshore development on emergent and floating-leaf vegetation abundance. North Am J Fish Manage 21:46–61

Strand RF (1986) Identification of principal spawning areas and seasonal distribution and movements of muskellunge in Leech Lake Minnesota. American Fisheries Society Special Publication 15:62–73 Bethesda, Maryland

Werner RG, Klindt R, Jonckheere B (1996) Vegetative characteristics of muskellunge (*Esox masquinongy*) spawning and nursery habitat in the 1000 Islands section of the St. Lawrence River. Great Lakes Res Rev 2:29–35

Winter JD (1983) Underwater biotelemetry. In: Nielsen LA, Johnson DL (eds) Fisheries Techniques. American Fisheries Society, Bethesda, Maryland pp 371–395

Zorn SA, Margenau TL, Diana JS, Edwards CJ (1998) The influence of spawning habitat on natural reproduction of muskellunge in Wisconsin. Trans Am Fish Soc 127:995–1005

Environ Biol Fish (2007) 79:111–123
DOI 10.1007/s10641-006-9091-7

CROSSMAN

Development, implementation, and evaluation of an international muskellunge management strategy for the upper St Lawrence River

John M. Farrell · Rodger M. Klindt · John M. Casselman · Steven R. LaPan · Robert G. Werner · Albert Schiavone

Received: 9 September 2005 / Accepted: 20 June 2006 / Published online: 25 July 2006

Abstract The muskellunge, *Esox masquinongy*, fishery in the St Lawrence River is believed to have declined significantly from historical levels and reached critically low levels during the 1970s. Over-exploitation caused by liberal angling regulations, and loss and alteration of critical spawning and nursery habitat probably contributed to this decline. In 1980, a St Lawrence River Muskellunge Management Work Group comprising resource managers and several advisors, including E.J. Crossman, to whom this symposium is dedicated, was created to address research and management needs. A trophy muskellunge management strategy was implemented including more restrictive harvest regulations, public education promoting "catch and release", and protection of spawning and nursery habitats. Age and growth information obtained from cleithra analysis indicated the need for increased size limits to adequately protect spawning stocks. Research efforts have developed a biological information base and monitoring tools to guide management decisions and evaluate responses. Over 100 spawning and nursery locations have been identified in US and Canadian waters leading to improved protection of critical habitats. An angler diary program shows a decline in the number of fish being harvested and a local muskellunge release award program implemented in 1987 has logged over 1000 releases of fish at least 44″ in length. Adult muskellunge monitoring in eleven spawning areas revealed an increase in mean total length of over 63 mm (>2.5 inches) after the regulation changes. Monitoring of age-0 muskellunge by use of seining surveys (1997–2005) indicates consistent reproductive success with the potential for several strong year-classes. Improvements in the muskellunge population and fishery are attributed to the progressive management action and a united community response.

Keywords Muskellunge · Management · Monitoring · St Lawrence River

J. M. Farrell (✉) · R. G. Werner
Department of Environmental Forest Biology, State University of New York College of Environmental Science and Forestry, Illick Hall, 1 Forestry Dr., Syracuse, NY 13210, USA
e-mail: jmfarrell@esf.edu

R. M. Klindt · A. Schiavone
New York State Department of Environmental Conservation, Region 6, Bureau of Fish Wildlife and Marine Resources, State Office Bldg., Watertown, NY 13601, USA

J. M. Casselman
Department of Biology, Queen's University, 2406 Biosciences Complex, K7L 3N6 Kingston, ON, Canada

S. R. LaPan
New York State Department of Environmental Conservation, Cape Vincent Fisheries Station, Cape Vincent, NY 13618, USA

Introduction

The upper St Lawrence River provides an important self-sustaining population of Great Lakes-strain muskellunge, *Esox masquinongy* (Mitchell). Its large size and the difficulty of catching a mature fish make the muskellunge one of the most revered and sought after freshwater game species in North America. The St Lawrence River muskellunge is both an important ecological and an economic resource to the State of New York and the Province of Ontario, but its value goes well beyond the fishery alone, because its presence is part of the history and present culture of maritime communities.

Resource managers from the New York State Department of Environmental Conservation (NYSDEC) and the Ontario Ministry of Natural Resources (OMNR) recognized in the late 1970s that information for making management decisions about muskellunge was lacking. The first comprehensive plan for the management of muskellunge in the St Lawrence River was published in 1980 (Panek 1980). The goals of the original management plan for the St Lawrence River muskellunge population were, and continue to be: *To perpetuate the muskellunge as a viable, self-sustaining component of the fish community in the St Lawrence River, and to provide a quality trophy fishery.*

The plan called for formulation of an International St Lawrence River Muskellunge Management Work Group, created within the Lake Ontario Committee of the Great Lakes Fishery Commission. This Work Group comprises fisheries research advisors from the SUNY College of Environmental Science and Forestry, and fisheries biologists from the NYSDEC and the OMNR. Responsibilities include identification of research needs and coordination of a cooperative research and management effort to protect and enhance the St Lawrence River muskellunge population and sport fishery.

A "Phase II Strategic Plan" (LaPan and Penney 1991) outlined research achievements from 1980 to 1990 and set directives for continued research on muskellunge behavior and biology for the next decade. Phase II objectives and tactics addressed four primary areas: habitat protection, population quantification, standardization of international regulations, and restoration of stocks. Identification of spawning, nursery, and sub-adult areas was regarded a key element in the protection of critical habitats and for understanding of muskellunge reproductive processes. Objectives and strategies of the muskellunge management plan were again updated for "Phase III" (Farrell et al. 2003) and continue to focus on habitat identification and protection, population monitoring, conservation education, and guiding future research activities.

We review the process that led to the development and implementation of new management strategies and evaluate what has been learned about the biology of upper St Lawrence River muskellunge. Second, databases are used to assess if management actions have affected the muskellunge population and fishery. Consideration is given to future management needs and lessons learned for guiding this adaptive approach.

Population decline

The quality of the St Lawrence River muskellunge fishery has been in question since ardent anglers and guides first voiced their concerns in the 1940s. Concerns about muskellunge population status occurred at a time when native self-sustaining populations were declining or being lost throughout the range (Trautman 1981; Dombeck et al. 1986). Although no fishery data existed for the St Lawrence River muskellunge population before these complaints, attempts to obtain brood stock to enhance the muskellunge population in the 1950s suggest there was a problem (Anon 1953).

After the capture of the world record muskellunge in 1957 in the St Lawrence River, interest in the fishery probably increased. Fishing pressure on popular spots was high and white flags displayed upon re-entry to a local port designated that a muskellunge was on board and signified the prevailing attitude that the fishery was an inexhaustible resource. Between 1969 and 1977, New York diary participants required 32 h to capture a legal sized muskellunge (Panek 1980). Data from the Clayton Muskellunge Derby (1969–1978) showed a decrease in the mean size of muskellunge, a 25% harvest of immature fish, and a 90% overall harvest, with a consistent decline in numbers of fish entered (LaPan and Penney 1991). The derby was abandoned in the late 1970s because of a lack of interest, apparently because of low catch rates. Major crude oil spills in the river in 1973 and 1976, coupled with water-quality issues and proposals for winter navigation in St Lawrence Seaway, bolstered concern for the muskellunge fishery and the environment as a whole. Despite the absence of data, there was a general feeling that the population was in poor condition and potentially near collapse when the first management plan was conceived. Local attitudes fostered distrust of government management agencies and muskellunge research. Hasse (1976) reported difficulty in obtaining reliable catch and effort information from angler cooperators. The lack of basic biological and fishery data was cited as the single largest problem facing management of the muskellunge fishery (Panek 1980).

Evolution of a muskellunge management strategy

Changes in size limits and regulations

Management of St Lawrence River muskellunge in New York State can be traced to 1909 and has since followed a trend from extremely liberal regulations, with no biological basis, to a more restrictive plan with a trophy-management strategy. From 1909 until 1960 muskellunge could be harvested at a size of 610 mm (24 inches), with no limit on the number harvested. In 1961–1962, the size limit was increased to 711 mm (28 inches) and from 1963 until 1977 the creel limit was 2 fish per day. On the basis of recommendations by Hasse (1976), the size limit was increased to 914 mm (36 inches) in 1978 to allow females a minimum 1-year opportunity to spawn. The creel limit was also reduced to one fish.

On the basis of spring trapnet catches during spawning (1990–2000), female muskellunge captured at 864–965 mm (34–38 inches) were sexually mature. Back calculated length at age from cleithra of "trophy" muskellunge obtained from local taxidermists indicated ages corresponding to this size range would be from 5 to 7. It was clear that the 914 mm (36 inch) size limit was inadequate for protection of muskellunge to promote population recruitment. After recommendations by the Muskellunge Work Group, the size limit was increased to 1118 mm (44 inches) in 1986 for New York waters and in 1991 for Ontario to allow females greater opportunity to spawn before being available for exploitation.

Size limits were re-evaluated on the basis of muskellunge growth information (Casselman et al. 1999). Three separate growth trajectories were estimated by use of von Bertalanffy growth models, mean lengths at age, and associated 99% confidence limits. Mean length at age was assumed to represent average growth potential, and confidence limits were assumed to represent fast (upper CL) and slow (lower CL) growth potential (Fig. 1). Individual growth models were fit using each of these data series. The minimum ultimate length estimate of 1268 mm (49.9 inches) represented the von Bertalanffy model fit for the lower 99% CL for mean lengths observed at each age. Age predictions derived from the reciprocal von Bertalanffy model suggested the youngest females harvested would be between 10 and 19 years old, substantially older than the age at maturity previously used for management (5–7 years).

In the fall of 2002, a 1219 mm (48 inch) size limit, approaching recommendations based on growth potential, was enacted for both Ontario and New York waters of the St Lawrence River. Growth trajectories for the males indicated a minimum ultimate size limit of 1041 mm (41 inches). Managers acknowledged that male muskellunge rarely reach the size limit, and hence exploitation rates would be low.

Changes in harvest philosophy

There has been a nationwide trend in the adoption of the catch-and-release angling philosophy. For the St Lawrence River, a "volunteer" catch and release ethic can be traced to an incentive program developed by Save The River, Inc. Beginning in 1987, a release award was offered via a signed affidavit stating that an angler had released a legal sized muskellunge in a proper

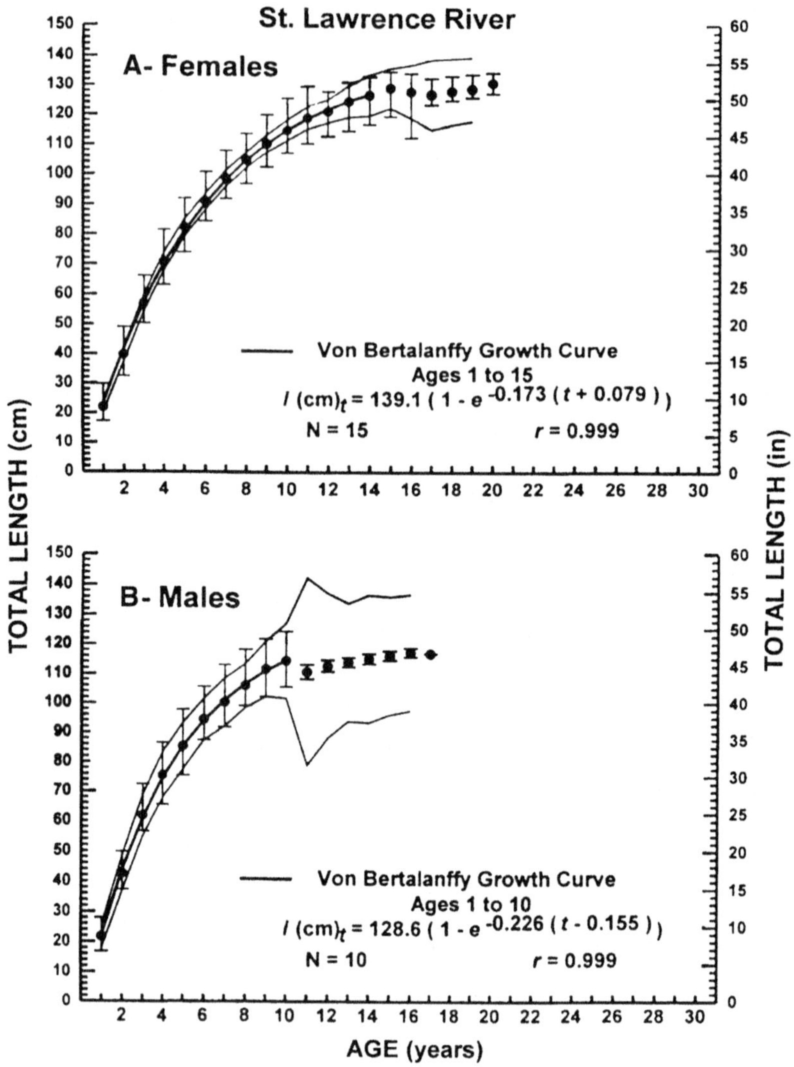

Fig. 1 Age-specific total length (cm) of (**A**) female and (**B**) male muskellunge sampled from the Thousand Islands section of the St Lawrence River (Casselman et al. 1999). Means (back-calculated) are indicated by *closed circles*, ranges by *bars*, and 95% confidence limits by *thin lines*. Von Bertalanffy growth curves (*thick lines*) are also provided as equations with number of ages used

manner. The angler received a limited edition rendition muskellunge print by a popular regional artist, Michael Ringer. Ringer has produced a total of three muskellunge prints, with the latest released in 2005, to assist in maintenance of the vitality of the catch and release program. Since inception of the program, nearly 1000 prints have been issued. Data contained in the affidavits include information on the size and location of each catch, and have been used to contact anglers for mail surveys and an angler diary program.

Identification and characterization of spawning and nursery habitats

In addition to the more restrictive harvest regulations, the muskellunge management strategy implemented by NYSDEC and OMNR has led to greater protection of spawning and nursery habitats. Many sites (n=23) have been identified as both spawning and nursery areas, supporting the belief that muskellunge spawn and develop in the same general location. In an international effort, 103 muskellunge nursery locations (69 US and 34 Canadian) have been identified in the upper St Lawrence River (Fig. 2). Sites are clustered in areas that have a relatively high abundance of shallow littoral habitats.

Critical habitats were discovered by trapnetting and radiotracking spawning adults (LaPan et al. 1996) and by extensive seining surveys in potential nursery areas (Werner et al. 1990). Spawning site fidelity was observed for radiotracked muskellunge through returns to locations

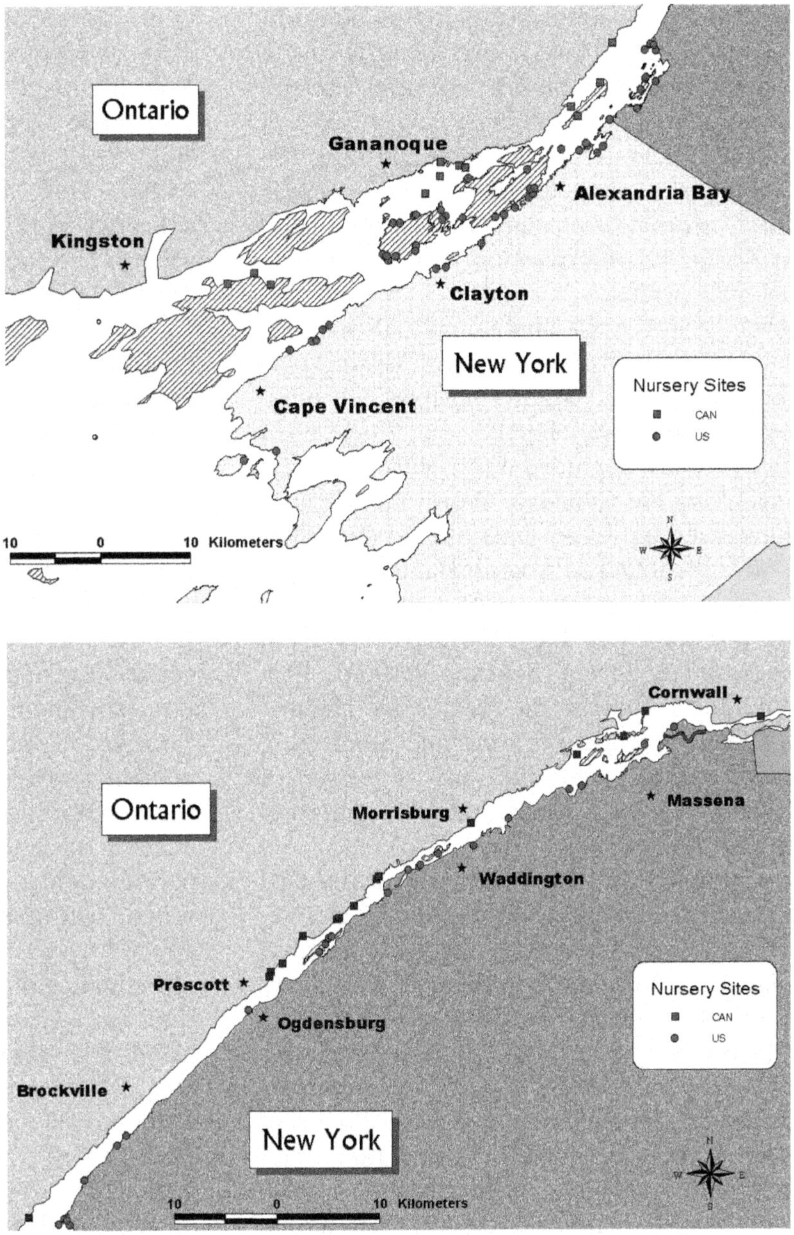

Fig. 2 Locations of muskellunge nursery areas in the International Eastern Lake Ontario and the St Lawrence River from Cornwall, Ontario, to Cape Vincent, New York. All muskellunge nursery areas were identified by use of seining surveys as a major initiative of the International Muskellunge Management Plan

over two successive years. Subsequent data on tagging and recapture of trapnetted spawning adults corroborates this finding. Of 33 fish tagged and recaptured during spawning over many years, all were recaptured at the location of original tagging. Crossman (1990) reported similar findings of muskellunge spawning site fidelity for Nogies Creek, Ontario. Whether this behavior represents a natal homing instinct remains uncertain.

The potential for natal homing was the basis for a population enhancement effort in the 1990s. Site-specific fry and fingerling stocking was conducted "restore" spawning locations that were perceived as unproductive, yet had suitable habitat characteristics (Werner et al. 1996; Farrell and Werner 1999). After stocking, comparison of the "natural" and "restoration" bays revealed no significant differences in survival (0.704% in restoration bays compared with0.678% in natural sites) or density (natural, 18.8 young-of-year (YOY) ha^{-1}; restoration 20.5 YOY ha^{-1}). Natural reproduction was eventually detected in all study bays and revealed that some reproduction was occurring even in sites labeled as non-productive. Stocked fry (19–24 mm), identified with thermal and chemical otolith markers, contributed substantially to abundance (over 50% of sampled juveniles) despite low survival rates (0–3%). Stocking of early-life stages seems to be a promising restoration technique where quality habitat exists, but the question of natal homing remains unanswered.

Knowledge of muskellunge critical habitats and their locations assists managers issuing permits for development. Many spawning and nursery areas are being developed; others have been filled or otherwise degraded. Over 50% of coastal wetlands along the Canadian shore of Lake Ontario have been lost or severely altered (Whillans 1982). Sites identified as spawning and/or nursery areas now receive greater protection under the Canada Fisheries Act and have added value in ranking as a Significant Coastal Wetlands Fish and Wildlife Habitat by the New York Department of State. Site identification and monitoring have also increased education and awareness of the importance of aquatic habitat and its protection at a local level.

Much progress has been achieved with regard to the need for biological information to characterize muskellunge spawning, early life history, and related critical habitats, including the physical, chemical, biological, and land-use characteristics associated with muskellunge spawning success (Diana et al. 2006). Muskellunge typically have a protracted spawning run in the St Lawrence River from early May to mid-June. The presence of muskellunge on spawning grounds, based on trapnet captures of over 280 adults (from 1990 to 2003), was observed between 26 April and June 13. Spawning occurred at different times in spring, in different bays, because of temperature variations. The main channel of the St Lawrence is very slow to warm in springtime and spawning runs are often later in spawning sites exposed to this cool water; other more sheltered locations often warm earlier and early spawning runs are observed.

Studies using the capture of naturally spawned eggs to evaluate spawning distribution and habitat use also yielded data on egg fertilization, viability, and survival rates (Farrell 1991, 2001; Farrell et al. 1996). Collections of naturally spawned eggs at Point Marguerite Marsh, near Alexandria Bay, NY, occurred from 13 May to 12 June 1989 (Farrell et al. 1996). Water temperatures ranged from 7 to 17°C and spawning peaked at 10–13°C. In Rose Bay, near Cape Vincent, NY, eggs were collected from 23 May to 22 June 1994, and 23 May to 22 June 1995. Water temperatures during the 1995 muskellunge spawning run ranged from 13.2 to 18.1°C. Similar ranges are reported in the literature (Scott and Crossman 1973) and are useful for describing spawning periods.

Recent laboratory experiments have revealed the exponential relationship between increased rates of egg and larval development with increasing temperature (Farrell and Toner 2003). For example, equations predict that at 15°C, eggs would require 10 days to hatch and an additional 13.3 days for larvae swim-up. Water temperature variation in spawning sites has been used to predict spawning times and developmental rates for a variety of purposes, including physiological modeling (Farrell 1998). Similar models have been created for northern pike to predict reproductive success outcomes given long-term temperature data (Farrell et al. 2006) and varying water level management scenarios, and could be applied to muskellunge.

Estimates of natural egg fertility for individual bays varied from 76% (1994) to 97.4% (1995); viability at time of collection was 67% (1989), 68.4% (1994), and 92.1% (1995) (Farrell et al. 1996; Farrell 2001). Survival of wild muskellunge from egg to fall juvenile in 1994 (0.063%) and 1995 (0.183%) was adequate to produce relatively high YOY density. Natural viability rates could serve as an indicator of spawning habitat quality in terms of environmental conditions such as water temperature and dissolved oxygen concentrations. Dombeck (1984) and Zorn et al. (1998) demonstrated how low dissolved-oxygen levels at the spawning substrate–water interface could cause widespread egg losses. Natural coastal wetland processes and biochemical pathways affecting DO microstratification, and the effects of localized eutrophication and flow interruption, are a potential major detriment to muskellunge habitat quality. Recent studies of wetland tributary spawning and nursery in the region revealed extensive use by northern pike, but no use by muskellunge (Farrell and Bosworth 2003). The upper river muskellunge spawning distribution is usually restricted to bays and coastal marshes in shallow waters <1.5 m deep, although a few deep-spawned eggs were found at Rose Bay at a depth of 2.8 m (Farrell 2001). In many upper St Lawrence River tributaries peat accumulation is high, and occurrences of low DO during the muskellunge spawning period are probably more frequent, possibly making the habitat unsuitable. Other tributaries in the Lake St Lawrence Region (e.g. the Grasse, Oswegatchie, St. Regis, and Deer rivers) are more lotic, and contain muskellunge spawning populations, but little research has been conducted in these areas.

Muskellunge egg distribution after spawning has no statistical relationship with a specific set of vegetation variables, but vegetation has not been a limiting feature within the habitats studied. Muskellunge spawn over a variety of new submersed and emergent vegetation growth, and over several substrate types including those high in sand, silt, and organic content (Farrell 1991; Farrell et al. 1996). Determining patterns of spawning distribution in relation to substrate DO levels and subsequent survival may be useful in assessment of habitat quality.

Because YOY muskellunge are found in shallow littoral habitats <1.5 m deep, mean water depths for successful seine hauls (≥1 YOY muskellunge captured) within nursery habitats were significantly shallower (mean=0.65 m) than for hauls with no catch (mean=0.72 m) and a reduction in YOY catch occurred with increasing water depth (Farrell and Werner 1999).

Data from the sampling program in Ontario waters revealed that a variety of submersed and emergent aquatic vegetation types were present in muskellunge nursery locations, but wild celery, *Vallisneria americanus*, was most prevalent (found in 95% of the locations), followed by coontail, *Ceratophyllum demersum*, bulrushes, *Scirpus* sp., and *Myriophyllum* species (Bendig 1996).

Findings were similar in a study of the vegetative characteristics of muskellunge nursery habitat on the American side of the river (Werner et al. 1996). The presence of mixed aquatic vegetation with heights that approach the water's surface, typical of the shallow littoral environment, has been shown to be an important habitat for YOY muskellunge during the summer nursery period (Jonckheere 1994). Vegetation identified included 17 different genera including muskgrass, *Chara vulgaris*, milfoil, *Myriophyllum sp.*, waterweed, *Elodea Canadensis*, pondweeds, *Potomogeton sp.*, and wild celery, *Vallisneria americanus* (Werner et al. 1996).

Estimates of stem densities of these plant types indicated that muskgrass densities were significantly lower where muskellunge YOY were present. This finding contrasts with Dombeck (1986) and Strand (1986) who reported positive associations between muskellunge nursery and muskgrass presence. A study by Clapsadl (1993), however, suggested that dense mats of muskgrass might negatively affect survival of muskellunge eggs and larvae incubated on muskgrass, because of low dissolved oxygen or, perhaps, chemical toxicity. As a mat-forming macroalgae, muskgrass forms dense monocultures in many St Lawrence bays. Dominance by muskgrass probably inhibits other submersed aquatic plant growth, resulting in a reduction in the quality of a nursery habitat (an intermediate density of mixed vegetation that approaches the water surface).

In nursery sites vegetative coverage increased from 9% (0–10 m offshore) to 59–73% (20–40 m offshore) because of a transition from emergent to submergent vegetation zones (Werner et al. 1996). Vegetation cover within seine hauls was high (77–89%) where YOY muskellunge were present (Jonckheere 1994). Similarly, submersed vegetation height was greater in successful seine hauls for YOY muskellunge than in unsuccessful hauls. In a recent analysis of muskellunge nursery habitat in the St Lawrence River, vegetation data were incorporated with fish community variables and prey abundance over two seining periods, July and August (Murry and Farrell 2006). Similar to past findings, YOY muskellunge were captured in areas of intermediate vegetation coverage. A shift in vegetation type used, from fine-leaved submersed vegetation (i.e. *Potomogeton pusillus* and *P. pectinatus*) to broad-leaved types (i.e. *Vallisineria americana*, *Alisma sp.*, and *P. richardsonia*), was proportional to their relative abundance between the two seining periods.

Prey abundance was also a highly predictive variable explaining muskellunge presence. At locations where YOY muskellunge were captured in seine hauls, densities of cyprinids were three times greater, and numbers of tessellated darters and killifish were twice as high as in hauls without muskellunge. These prey species constitute most of the diet of YOY muskellunge (Farrell 1998).

Characterization of St Lawrence River muskellunge nursery habitat has led to important baseline information useful for monitoring annual YOY abundance and the effects of environmental change (e.g. climate change, habitat dynamics, invasive species, and water-level management). Because wetland loss and degradation is widespread, maintenance of quality habitat is critical to muskellunge sustainability. Building an understanding of habitat characteristics in relation to reproduction success is, therefore, an important management consideration.

Muskellunge population responses to management

A mail survey completed by St Lawrence River anglers who had purchased non-resident muskellunge stamps (required for angling) at Hill Island, Ontario, was conducted in 1990 (LaPan and Schiavone 1991). From these 639 license sales, 285 surveys were completed and returned, and 200 (69%) of these anglers indicated they had fished for muskellunge in 1989. At that time, the muskellunge catch rate was 0.038 fish per hour (or one per 26 hours fished) for a total of 290 fish landed. Catch rates of legal-sized fish in New York (1118 mm, 44 inches) were 0.011 fish per hour and the mean length of the catch was 991 mm (39 inches). Harvest was reported at 10% of all fish caught.

This survey was repeated eight years later in 1998 using a combined list from the 1992 non-resident muskellunge stamp list (the last year of the requirement), the NYSDEC statewide angler survey, lists compiled of active guides, and the Save The River Muskellunge Release Program. The survey questionnaire was identical and of 305 surveys completed, 167 respondents indicated they fished for muskellunge. The muskellunge catch was 242 fish with a catch rate of 0.037 fish per hour (one per 27 hours fished). Results were nearly identical with those of the 1989 survey. Legal-sized catch rates, however, were slightly higher at 0.015 fish per hour and the mean length of the catch was also greater at 1048 mm (41.25 inches). Despite the greater sizes of fish caught, harvest rates were lower, at 7.9%.

As expected, the results change considerably if only data from professional fishing guides were considered. In 1989, four respondents accounted for 20% of the total catch with a 0.10 fish per hour catch rate. For 1998, 20 guides caught over 1/3 of the total catch with a 0.11 fish per hour rate. Harvest by guides was higher than for other anglers, but was reduced from 15.3% in 1989 to 9.8% for 1998.

An angler diary program of several guides and serious anglers has been maintained for the Thousand Island Region since 1997 in an attempt to monitor muskellunge fishery trends. The catch per effort data are similar to those at by guides for the mail surveys described above, ranging from 0.026 to 0.118 fish per hour (Table 1). Catch rates seem to have peaked in 2000 and 2001. Subsequent levels were well below 50% of the peak catch and corresponded to a doubling of effort. Mean length of the catch, however, has steadily increased and peak catch rates seem to have been related to the capture of relatively smaller muskellunge. The largest fish of 523 muskellunge captured was nearly 1499 mm (59 inches). Kerr (2006) reported a St Lawrence River angler diary recorded muskellunge of 1524 mm (60 inches) caught in 2004, approaching the estimated maximum female growth potential of 1549 mm (61 inches; Casselman et al. 1999).

Length–frequency histograms from the angler diary program catch and spring trapnet surveys (1990–2000) reveal a similar increase in catch in the 914–940 mm (36–37 inch) size interval (Fig. 3). Smaller fish are not fully represented either in angler catch or in the spawning population.

Harvest rates of muskellunge reflect a remarkable change in the philosophy of anglers with regard to exploitation of St Lawrence River muskellunge. During the Clayton Muskellunge Derby, 90% of fish captured in 1975 were

Table 1 St Lawrence River muskellunge angler diary program data summarized for 1997–2004

	1997	1998	1999	2000	2001	2002	2003	2004
Participants	5	6	4	7	7	5	4	5
Angler hours	468	564	450	899	2344	1445	1542	1441
No. of musky caught	46	51	53	102	111	37	61	62
CPUE (fish per hour)	0.098	0.09	0.118	0.113	0.047	0.026	0.039	0.043
Number harvested	19	15	10	7	3	0	0	0
Number released (%)	59	73	81	93	97.3	100	100	100
Mean length	45.3	44.7	43.7	40.5	45.6	46.3	46.1	46.2
Length range	30–54	36–58	36–50	32–56	30–59	33–54	30–57	36–58.5

Program participants are primarily professional guides and many of the individuals remain the same each year. Lengths are given in inches

CPUE is catch-per-unit-effort

harvested, with over 1000 fish removed from the population from 1969 to 1978. Hasse (1976) reported an 87% harvest (20 of 23 fish logged) in a voluntary creel program in 1975. Harvest rates have clearly declined in recent years with an estimated harvest of 10% in the 1989 mail survey and 7.9% in 1998 survey, and a declining trend from 41% in 1997 to 0% harvest for 2002–2005 during the angler diary program (Table 1).

Comparison of the length frequency histogram for both male and female muskellunge captured over two extended periods of spring trapnetting (1983–1991 and 1992–2000) in the upper river indicates a shift in the size distribution from smaller to larger individuals (Fig. 4). Mean length of male muskellunge has increased 71 mm (2.8 inches) from 1014 mm, SD=89 (39.9 inches, SD=3.5) to 1085 mm, SD=109 (42.7 inches, SD=4.3) between the periods (t-test, df=140, $P<0.0001$). The increase in female muskellunge

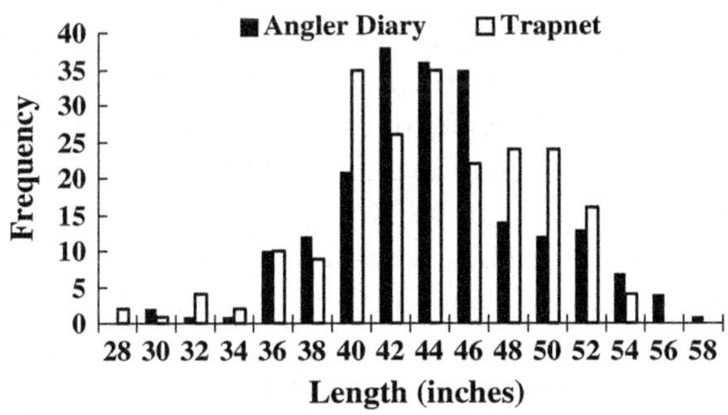

Fig. 3 Length–frequency histogram of male and female muskellunge captured in the New York 1997–2001 St Lawrence River angler diary program and in a New York trapnetting survey (1990–2000)

mean length increased 68.6 mm (2.4 inches) from 1163 mm, SD=134 (45.8 inches, SD=5.3) to 1227 mm SD=119 (48.3 inches, SD=4.7). The male distribution shows a normal bell-shaped curve whereas females are skewed toward larger individuals. The changes in the size distributions may be because of greater muskellunge release rates associated with increasing size limits of 1118 mm (44 inches) in New York in 1986 and Ontario in 1991) and the adoption of voluntary release. Unfortunately, no corresponding age information was available to assess changes in age distribution related to the management changes. Similar length distribution responses, attributed to increased size restriction and voluntary release, have been reported for Lake of the Woods, Ontario (Mosindy 1996) and Bone Lake Wisconsin (Cornelius and Margenau 1999). Catch rate increases, for example those observed at Lake St. Clair and believed to be related to multiple captures of released muskellunge (MacLennan 1996), were not apparent in the St Lawrence River.

Responses of muskellunge reproductive success

A muskellunge YOY monitoring program, with standardized effort, in eleven upper St Lawrence River nursery bays has been in effect from 1997 to 2005. Surveys are completed each year with a July fine-mesh seining procedure and a large-mesh seine used in August. A catch-per-unit-effort (CPUE) relationship between the surveys is apparent, except for 2000 when August muskellunge CPUE was much less than expected

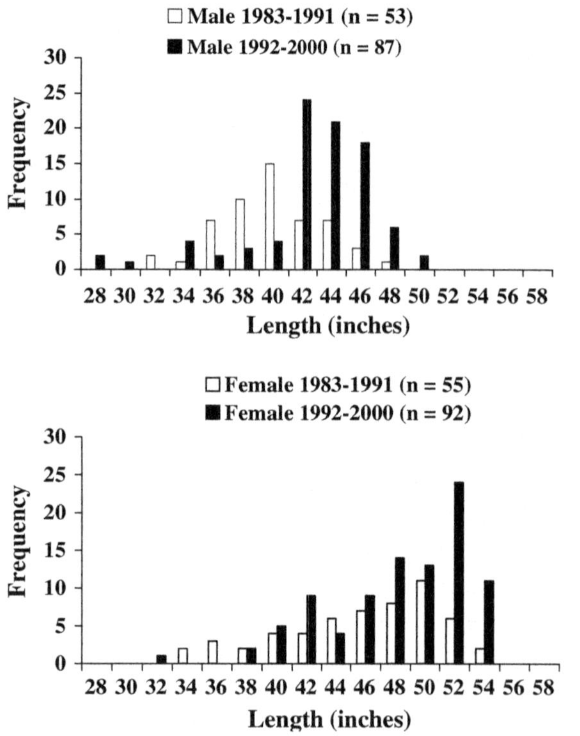

Fig. 4 Length–frequency histogram for female (**top**) and male (**bottom**) muskellunge captured in trapnet sets in New York waters of the Thousand Islands section of the St Lawrence River during spring spawning runs over two eight-year periods (1983–1991 and 1992–2000). Size-limit restrictions changed from 36 to 44 inches in New York waters in 1986 and in Ontario waters in 1991

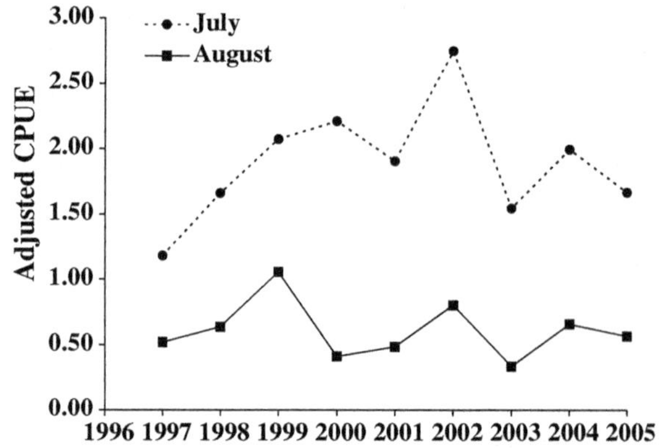

Fig. 5 Catch per unit effort of YOY muskellunge captured in standardized seine hauls in eleven upper St Lawrence River nursery sites from 1997 to 2005. A 30-foot fine-mesh is used in July 15–31 and a 60-foot large-mesh is used from August 15 to 31. The fine-mesh seine CPUE was doubled to standardize the area swept by the two pieces of mesh

(Fig. 5). Although standardized data are not available before implementation of management actions, trends in YOY muskellunge CPUE have indicated consistent reproductive success. High YOY abundance observed in 1999, 2002, and 2004 was indicative of potentially strong year classes.

Research and management needs and priorities

Ongoing efforts will continue to identify nursery and spawning habitats, monitor adult and YOY populations, work and with anglers to develop long-term databases to evaluate impacts of fishery management, and monitor muskellunge populations. Efforts to foster conservation of muskellunge, by education, remain integral to successful management.

Identification of all muskellunge spawning and nursery habitats, including sites in Lake Ontario, remains a priority. Many locations remain to be identified, especially on the Canadian part of the river. OMNR has increased muskellunge nursery habitat identification efforts in 2005 and seven new sites were recently located (Lake 2005).

The quality of the muskellunge fishery will continue to be monitored using angler-derived catch data, but more information is needed to supplement the existing program. The newly established Muskies Inc., Gananoque, Ontario Chapter, has established Muskellunge Angler Logs and data will be made available to the Esocid Working Group. Additional efforts should be made to expand the program geographically to include all key sportfishing locations in the study area. Standardized spring trapnetting occurs every third year to monitor the health of adult muskellunge during spawning, develop an index of abundance, and evaluate spawning conditions, size structure, and spawning site fidelity.

More research is needed to assess niche overlap of northern pike and muskellunge in nursery areas. Spawning habitat segregation of northern pike and muskellunge is well understood, but the differences in nursery habitat among the esocids are not well known. Well-designed field experiments may elucidate the mechanisms of esocid interaction. This information will be very important for the planning and implementation of measures to enhance the northern pike population while preventing negative impacts on muskellunge YOY production.

Models guiding habitat restoration efforts are proposed as an important tool in enhancing or restoring muskellunge reproduction at sites that have been degraded by human activity and invasive plant species. Information collected from monitoring sites should be used to develop statistical models to predict habitat factors relating to the presence or absence, and abundance, of YOY. An inventory of potential restoration/enhancement locations could be developed using modeling tools and field sampling. Factors limiting reproductive success can be isolated and targeted as habitat restoration/enhancement priorities.

It is imperative to acquire better understanding of how a variety of environmental factors (i.e. temperature, water levels, habitat, and biotic interactions) affect muskellunge recruitment processes. Models for spawning and nursery habitat, and for maximization of YOY growth, must be developed. The models should be developed using long-term field data from nursery sites and experimental data to predict which conditions promote strong year-classes and maintain high-quality habitat. Models are currently being developed to assess environmental conditions important to muskellunge growth using data from the OMNR Cleithrum Project (Casselman et al. 1999; Robinson and Casselman 2006a, b). Continued collection of cleithra bones from taxidermists is imperative to further development of the models. Both approaches will yield useful information affecting future management considerations.

Despite management efforts, diseases of muskellunge have had devastating effects on populations in other waters such as Chautauqua Lake, New York. Muskellunge in Lake St. Clair have recently been infected by *Piscirickettsia*, and viral hemorrhagic septicemia (VHSV) and its effects on the population are poorly understood. The VHSV virus has recently been detected in freshwater drum, *Aplodinotus grunniens*, in eastern Lake Ontario and assays are needed to detect if it spreads to the muskellunge population. The bacterium *Aeromonas sobria* is suspected of being related to a major muskellunge mortality event during June 2005 (Paul Bowser, Cornell University College of Veterinary Medicine, personal communication). Unusually high water temperatures in the St Lawrence River, an 8°C increase during a 2-week period from the last week of May to June 13, are believed to have stressed spawning adult muskellunge and triggered the mortality. Over 70 dead muskellunge were observed by OMNR, NYSDEC, and ESF personnel, with many additional fish reported in parts of Eastern Lake Ontario and the entire Thousand Islands Region.

Management of the St Lawrence River muskellunge population continues to be an adaptive process, building on what is learned while maintaining a commitment to long-term monitoring. Implementing strategies by incorporating new technologies, including statistical and modeling techniques, will enable future advancement of knowledge. Social, economic, and political considerations must be balanced with the ecological need to sustain critical habitats. The current improvements observed in the muskellunge population and fishery are only a beginning; further enhancement through habitat and population restoration are a greater challenge.

Acknowledgements The primary source of funding was through the Federal Aid in Sportfish Restoration FA-5-R and FA-48-R. We thank Doug Stang, Patrick Festa, and Steve Hurst of the NYS Department of Environmental Conservation for their support of and comments on this research. We also thank members of the Esocid Working Group for helping to guide our research and management efforts. We are indebted to the staff of the Thousand Islands Biological Station, including Brent Murry, Kristen Hawley, Tom Hughes, Lea Calhoun, Sarah Walsh, Sue Sabik, and previous employees and countless volunteers who have assisted with data collection. We thank K. Farrell for reviews of this manuscript. This work is a contribution of the Thousand Islands Biological Station and is dedicated to the memory of E.J. Crossman.

References

Anon (1953) Conservation Department Forty-third Annual Report for the year 1952. Legislative Document (1953), No. 27

Bendig A (1996) 1994 St. Lawrence River muskellunge nursery habitat inventory. In: Kerr SJ, Olver CH (eds) Managing Muskies in the 90's. Workshop proceedings. OMNR, Southern Region Science and Technology Transfer Unit WP-007, pp 83–91

Casselman JM, Robinson C, Crossman EJ (1999) Growth and ultimate length of muskellunge from Ontario water bodies. North Am J Fish Manage 19:271–290

Clapsadl M (1993) An examination of habitat characteristics influencing the survival of stocked muskellunge (*Esox masquinongy*) in four St. Lawrence River embayments. Master's Thesis, State University of New York, College of Environmental Science and Forestry, Syracuse, New York

Cornelius RR, Margenau TL (1999) Effect of length limits on muskellunge in Bone Lake, Wisconsin. North Am J Fish Manage 19:300–308

Crossman EJ (1990) Reproductive homing in muskellunge, (*Esox masquinongy*). Can J Fish Aquat Sci 47:1803–1812

Diana JS, Rust A, Zorn S, Margenau T (2006) Human influences on natural spawning and reproductive success of muskellunge. Environ Biol Fish 00:000–000 (in press)

Dombeck MP, Menzel BW, Hinz PN (1984) Muskellunge spawning habitat and reproductive success. Trans Am Fish Soc 113:205–216

Dombeck MP, Menzel BW, Hinz PN (1986) Natural reproduction in midwestern lakes. Am Fish Soc Special Publ 15:122–134

Farrell JM (1991) Spawning ecology of sympatric northern pike and muskellunge in the St. Lawrence River. Master's Thesis, State University of New York, College of Environmental Science and Forestry, Syracuse, New York

Farrell JM (1998) Population ecology of sympatric age-0 northern pike and muskellunge in the St. Lawrence River. Doctoral Thesis, State University of New York, College of Environmental Science and Forestry, Syracuse, New York

Farrell JM, Werner RG, LaPan SR, Claypoole KA (1996) Egg distribution and spawning habitat of northern pike and muskellunge in a St. Lawrence River marsh, New York. Trans Am Fish Soc 125:127–131

Farrell JM, Werner RG (1999) Distribution, abundance, and survival of age-0 muskellunge in Upper St. Lawrence River nursery bays. North Am J Fish Manage 19:310–321

Farrell JM (2001) Reproductive success of sympatric northern pike and muskellunge in an Upper St. Lawrence River bay. Trans Am Fish Soc 130:796–808

Farrell JM, Toner JA (2003) Fish recruitment: evaluation of hydrologic management effects on northern pike and muskellunge performance in Lake Ontario and the upper St. Lawrence River. Lake Ontario-St. Lawrence River Water Level Study Year 2 Report, International Joint Commission, 15 pp

Farrell JM, Bosworth A (2003) Use of spawning marshes in rehabilitation of St. Lawrence River northern Pike and muskellunge populations. Final Report submitted in the New York State Great Lakes Protection Fund, 31 pp

Farrell JM, Klindt R, Casselman JM (2003) Update of the strategic plan for management of St. Lawrence River muskellunge population and sportfishery. NYS Department of Environmental Conservation, Albany, NY, p 40

Farrell, JM, Mead JV, Murry BV (2006) Protracted spawning of St. Lawrence River northern pike (*Esox lucius*): simulated effects on survival, growth, and production. Ecol Freshwater Fish 00:000–000 (in press)

Hasse JJ (1976) A comparison of the St. Lawrence River muskellunge fishery to other muskellunge waters with a recommendation on regulation changes. NYSDEC report, Region 6 Sub-Office, Utica p 12

Jonckheere BV (1994) Production and survival of juvenile muskellunge (*Esox masquinongy*) in the St. Lawrence River, and the habitat characteristics affecting distribution. Master's Thesis, State University of New York, College of Environmental Science and Forestry, Syracuse, New York

Kerr SJ (2006) Characteristics of Ontario muskellunge (*Esox masquinongy*) fisheries based on volunteer angler diary information. Environ Biol Fish 00:000–000 (in press)

Lake C (2005) St. Lawrence River muskellunge nursery habitat inventory. Ontario Ministry of Natural Resources, Lake Ontario Fisheries Unit, Glenora Fisheries Station, Picton, Ontario. Report number LOA 05.08

LaPan SR, Penney L (1991) Strategic plan for management of the muskellunge population and sportfishery of the St. Lawrence River Phase 11:1991–2000. New York State Department of Environmental Conservation, Albany, New York, 51 pp

LaPan SR, Schiavone A (1991) Angler survey, St. Lawrence River muskellunge. New York State Department of Environmental Conservation, Region 6, Watertown, New York, 18 pp

LaPan SR, Schiavone A, Werner RG (1996) Spawning and post-spawning movements of the St. Lawrence River muskellunge (*Esox masquinongy*). In: Kerr SJ, Olver CH (eds) Managing Muskies in the 90's. Workshop proceedings. OMNR, Southern Region Science and Technology Transfer Unit WP-007, pp 73–82

MacLennan D (1996) Changes in the muskellunge fishery and population of Lake St. Clair after an increase in the minimum size limit. In: Kerr, SJ, Olver CH (eds) Managing muskies in the '90s. Workshop Proceedings. Ontario Ministry of Natural Resources, Southern Region Science, Technology Transfer Unit, WP-007, pp. 9–28

Mosindy T (1996) Recent trends in the Lake of the Woods muskellunge fishery. In: Kerr SJ, Olver CH (eds) Managing muskies in the '90s. Workshop Proceedings. Ontario Ministry of Natural Resources, Southern Region Science and Technology Transfer Unit, WP-007, pp 105–114

Murry BM, Farrell JM (2006) Quantification of native Muskellunge Nursery Habitat in the Upper St. Lawrence River: influence of body size, fish community composition, and vegetation structure. Environ Biol Fish 00:000–000 (in press)

Panek FM (1980) Strategic plan for management of the muskellunge population and sport fishery of the St. Lawrence River. NYS Department of Environmental Conservation, Watertown, NY

Robinson C, Casselman JM (2006a) Historical trends in body growth of five Ontario muskellunge populations. Environ Biol Fish 00:000–000

Robinson C, Casselman JM (2006b) Factors affecting historical year-class strengths of trophy muskellunge populations. Environ Biol Fish 00:000–000

Scott WB, Crossman EJ (1973) Freshwater fishes of Canada. Bulletin 184. Fisheries Research Board of Canada, Ottawa

Strand RF (1986) Identification of principal spawning areas and seasonal distribution and movements of muskellunge in Leech Lake, Minnesota. Am Fish Soc Spl Publ 15:62–73

Trautman MB (1981) The fishes of Ohio, revised edition. Ohio State University Press, Ames, Iowa

Werner RG, Klindt R, Jonckheere BV (1996) Vegetative characteristics of muskellunge (*Esox masquinongy*) spawning and nursery habitat in the 1000 Islands section of the St. Lawrence River. Great Lakes Res Rev 2:29–35

Werner RG, LaPan SR, Klindt R, Farrell JM (1990) Identification of muskellunge spawning and nursery habitat. St. Lawrence River Muskellunge Investigations Phase I Federal Aid in Sportfish Restoration Final Report. NYS Department of Environmental Conservation Publication, Albany NY, p 47

Whillans TH (1982) Changes in marsh area along the Canadian shore of Lake Ontario. J Great Lakes Res 8:570–577

Zorn SA, Margenau TL, Diana JS, Edwards CJ (1998) The influence of spawning habitat on natural reproduction of muskellunge in Wisconsin. Trans Am Fish Soc 127:995–1005

Environ Biol Fish (2007) 79:125–136
DOI 10.1007/s10641-006-9168-3

SPECIAL ISSUE

An examination of Minnesota's muskellunge waters

Jerry A. Younk · Donald L. Pereira

Received: 7 October 2005 / Accepted: 11 October 2006 / Published online: 30 November 2006

Synopsis We examined Minnesota's muskellunge, *Esox masquinongy*, populations using various data sets including spring trap net assessments and angler diary surveys. We confronted a potentially problematic data set that had a large number of zero catches by fitting logistic models to the presence/absence data set and linear models to the subset of data that included catches with one fish or more. Currently, 105 lakes have been identified as muskellunge waters, of which 61 lakes have been created and maintained by stocking. Although the proportion of successful anglers has increased over time, catch rates have remained the same. It appears from analysis of trap net catches that the abundance of 102 cm (40 in) and larger muskellunge has been increasing over time. Both trap net and angler data provide some indications that size of muskellunge caught has also increased over time. Angler-harvested muskellunge averaged 11 years of age and 115 cm (45 in) total length. For Minnesota waters, muskellunge ultimate length averaged 134 cm (53 in) for females and 119 cm (47 in) for males. All evidence indicates a successful management program.

Keywords Muskellunge · Management · Population trends · von Bertalanffy growth parameters

J. A. Younk (✉)
Division of Fish and Wildlife, Minnesota Department of Natural Resources, 2114 Bemidji Avenue, Bemidji, Minnesota 56601, USA
e-mail: jerry.younk@dnr.state.mn.us

D. L. Pereira
Division of Fish and Wildlife, Minnesota Department of Natural Resources, 500 Lafayette Road, St. Paul, Minnesota 55155-4020, USA

Introduction

The muskellunge, *Esox masquinongy*, is regarded as a prized game fish and is the largest of the esocids found in Minnesota. Minnesota's Muskellunge Long Range Plan (MNDNR 1994) established a goal of managing natural and introduced populations of muskellunge for a range of quality angling experiences, while maintaining trophy opportunities and preserving genetic integrity. However, as muskellunge-angling popularity increases, so does the demand for more muskellunge waters and larger minimum length limits.

Muskellunge size is perceived as the key component of a quality fishery, and the use of harvest regulations (i.e., bag and length limits, restricted seasons) are viewed as tools to improve fishing quality. Leitch and Baltezore (1987) illustrated the importance of trophy angling opportunities from an angler attitude survey of Minnesota anglers. Cunningham and Anderson (1992) found that anglers typically associated with chartered fishing organizations prefer quality-sized fish and support increased regulatory restrictions. Anglers fishing muskellunge in Wisconsin defined a trophy as at least 102 cm (40 in), but preferably greater than 127 cm (50 in) in total length (Margenau and Petchenik 2004). The practice of catch and release is also testimony to the significance size plays in muskellunge angling. More than just an ethic, catch and release has been viewed as a tool in managing a trophy muskellunge resource (MNDNR 1994, Simonson and Hewett 1999).

Evidence of increased exploitation coupled with changes in population size structure was documented for muskellunge in the Park Rapids area over a 58-year period (Olson and Cunningham 1989). The historical qualities of these fisheries have not been restored, although stocking and length restrictions were applied as corrective measures. Hanson (1986) attributed limited trophy potential and poor quality size structure in some Wisconsin lakes to exploitation. Highly variable muskellunge population characteristics were found in these eight study lakes. In contrast, a 36-year experiment with liberal angling regulations in Escanaba Lake failed to alter muskellunge population trends as theoretically predicted (Hoff and Serns 1986). It is apparent that exploitation as well as other factors have played a role in changing the quality size structure of muskellunge populations in some waters. These cases exemplify the need to more closely examine Minnesota's muskellunge waters.

Although muskellunge are one of the few fish in Minnesota managed exclusively for trophy purposes, the definition of a trophy muskellunge is as diverse as the experience of the anglers who pursue this fish. As angler experience and catch (numbers and size) increase, so does their perspective of what constitutes a trophy fish (Wingate 1986). Confounding this issue is the inherent difference between biological and social definitions of trophy management. Both resource managers and muskellunge anglers must recognize the biological limitations and associated social issues accompanying muskellunge management, and react accordingly.

Minnesota's muskellunge program has primarily relied on two management practices: (1) population manipulation (stocking) and (2) harvest restrictions (regulations). A third management practice, although not directly targeting muskellunge populations, involves habitat protection implemented by the Minnesota Department of Natural Resources (MNDNR), Aquatic Plant Management Unit. The major change to stocking that began in the 1950s involved discontinuing the use of Shoepack strain and replacing it with the Mississippi (Leech Lake) strain in the 1980s (Younk and Strand 1992, Wingate and Younk in press). Strain, as used in this paper, means fish that were from or descended from their respective source populations. A progressive increase in minimum length limit also occurred statewide between 1986 and 1993, resulting in our present statewide minimum size regulation of 102 cm (40 in) (Younk and Pereira 2003, Wingate and Younk in press). Exceptions to this regulation include a minimum length limit of 76 cm (30 in) in Shoepack Lake and 122 cm (48 in) in 7 brood stock lakes.

With the numerous changes that have occurred over the past 25–30 years came an opportunity to begin examining the status of the state's muskellunge populations. The primary objective of this study was to determine if any changes in length and relative abundance has occurred since the MNDNR instituted its muskellunge management program. Specifically we use trap net assessments and angler diary surveys to compare native and introduced (stocked) waters by examining catch rates, size of fish caught, and angler success. Secondary objectives were to determine age and length of harvested muskellunge and trophy potential of Minnesota's muskellunge waters using cleithra. Of particular interest were populations in stocked waters, which in the past received numerous angler complaints about the lack of quality (trophy) size individuals.

Material and methods

Minnesota's muskellunge program is relatively young and does not have the advantage of a large database to provide management direction. Spring trap net assessments have been the primary means of collecting population information since 1976. Assessments followed trap netting procedures first described by Scidmore (1970) and later revised by MNDNR (1993). Trap nets were fished daily from after ice-out (5–14 days) until after peak spawning activity. A total of 210 spring trap net assessments conducted on 47 lakes were available for analysis, indicating that 55% of the muskellunge waters have been assessed. The distribution of these assessments compared favorably with the statewide number of muskellunge waters in each size group (Fig. 1). The sampled waters represent the core of muskellunge lakes that are routinely assessed as needed to develop management plans for the species.

Both angler diary and taxidermist information were collected with the help of volunteers. The amount of data collected varied between lakes and among volunteers. The voluntary angler reporting system used diaries to obtain information on muskellunge angling trips from 1996 to 1998. The diary design was similar to the 1986–89 "Project Muskie" angler diary (Younk and Cook 1992). Angler diary data was available for 44 muskellunge lakes, of which 89% coincided with trap net assessed waters. Taxidermist and angler cooperation along with the amount of information collected from these sources were limiting factors in performing lake specific analyses.

Confounding factors that complicate analysis and interpretation of trap net and diary data include changes in sampling gear, minimum size regulations, and stocking. Sampling gear used during netting assessments included large (1.5 × 1.8 m (5 × 6 ft) frame) and small (0.9 × 1.8 m (3 × 6 ft) frame) trap nets. With a few exceptions, most lakes during early assessments were sampled using small trap nets. Starting in 1999, large trap nets became the standard assessment net. Three regulation changes occurred over a 10-year period. Although we could not partition individual regulation effects, these effects are considered collectively over the time periods analyzed. The stocking variable we used only separates introduced (stocked) lakes from native lakes. Stocked waters include some lakes where muskellunge existed naturally but showed declines in abundance and were augmented by stocking. A more detailed assessment of stocking would have required additional information about stocking rates, frequency, size of fish stocked, and strain.

Analytical methods

Both angler diary and trap net CPUE data contained large numbers of zero values (i.e., anglers that were not successful in catching at least one fish, and trap net assessments where no fish were captured), and thus could not be subjected to standard parametric procedures based on normal or lognormal distribution. For these data, we applied analytical procedures

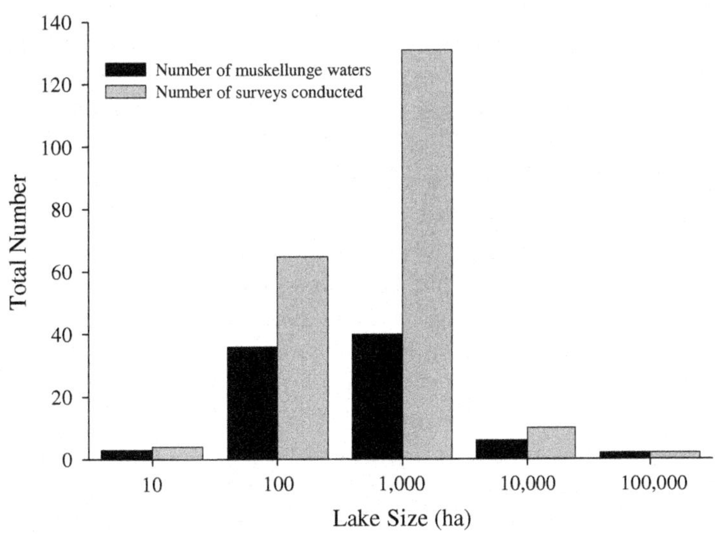

Fig. 1 Distribution of waters where spring trap net assessments were conducted from 1976–2005 in relation to the 105 muskellunge waters in Minnesota

appropriate for contending with the large number of zero observations. To achieve this, we first categorized each observation (i.e., fishing trip or trap net assessment) as successful (caught at least one fish), or not (no fish were caught). These binomial data were then analyzed using logistic models. We then used a subset of both angler diary and trap net data that were successful. These data were analyzed with standard general linear model procedures, where the dependent variable was $\log_e$ CPUE. Where appropriate, we applied Tukey's HSD with a Type I error rate of 5% to identify homogenous subsets for the relevant independent variables and examined residuals to assess model fit.

When applying linear model analyses for both length and CPUE data, we did not use standard model selection procedures such as best subset regression because several independent variables were confounded over time. For trap net data, we had to separate analyses by gear type, since the dimensions and design of trap nets changed over time. Year was a continuous variable when testing trends over time with trap net data. We did this analysis for three length classes of fish: fish ≥ 64 cm (25 in), 76 cm (30 in) and 102 cm (40 in). For testing trends over time with angler diary data, time was arranged as a categorical variable with an early period from 1986 to 1989 and a later period from 1996 to 1998. In addition to time, another independent variable for angler diary data included lake status (native or stocked). We did this analysis for all reported fish caught, and for fish greater than 102 cm (40 in). We performed all analyses of length and CPUE data for both angler diary and trap net assessments using JMP software (SAS Institute 2002).

Length data from angler diaries and trap net assessments were analyzed with standard parametric procedures, including analysis of variance (ANOVA) followed by multiple comparison of means using Tukey's HSD with a Type I error rate of 5%. Length data from trap nets were not available for individual fish, so we used mean length per assessment with reciprocal variance weighting, where the mean and variance were provided in survey reports for all fish sampled within an annual assessment.

We used cleithra as the primary structure for analysis of age and growth. The analysis of cleithra has provided a direction for defining trophy muskellunge growth parameters (Casselman and Crossman 1986). Cleithra from taxidermists were supplemented with cleithra from other sources including lake assessments to increase sample size. Population characteristics described include mean age and size of harvest, and age and size frequency distributions. Additional length-based analysis (Pauly 1984) used the von Bertalanffy growth formula (VBGF) fit with a nonlinear least squares method (Prager et al. 1989) for determining the asymptotic length (L_∞). We used an unpaired *t*-test to determine whether individual L_∞ estimates differed between native and introduced waters.

Results and discussion

Distribution of Minnesota muskellunge waters

A key management issue in Minnesota is to increase the number of muskellunge angling opportunities by expanding the number of lakes managed for muskellunge. One hundred five lakes with a combined area of 196,800 ha (486,302 ac), and six river systems have been identified as muskellunge waters (including hybrid muskellunge). Minnesota muskellunge waters are present in all three major drainage basins: Hudson Bay; Mississippi River; and Lake Superior. The majority of these waters are found in the north-central region and the Minneapolis-St. Paul metropolitan area, although angling opportunities exist in all fisheries management regions of the state (Fig. 2). Forty-four lakes and all six rivers are recognized as native muskellunge waters. In addition, muskellunge are present in the following border waters with Canada and Wisconsin: Lake of the Woods; Rainy River; Rainy Lake; St. Louis River Estuary; and St. Croix River. Introduced populations have been developed statewide, and are maintained by a stocking program. The stocking program has created and maintains 100,035 ha (247,192 ac, 41 lakes) of muskellunge waters and 1766 ha (4364 ac, 20 lakes) of muskellunge hybrid waters.

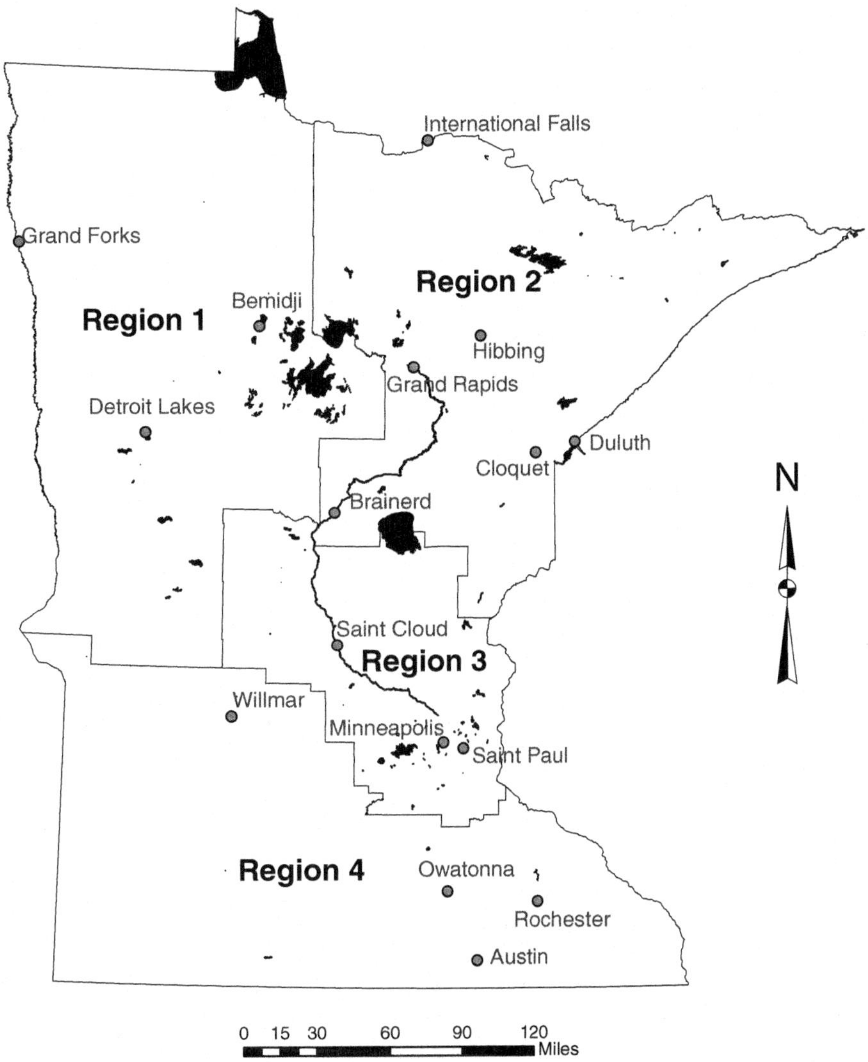

Fig. 2 Statewide distribution of muskellunge waters. Lines represent the four regional fisheries management boundaries

Although hybrid muskellunge occur naturally in some waters, managed hybrid lakes are maintained by stocking, and are located only in the Minneapolis-St. Paul metropolitan area.

Analysis of relative abundance

The following results were based on independent analysis of two data sets (diary and trap net assessments), and then compared to assess whether they led to similar conclusions. For logistic regression of binomial data from trap net assessments, year was the only significant effect and only for the 102 cm (40 in) length class (Table 1A). This indicated that the odds of catching at least one fish 102 cm (40 in) or larger in a trap net assessment has increased over time (Fig. 3). Muskellunge 127 cm (50 in) and larger started to appear more frequently in trap net assessments starting in 1992. About 57% of muskellunge waters had at least one fish 127 cm (50 in) or longer present in trap net assessments conducted between 1997 and 2005.

There was also evidence of a trend of increasing abundance over time for all three-length classes when we applied ANOVA to CPUE from successful trap net assessments (Table 1B). For the 102 cm (40 in) data set, both the Year and Gear main effects were significant with no significant interaction detected. This indicates that both the large and small trap nets showed an increase in CPUE over time, however the large

Table 1 Analysis of CPUE data from trap net assessments: (A) Logistic regression for binomial data (assessments where muskellunge were present or absent); (B) ANOVA of Log_eCPUE for successful trap net assessments

Size of fish cm (in)	Source	df	Chi square	Prob. > Chi square	
(A)					
64 (25)	Year	1	2.932	0.0868	
	Gear	1	0.050	0.8235	
76 (30)	Year	1	2.086	0.1486	
	Gear	1	0.301	0.5834	
102 (40)	Year	1	14.066	0.0002	
	Gear	1	0.430	0.5120	
	Year*Gear	1	0.028	0.8674	
Size of fish cm (in)	Source	df	Sum of squares	Mean square	Prob. > *F*
(B)					
64 (25)	Year	1	6.0	6.0	0.0406
	Gear	1	0.9	0.9	0.4272
	Year*Gear	1	9.0	9.0	0.0125
	Error	135	190.4	1.7	
	Corrected Total	138	219.0		
76 (30)	Year	1	14.1	14.1	0.0017
	Gear	1	0.5	0.5	0.5599
	Year*Gear	1	3.9	3.9	0.0920
	Error	134	183.5	1.4	
	Corrected Total	137	221.9		
102 (40)	Year	1	26.2	26.2	<0.0001
	Gear	1	4.4	4.4	0.0356
	Year*Gear	1	0.3	0.3	0.5514
	Error	106	102.2	1.0	
	Corrected Total	109	171.9		

nets showed a faster rate of increase. Year was also significant for fish 76 cm (30 in) and larger, providing additional evidence for an increasing trend in abundance of larger fish (Table 1B). The only significant interaction occurred for the 64 cm (25 in) length class (Table 1B). While large trap nets showed an increase in CPUE over time, small trap nets showed a decrease, although not significant. This may reflect some of the changes that have occurred over the past 25–30 years in Minnesota's muskellunge stocking program. Small trap nets were used in spring assessments until 1999 and were considered efficient at catching muskellunge during these earlier assessment periods. However, both types of analyses that we applied to the trap net assessment data (logistic regression of presence/absence binomial data and ANOVA of Log_e CPUE) indicate that the abundance of larger muskellunge has increased over time. By including fish greater than or equal to 64 cm (25 in) in our analysis, we may be seeing the first evidence of maturing populations dominated by larger individuals, and thus a reduction in relative abundance of smaller fish in the population. This may also reflect the number of muskellunge stocked over time, with total numbers of stocked reaching a maximum from the late 1980s to early 1990s, followed by a slight decline (Younk and Pereira 2003, Wingate and Younk in

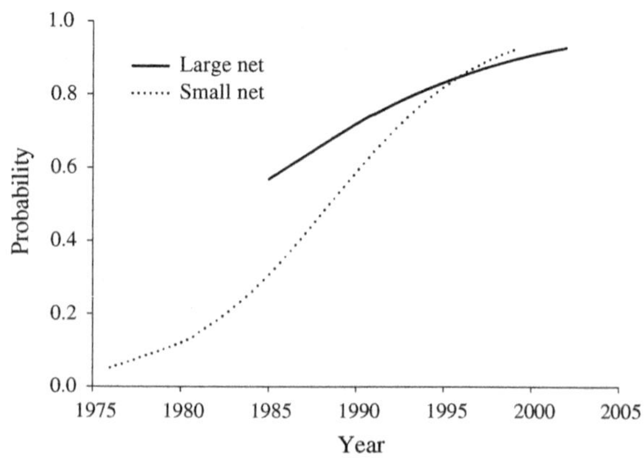

Fig. 3 The probability of catching at least one muskellunge 102 cm (40 in) or larger in a trap net assessment. Since the gear effect was not significant, the two lines for small and large nets are not significantly different

press). Increased abundance of small fish during the stocking peak followed by the decline in stocking may explain a declining CPUE trend in the small nets when analyzing all fish 64 cm (25 in) and larger.

Angler catch data also provided evidence of an improving muskellunge fishery over time. The proportion of successful anglers increased from the early to late time period for both length classes of fish (Table 2A; Fig. 4). However, anglers fishing stocked lakes were more successful (from 28.0% early period to 29.8% late period) than those fishing native waters (from 16.6 % early period to 20.4% late period), but only for all fish caught. The greatest increase in percent of successful anglers from the early to late period for all fish caught occurred in native waters (22.9% vs. 6.4% in stocked waters). In contrast, for fish 102 cm (40 in) and larger in length, stocked waters exhibited the greatest percent increase in successful anglers from the early to late period (322.7% vs. 89.8% native waters). This produced a significant interaction for the 102 cm (40 in) length class (Table 2A). The probability of an angler catching a fish 102 cm (40 in) and larger was greater in native waters than stocked waters in the early period only (Fig. 4). During the late time period, anglers were equally successful in catching at least one fish 102 cm (40 in) or larger, regardless of which waters (native or stocked) they fished (Fig. 4), and again may be related to the temporal trend in total numbers of fish stocked. This may also represent additional evidence that abundance of larger muskellunge has improved over time.

The magnitude of catch rates for all fish differed between time periods within lake type (native or stocked), as suggested by a strong time by stocked interaction ($F = 16.70$; df = 1; $P < 0.0001$; Table 2B). Angler catch rates did not change significantly in stocked waters between the two time periods for all fish caught (0.185 fish per hour for early period and 0.172 fish per hour for later period), but did increase significantly for native waters (from 0.099 fish per hour during early period up to 0.122 fish per hour during late period). However, stocked waters had higher catch rates than native waters for fish 102 cm (40 in) and larger (0.148 vs. 0.098 fish per hour). No change in catch rates was found between time periods for fish 102 cm (40 in) and larger (Table 2B; early period 0.120 vs. late

Table 2 Analysis of CPUE data from angler diaries: (A) Nominal logistic fit for binomial data (successful and unsuccessful anglers); (B) ANOVA of Log_eCPUE for successful anglers only

Size of fish cm (in)	Source	df	Chi square	Prob. > Chi square	
(A)					
All Fish	Time	1	10.6	0.0011	
	Stocked	1	125.2	<0.0001	
	Time*Stocked	1	2.4	0.1253	
Fish ≥ 102 (40)	Time	1	88.8	<0.0001	
	Stocked	1	11.9	0.0006	
	Time*Stocked	1	11.9	0.0005	
Size of fish cm (in)	Source	df	Sum of squares	Mean square	Prob. > F
(B)					
All Fish	Time	1	2.2	2.2	0.0499
	Stocked	1	112.8	112.8	<0.0001
	Time*Stocked	1	9.6	9.6	<0.0001
	Error	3287	1889.3	0.6	
	Corrected Total	3290	2024.5		
Fish ≥ 102 (40)	Time	1	0.004	0.004	0.9241
	Stocked	1	13.2	13.2	<0.0001
	Time*Stocked	1	0.7	0.7	0.2620
	Error	992	546.1	0.6	
	Corrected Total	995	574.0		

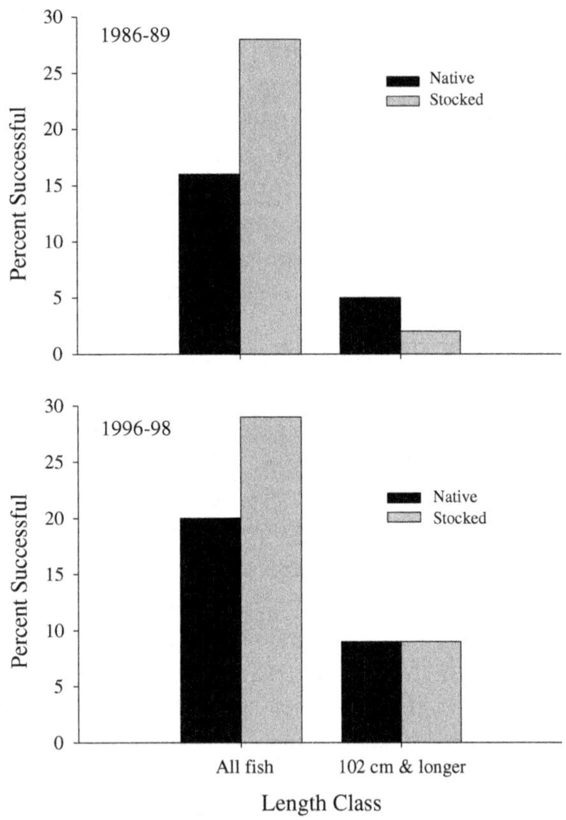

Fig. 4 The proportion of successful anglers during early and late time periods of the angler diary project and for native and stocked waters

period 0.121 fish per hour). We conclude that there is little evidence for any change in catch rates by successful anglers between the two time periods, with the exception of all fish in native waters.

Although the proportion of successful anglers increased over time, we found inconsistent evidence for an increase in catch rates over time. The differences between these two results may reflect the nature of muskellunge angling, where very few anglers catch more than one fish per trip, thus significantly hindering our ability to detect changes over time in catch rates by successful anglers. We could also attribute these differences to a limited resource being pursued by an ever-increasing number of anglers. Based on increasing license sales (Cook et al. 1997), we assume that the number of muskellunge anglers has also increased over time. Coupled with the increase in fishable muskellunge waters, it appears that the catch is distributed over more anglers, thus catch rate has not changed. Simonson and Hewett (1999) reported that angling effort targeting muskellunge in six Wisconsin lakes during the 1990s was higher than during the 1980s. During these same two time periods catch rates were similar and harvest rates declined, but total muskellunge catch remained the same.

Analysis of length data

Analysis of length data from angler diaries indicates that size of muskellunge caught by anglers has increased over time (Table 3A). Mean length of muskellunge reported in angler diaries increased from 87 cm (34 in) in the early period to 93 cm (36 in) in the later period. A similar trend can be seen when examining pooled

Table 3 Analysis of variance for length data from (A) angler diaries and (B) trap nets

Source	df	Sum of Squares	Mean Square	Prob. > F
(A)				
Time	1	3465.5	3465.5	<0.0001
Stocked	1	2800.2	2800.2	<0.0001
Time*Stocked	1	125.1	125.1	0.0745
Error	4315	169558.4	39.3	
	4318	174792.2		
Source	df	Sum of Squares	Mean Square	Prob. > F
(B)				
Year	1	1439.1	1439.1	<0.0001
Gear	1	1.3	1.3	0.8853
Year*Gear	1	1067.9	1067.9	<0.0001
Error	375	23138.5	61.7	
Corrected Total	378	29420.6		

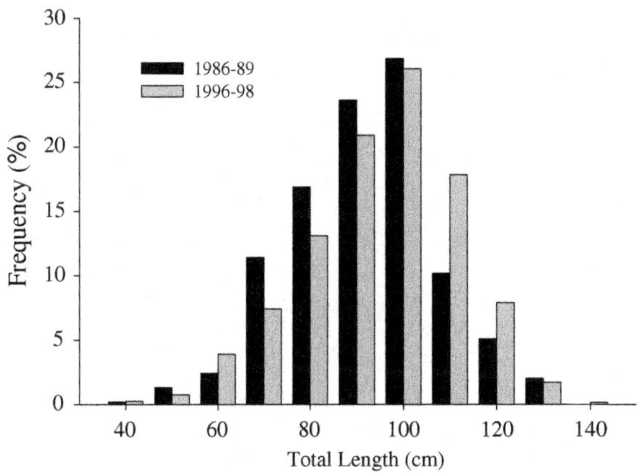

Fig. 5 Length frequency distributions of muskellunge caught by anglers during the early (1986–89) and late (1996–98) diary periods. Data was pooled for all lakes in each period

length data between early and late periods (Fig. 5). Although the mode is similar, anglers reported a higher proportion of fish 110 cm (43 in) and longer. We also found that anglers caught significantly longer fish in native (92 cm; 36 in) vs. stocked (87 cm; 34 in) waters ($F = 71.2$; df = 1; $P < 0 .0001$).

Length data from trap nets were analyzed separately for small and large nets. No difference was found in mean length of fish caught by either small (95 cm; 37 in) or large (95 cm; 37 in) trap nets. However, mean length in trap net assessments increased over time in both gear types, but the magnitude of increase was greater for small trap nets, which is the cause of the significant interaction (Table 3B). This trend remained the same even when we used subsets of data to reflect only those years when both gears were deployed. One reason for these inconsistent results may be related to the initial unfamiliarity with the deployment of large trap nets. Site selection and net deployment procedures needed to be modified because of the larger frame and hoops. Net deployment was one of several factors listed by Hubert (1996) as having an influence on catch. However, change in minimum size regulation from 91 to 102 cm (36–40 in) in 1993 is the more likely reason for these differences. Much of the effect that this regulation change had on the size structure of the population likely occurred during the 6-year period prior to the full-scale use of large trap nets.

Our results indicate consistency between trap net assessments and angler diary data. However, we urge caution in how inferences may be applied or interpreted. For example, we can't apply inferences to all muskellunge anglers because participants were a distinct subset of the statewide muskellunge angling population and were not directed in any way on how or where to fish. Thus, it may not reflect behaviors of all muskellunge anglers. There are similar restrictions with the trap net assessment data. The MNDNR currently does not have a clear sampling design for trap net assessments, and managers often conduct assessments based on local information needs (i.e., lake management plans), or to address issues that may arise from external pressures. However, trap net assessment data represents a majority (55%) of the state's muskellunge waters. Furthermore, similar results from these two disparate data sources indicate that Minnesota's muskellunge waters are providing larger fish while maintaining angler catch rates over the time period evaluated.

Age and growth parameters

A total of 564 muskellunge cleithra were collected; however, not all were available for use in analyzing growth parameters. Taxidermist samples ($N = 389$) accounted for 69% of the cleithra aged, with the remaining samples ($N = 175$) coming from other sources. Muskellunge from taxidermists ranged from 4 to 22 years and averaged 11 years old (Fig. 6a). Although females were 1.2 years older than males on average, results showed a modal age of 9 years for females and 11 years for males. Forty-two percent of the females and 23% of the males caught by anglers exceeded 11 years of age (Fig. 5a). Casselman and Crossman (1986) reported a modal age of trophy muskellunge from Ontario waters of 11 years and from U. S. waters of 8 years. The oldest aged fish from Ontario and U. S. waters was 30 and 23 years, respectively (Casselman and Crossman 1986). When cleithra from all Minnesota sources were combined, a broader range of

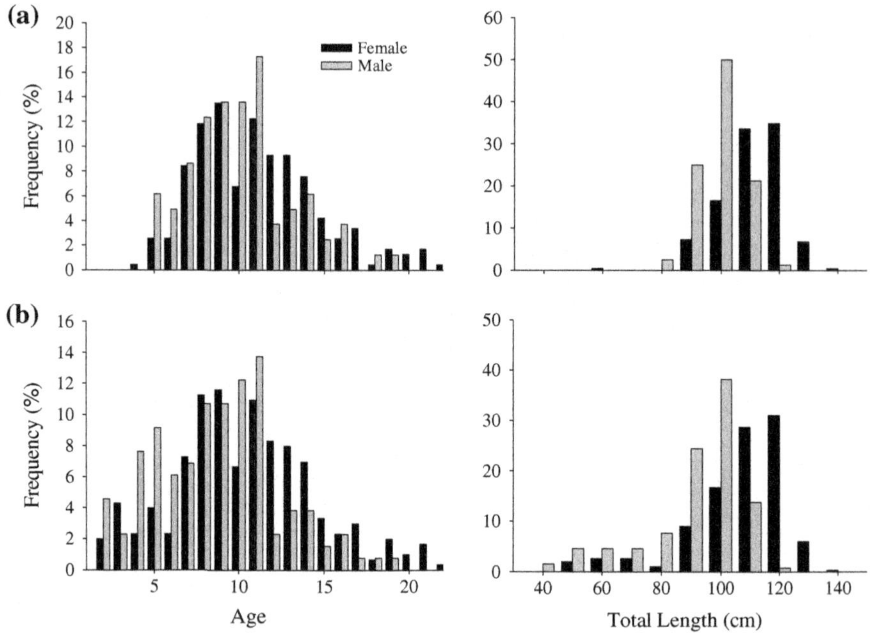

Fig. 6 Length and age frequency distribution of (a) muskellunge from taxidermists and also (b) combined with muskellunge collected from other sources

ages (range 1–22) was available, but skewed towards younger fish (Fig. 6b).

Minimum size regulations, coupled with angler preferences and harvest ethics, probably have the greatest influences on the harvest of trophy muskellunge. Muskellunge reported from taxidermists averaged 115 cm (45 in) with a modal length-class of 120 cm (47 in) (Fig. 6a). Sex ratio of taxidermist reported muskellunge was skewed toward females (2.9:1), as would be expected based on anglers targeting the largest individuals, which are typically females. This disproportionate number of females reflects the selective harvest of the larger individuals by anglers. Reports of skewed sex ratios of angler-harvested muskellunge have ranged from 2.7:1 (Casselman et al. 1999) to 6.3:1 (Casselman and Crossman 1986). About 7% of the females and 0% of the males exceeded the 120 cm (47 in) length-class (Fig. 6a). The majority (51%) of females were from 100 (39 in) to 120 cm (47 in) in length, while 52% of the males were from the 100 cm (39 in) length-class. Other sources of muskellunge provided an additional 173 samples, resulting in a broader length distribution (Fig. 6b). Considering that the majority of fish from other sources were found dead along lakeshores, this information could provide some insight into the size of fish associated with either natural or hooking mortality.

A total of 360 cleithra were available for calculating von Bertalanffy growth estimates, with 81% collected from taxidermists. Estimates of ultimate lengths from individual muskellunge ranged from 97 (38 in) to 156 cm (61 in) and averaged 130 cm (51 in) (Fig. 7a). Estimates of female ultimate length averaged 134 cm (53 in) while males averaged 119 cm (47 in). Box plots also seem to indicate that the majority of fish identified as unknown sex are female. Although the majority (79%) of these samples came from Leech and Cass lakes (native waters) the remaining samples came from a number of different waters representing both native and introduced populations. The difference of 3 cm (1.1 in) in average ultimate lengths between native and introduced populations was not statistically significant ($t = 1.42$; df = 358; $P = 0.15$; Fig. 7b), indicating that both native and introduced waters have the potential to produce trophy fish. For Ontario muskellunge waters, average ultimate length ranged from 81 (32 in) to 140 cm (55 in) for females and 71 (28 in) to 116 cm (46 in) for males (Casselman et al. 1999). Projected growth estimates for Minnesota female muskellunge ranged from 100 to 154 cm (39–61 in) in total length, comparing favorably with Ontario waters. The two most frequently sampled native lakes in the database had average

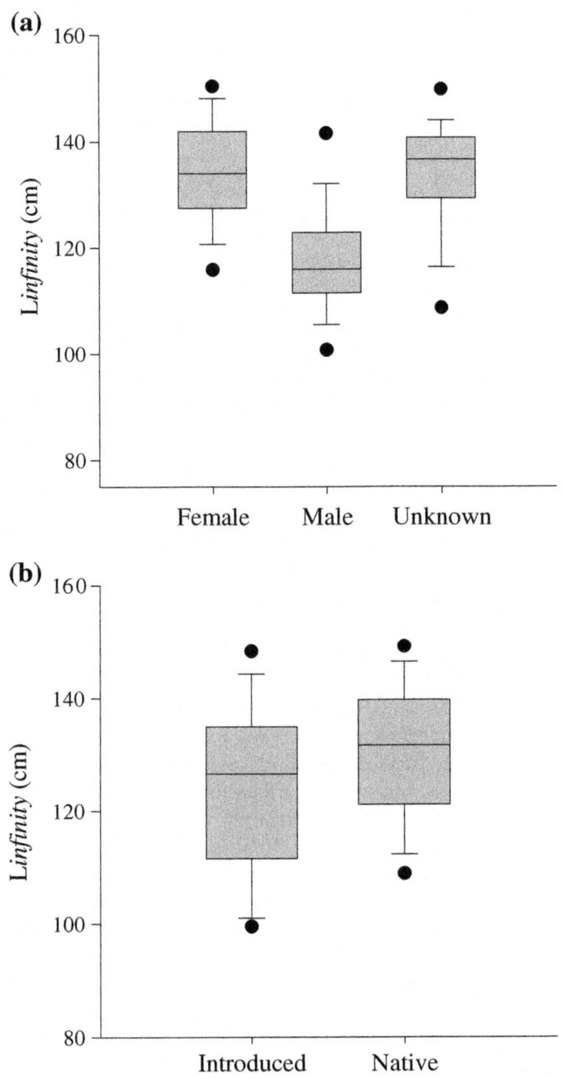

Fig. 7 Box plots of von Bertalanffy growth parameter L_∞ (cm) for (a) male and female muskellunge and (b) for native versus introduced (stocked) waters. The box defines the 75th and 25th percentiles, line within the box marks the median, whiskers above and below the box indicates the 90th and 10th percentiles. Outliers represented by dark dots

female ultimate lengths of 137 cm (54 in; Leech) and 132 cm (52 in; Cass).

Summary

Muskellunge angling opportunities in Minnesota have been increased and enhanced through the establishment of new populations, supplemental stocking and the use of conservative, but progressive regulations. Although we were unable to address specific cause and effect relationships, Minnesota's muskellunge populations appear to have improved in quality over the past 25–30 years. Evaluation of data utilizing trap net and angler diary information suggests that both size and catch of muskellunge have increased over this time period. Angler catch rates have not improved over time, but the proportion of successful anglers has increased. Although voluntary catch and release has played a role in maintaining the quality of the muskellunge fishery, the majority of fish harvested were 11 years old or younger. Taking into account the maximum assessed age of 30 years in Ontario waters (Casselman et al. 1999) and 22 years in Minnesota waters, harvest remains directed at younger muskellunge. Growth information suggests that many of the state's muskellunge waters are capable of producing 140 cm (55 in) and larger fish. Overall, these results have shown that the management steps taken by the MNDNR has provided a quality muskellunge fishery over time. This is particularly true for those waters where Shoepack strain muskellunge were initially used early in the stocking program, but were subsequently stocked with Mississippi strain muskellunge because of concerns about the quality of populations with Shoepack heritage (MNDNR 1994, Wingate and Younk in press).

The overall number and acreage of muskellunge waters in Minnesota will continue to change as both introduced and historically important native muskellunge waters are evaluated. Other changes likely to occur in the future include minimum length regulations and stocking (rates and frequency). Any discussions about changes in minimum length regulations should reflect the addition of muskellunge growth potential as well as age and length of harvested fish. Future stocking discussions will need to address stocking rates and frequency based on the current status of population characteristics in these waters. However, there remain numerous information gaps that need to be addressed prior to making these types of management decisions. Standard fisheries sampling techniques and monitoring methods have failed to provide adequate information on muskellunge population characteristics (Strand 1986), thus there is a need to incorporate better sampling designs into muskellunge assessment programs. Lewis (1996) identified effective evaluation as a challenge for the future of muskellunge management in Ontario. These sampling designs must take into consideration both short-term and long-term data requirements.

Acknowledgments We thank all the Area Fisheries staff that provided historical and updated trap netting data. Reviews by J. Diana, M. A. Hanson, B. R. Herwig, T. Margenau, P. J. Wingate, and two anonymous reviewers improved the manuscript. Sanford Weisberg (University of Minnesota) and Larry Jacobson (National Marine Fisheries Service) provided statistical advice. Finally, a special thanks to all of the volunteer anglers and taxidermists for their cooperation during this study. This project was funded in part by the Federal Aid in Sport Fish Restoration (Dingell-Johnson) Program, Project F-26-R, Minnesota.

References

Casselman JM, Crossman EJ (1986) Size, age, and growth of trophy muskellunge and muskellunge-northern pike hybrids – the cleithrum project, 1979–1983. American Fisheries Society Special Publication 15:93–110

Casselman JM, Robinson CJ, Crossman EJ (1999) Growth and ultimate length of muskellunge from Ontario water bodies. North Am J Fish Manage 19:271–290

Cook MF, Younk JA, Schupp DH (1997) An indexed bibliography of creel surveys, fishing license sales, and recreational surface use of lakes and rivers in Minnesota. Minnesota Department of Natural Resources, Section of Fisheries Investigational Report 381, St. Paul

Cunningham PK, Anderson CS (1992) Opinions of angler groups and fisheries professionals in Minnesota. Minnesota Department of Natural Resources, Section of Fisheries Investigational Report 422, St. Paul

Hanson DA (1986) Population characteristics and angler use of muskellunge in eight northern Wisconsin lakes. Am Fish Soc Spec Publ 15:238–248

Hoff MH, Serns SL (1986) The muskellunge fishery of Escanaba Lake, Wisconsin under liberalized angling regulations, 1946–1981. Am Fish Soc Spec Publ 15:249–256

Hubert WA (1996) Passive capture techniques. In: Murphy BR, Willis DW (eds.), Fisheries techniques, 2nd ed. American Fisheries Society, Bethesda, Maryland. pp. 157–181

Leitch JA, Baltezore JF (1987) Attitudes of Minnesota anglers. Final Report to Minnesota Department of Natural Resources, Division of Fish and Wildlife, St. Paul

Lewis CA (1996) An overview of muskellunge management in Ontario. In: Kerr SJ, Olver CH (eds), Managing muskies in the '90s. Workshop Proceedings. Ontario Ministry of Natural Resources, Southern Region Science & Technology Transfer Unit Workshop Proceedings. pp 5–9

Margenau TL, Petchenik JB (2004) Social aspects of muskellunge management in Wisconsin. North Am J Fish Manage 24:82–93

MNDNR (Minnesota Department of Natural Resources). (1993) Manual of instructions for lake survey. Minnesota Department of Natural Resources, Section of Fisheries, Special Publication No. 147, St. Paul

MNDNR (Minnesota Department of Natural Resources). (1994) Fisheries long-range plan – muskellunge chapter: 1994–1999. Minnesota Department of Natural Resources, Section of Fisheries, St. Paul

Olson DE, Cunningham PK (1989) Sport-fisheries trends shown by an annual Minnesota fishing contest over a 58-year period. North Am J Fish Manage 9:287–297

Pauly D (1984) Fish population dynamics in tropical waters; a manual for use with programmable calculators. International Center for Living Aquatic Resources Management, ICLARM Studies and Reviews 8, Manila

Prager MH, Saila SB, Recksiek CW (1989) FISHPARM: a microcomputer program for parameter estimation of nonlinear models in fishery science, 2nd edition. Old Dominion University, Oceanography Technical Report 87–10, Norfolk, Virginia

SAS Insitute, Inc. (2002) JMP Statistics and graphics guide, Version 5. Cary, North Carolina

Scidmore WJ (1970) Manual of instructions for lake survey. Minnesota Department of Natural Resources, Section of Fisheries, Special Publication 1, St. Paul

Simonson TD, Hewett SW (1999) Trends in Wisconsin's muskellunge fishery. North Am J Fish Manage 19:291–299

Strand RF (1986) Review of technical contributions. Am Fish Soc Spec Publ 15:360–369

Wingate PJ (1986) Philosophy of muskellunge management. Am Fish Soc Spec Publ 15:199–202

Wingate PJ, Younk JA (in press) A recipe for successful muskellunge management – A Minnesota success story. Environ Biol Fishes 00:000–000

Younk JA, Cook MF (1992) Application of an angler diary for muskellunge, *Esox Masquinongy*. Minnesota Department of Natural Resources, Section of Fisheries Investigational Report 420, St. Paul

Younk JA, Strand RF (1992) Performance evaluation of four muskellunge *Esox masquinongy* strains in two Minnesota lakes. Minnesota Department of Natural Resources, Division of Fish and Wildlife Investigational Report 418. St. Paul

Younk JA, Pereira DL (2003) An examination of Minnesota's muskellunge waters. Minnesota Department of Natural Resources, Division of Fish and Wildlife Investigational Report 498. St. Paul

Environ Biol Fish (2007) 79:137–154
DOI 10.1007/s10641-007-9192-y

SPECIAL ISSUE CROSSMAN

Determining minimum ultimate size, setting size limits, and developing trophy standards and indices of comparable size for maintaining quality muskellunge (*Esox masquinongy*) populations and sports fisheries

John M. Casselman

Received: 8 August 2006 / Accepted: 10 December 2006 / Published online: 5 April 2007

Abstract Growth and ultimate size can provide important population insights and a sound biological basis for setting length limits, which can be the best single regulation for preventing overexploitation of muskellunge (*Esox masquinongy*) populations. A system was developed, using cleithral age and total length at age confidence limits (CL) data, to determine reproductive and growth potential (ultimate size) for calculating and setting increased size limits based on minimum reproductive size (upper 99% CL at age at first maturity + 2 year) and minimum ultimate size (MUS) calculated from the lower 99% CL—minimum ultimate size limit (MUSL). MUS also provides a trophy standard and an index of relative size for comparing trophy potential of individuals within and among populations. Guidelines are provided for determining minimum sample size (mean ± 95% confidence interval = 12 ± 4) and minimum age (8–10 ± 2.0 year) required to produce valid von Bertalanffy growth trajectories. MUS, MUSL, and trophy standards for both length and estimated weight are provided for female and male muskellunge from 14 Ontario sources. Mean MUS, or trophy standard, for females was 115 ± 10.3 cm (MUSL range 75–135) and 11.1 ± 2.6 kg (2.5–17.5) and for males was 95 ± 7.5 cm (66–110) and 6.1 ± 1.3 kg (1.9–9.2). These indices can precisely define growth and growth potential for muskellunge populations and individuals and can be used to better manage and maintain or improve the quality of muskellunge populations and fisheries.

Keywords Muskellunge · Size limit · Trophy standard · Minimum ultimate size · Growth index · Length · Weight · Age · Growth potential · L-infinity · 99% confidence limits

Dedicated to the late Dr. E.J. Crossman.

J. M. Casselman (✉)
Department of Biology, Queen's University, Biosciences Complex, 116 Barrie Street, Kingston, ON, CanadaK7L 3N6
e-mail: casselmj@biology.queensu.ca

Introduction

Growth of muskellunge, *Esox masquinongy*, varies widely across the range of the species. This has been well documented in recent years because accurate, validated age assessment has been developed for the species (Casselman et al. 1999). Differences in growth potential and ultimate size are well known to muskellunge anglers. In recent years, populations with high growth potential, producing large-bodied fish, have been heavily fished and are increasingly the focus of intense sport fisheries.

With the increasing practice of voluntary catch-and-release and its promotion by organized muskellunge angling groups (such as Muskies Inc. and Muskies Canada), there has been a substantial increase in size of fish caught and angling success. This has, to some extent, been facilitated by increasing length limits (commonly referred to as size limits) to restrict harvest and by more favourable environmental conditions (Robinson 2005) associated with warmer temperatures in the Great Lakes Basin (Casselman 2002).

In order to better manage populations and to provide diverse fishing opportunities, some agencies have implemented a range of size limits across their jurisdictions, depending upon whether the water body can produce small-, medium-, or large-bodied muskellunge (Casselman et al. 1999). In Ontario, more diverse growth-based size limits were implemented widely in the late 1980s (Ontario Ministry of Natural Resources 1986). Indeed, Hunt (1970) emphasized that a size limit, if appropriately applied, is the best single regulation for preventing overexploitation of fish populations.

Encouragement by organized muskellunge anglers to increase length limits necessitated a science-based method for determining desirable size, based on growth. Although age at first maturity seems to be quite consistent across the range of the species, growth and ultimate size seem much more variable (Casselman et al. 1999). This is probably related to productivity of the environment, abundance and size of available prey, and even size of the water body.

Genetics can also affect growth and growth potential of muskellunge. Koppleman and Philipp (1986) confirmed that there are distinct genetic stocks of muskellunge and emphasized that this should be taken into consideration when managing the species. Minnesota conducted performance evaluation of four muskellunge strains in two Minnesota lakes and confirmed that the Mississippi strain grew fastest and reached the largest size (Younk and Strand 1992). Margenau and Hanson (1996) documented that genetics affected fish size and growth of stocked muskellunge. This is strong evidence that genetics can affect growth potential and requires special consideration when stocking and managing for quality fisheries and overall trophy potential.

All of this emphasizes the need for a biologically based growth method for setting specific size limits. Initial discussions in the mid-1990s that led to new length-limit regulations for the province of Ontario in the late 1990s (Ontario Ministry of Natural Resources 1999) were initiated with a discussion of "favourite" length limits. This resulted in considerable debate that seemed irresolvable and fraught with controversy. It was apparent that what was needed was a universally applicable growth-based method for calculating length limits that would take into consideration population-specific growth potential.

Considerable data had been collected on size at age, mainly as a result of the success of The Cleithrum Project in assembling and archiving a large quantity of muskellunge data and cleithral samples from numerous water bodies throughout Ontario (Crossman and Casselman 1996). A grant from Muskies Inc. made it possible to extract a considerable amount of data from this archive (Robinson 2005). Detailed growth of muskellunge populations from 17 Ontario water bodies was assembled and analyzed (Casselman et al. 1999).

Detailed age and growth data that were available made it possible to examine size at age and growth rate quite precisely and to categorize specific populations, based on growth potential, producing small-, medium-, or large-bodied fish (Casselman et al. 1999). Size limits in the mid-1980s were grouped according to growth but did not take into consideration ultimate growth potential. These limits were initially set around age at first maturity + 2 years. It was generally assumed that reproduction was not totally successful at first maturity but that fish required several years to reach full reproductive potential. Early analyses of growth resulted in an increase in size limits but did not truly address the broad range of ultimate sizes, or growth potentials, that existed for populations across Ontario.

To develop a more refined set of length limits, it was necessary to examine growth rate and growth potential in more detail. Ideally, because of differential growth of the sexes, different limits would be needed for males and females. However, this required a level of complexity that anglers did not appear ready to accept at that time, even though external sex determination of live muskellunge was possible (LeBeau and Pageau 1989), using a method that had been developed for northern pike, *Esox lucius* (Casselman 1974). If female growth was used to set size limits, slower-growing males would be underexploited. Nevertheless, it seems appropriate to focus on female size limits since the reproductive capacity of the population is strongly influenced by egg production.

The general purpose of the present study was to develop a method and criteria for setting size limits based on linear growth potential, using growth trajectories and ultimate size, or L-infinity. The new upper size limit developed here uses $L\infty$, or ultimate length, calculated from the lower 99% confidence limit (CL) and is referred to as minimum ultimate size (MUS). MUS is a mathematical projection of the length that 99% of the fish would reach if they lived indefinitely. A number of other ultimate-size values could be considered; for example, maximum ultimate size (upper 99% CL) would be of interest to anglers seeking the largest possible individuals, but this is an extreme size that is rarely seen or reached. MUS is a more realistic size frequently seen and exceeded and is a better lower-level upper-length benchmark; theoretically, only one fish in 100 would not reach this MUS.

MUS has both a genetic (Koppleman and Philipp 1986; Margenau and Hanson 1996) and an environmental (Casselman et al. 1999; Robinson 2005) component, depending upon population and water body. It is best assessed empirically from actual growth data assembled for the population. It can also define trophy potential, or status, and not only categorizes the fish, the population, and the water body but can also be used to assess relative size of an individual within and among populations in terms of a trophy standard. This trophy standard can be used to describe a quite specific ultimate size (length or estimated weight) that could be defined as being equal to one trophy unit, and any fish could be compared to it on a proportional basis.

The specific purpose of this study was to analyze growth to determine MUS for both sexes and set a minimum ultimate trophy size limit (MUSL) and to develop trophy standards and indices for examining trophy size and status, using muskellunge populations in Ontario as examples. Muskellunge are rare fish and sample sizes are often small, so it would also be useful to determine just how few samples are required and how old the fish would need to be to calculate ultimate size accurately and consistently.

Materials and methods

Age and growth determination

Samples and data used in this study came from The Cleithrum Project (Crossman and Casselman 1996) and were primarily from fish provided by anglers and taxidermists. Age and growth were extracted from cleithra, using validated age-interpretation procedures (Casselman 1996; Casselman et al. 1999). The majority of the samples were from relatively large trophy muskellunge, defined by Casselman and Crossman (1986) as angled fish mounted or considered worthy of being mounted. Some samples from the Niagara River and Georgian Bay came from Ontario Ministry of Natural Resources (OMNR) assessment sampling but were from large fish that were of trophy size. The Nogies Creek sample came from a sanctuary population in central Ontario provided from research studies conducted by the late Dr. E. J. Crossman and was used to encompass the range in growth potential of the species. Additional samples were used from the French River, which flows into Georgian Bay. Recent samples submitted to The Cleithrum Project made it possible to include this new source. It included 17 females, maximum age 22 years, and five males, maximum age 11 years. Only samples for which sex was provided were used in the analyses. Total length, weight, and age of muskellunge used in this study are provided in Casselman et al. (1999).

To reconstruct growth, annual increments in cleithra were measured, using the Calcified Structure Age-Growth data Extraction Software

(CSAGES) (Casselman and Scott 2000) modified to incorporate electronic calipers for measuring and enumerating annuli (Robinson 2005). Cleithral length was converted to body length by using the body-cleithrum relationship in Casselman et al. (1986). For all populations, von Bertalanffy relationships were constructed for the mean and also for the upper and lower 99% CL. $L\infty$ for von Bertalanffy relationships for upper and lower 99% CL were used to estimate mathematically the maximum and minimum ultimate lengths. To provide a more complete comparison when too few samples were available to calculate an accurate von Bertalanffy relationship, the relationship for the population with the most similar size at age was selected, using concordance sum of squares (Casselman et al. 1999). When unrealistic maximum or minimum ultimate lengths were estimated, equal confidence intervals were used to provide more realistic estimates. This most frequently occurred in the case of upper 99% CL. No males were available for Lake of the Woods and Tweed water bodies, so for comparative purposes, male means and confidence limits were estimated from regressions of the average relationship between male and female growth (Casselman et al. 1999). All growth analyses were conducted on length, and weights, when presented, were estimated from length, using a standard length-weight relationship (Casselman and Crossman 1986).

Minimum sample size and age

Growth potential was estimated by using ultimate length calculated from the von Bertalanffy relationship. Estimates of $L\infty$ are greatly influenced by maximum age of the sample (Ontario Ministry of Natural Resources 1983), so additional analyses involving 10 populations were conducted to determine how old muskellunge needed to be in order to provide an $L\infty$ that fell within the confidence limits for the population. This provided the minimum age required to produce a statistically reliable mean von Bertalanffy growth curve. The age at which this mean ultimate size was reached was calculated and compared with minimum age. For this analysis, as was shown in a previous study (Casselman et al. 1999), at least six individuals were required at each age to calculate an accurate von Bertalanffy relationship.

In addition to age, the minimum number of samples required was considered an important variable in estimating an appropriate $L\infty$. A set of six different sequential removals was conducted for three populations (Indian Lake, Lake of the Woods, and Lake St. Clair) to determine the minimum number of samples needed. This also provided an estimate of the minimum age required to estimate $L\infty$ accurately.

Setting minimum ultimate size limits

A method was developed to determine minimum ultimate size limits based on female linear growth potential but also considering minimum reproductive length limit at the upper 99% CL at age seven (Fig. 1). Since the late 1980s in Ontario, reproductive minimum size limit has always been set at the age at which the majority of females were first mature (age five plus 2 years) to ensure experienced spawning before harvest. In females, this is age seven, and in males, it is age six, since males usually mature one year younger than females. The upper 99% CL at this age was chosen rather than mean size because it would reduce selective removal of faster-growing, young, mature muskellunge.

MUSL was MUS rounded down to the nearest cm (if limits were in inches, the same unit rounding was performed); this was done to make sure that rounding did not exclude more individuals. For a similar reason, reproductive size limit was rounded up to the nearest cm. Depending upon growth rate of the population, if reproductive size limit was greater than MUSL, then reproductive size was selected as the upper size limit. This occurred only in populations that had very rapid early growth and reached an asymptotic length early in life. If MUSL was greater than reproductive size, and this was usually the case, then MUSL remained the minimum upper size limit.

Trophy standards and indices of relative size and status

MUS also provides a trophy standard and index that is the mathematical projection of minimum

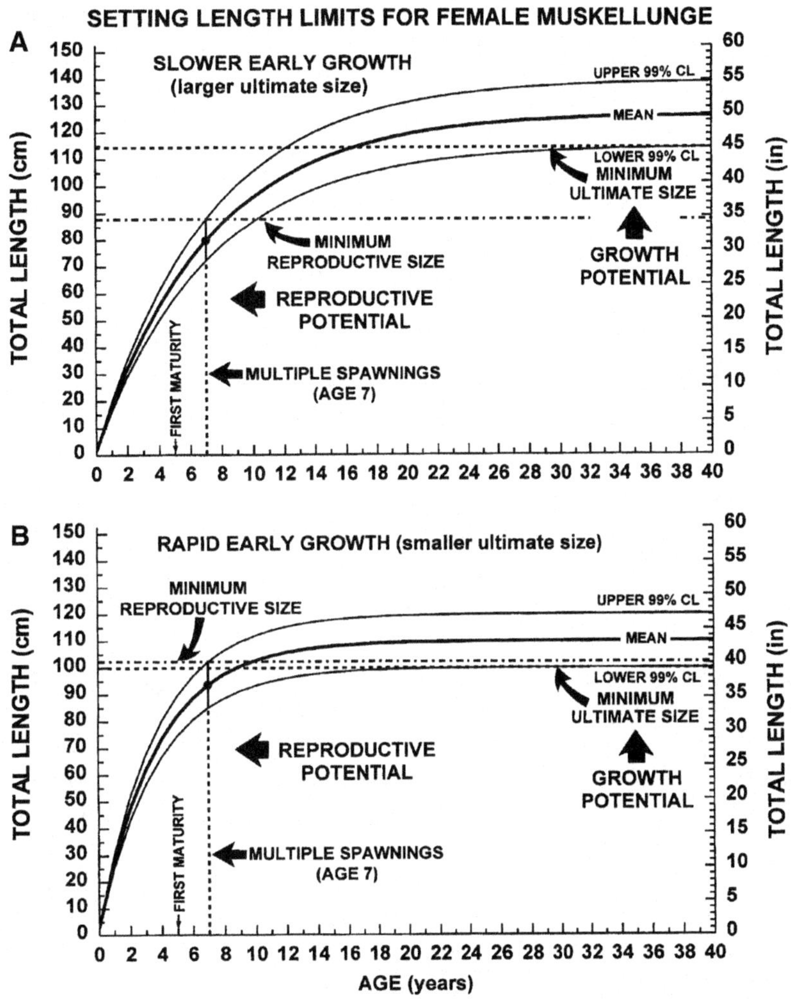

Fig. 1 Illustration of generalized size-at-age growth trajectories for female muskellunge with (**A**) slow early growth and large ultimate length and (**B**) rapid early growth and small ultimate length, indicating age and growth data important for setting size limits for muskellunge based on reproductive potential involving first maturity plus 2 years and growth potential involving ultimate length, calculated on the mean and 99% confidence limits, emphasizing the lower or minimum ultimate size

growth potential of the population, the theoretical size that 99% of individuals would reach if they lived indefinitely. MUS provides a trophy standard precisely equal to one trophy that other muskellunge can be compared with, using an index of relative size expressed as per cent, based on either length or weight, depending upon which criterion is chosen to define trophy. Mean or maximum ultimate length can also be expressed as a per cent of this minimum, or trophy, standard. Size varies considerably in relation to sex of a fish, so male and female standards were developed. If fish are released, external sex determination is necessary. However, if sex is unknown or not determined, then a sex-independent standard that equally weights males and females was developed and can be used. This eliminates the necessity to sex fish but is far less precise because it underestimates for females and overestimates for males.

Results

Minimum sample size and age

Calculation of the minimum age required to produce a mean von Bertalanffy curve and an $L\infty$ that fell within the 95% CL of the $L\infty$, using all ages, was examined for females from 10 water bodies (Table 1). Samples were fairly large (mean $N = 33.9$ fish) with a broad range of ages, producing a reliable overall $L\infty$. Mean minimum age required was 8.2 years, with 95% CL ranging from 6.2 to 10.2 years (Table 1). Mean age when overall $L\infty$ was reached was 29.5 years, and

Table 1 Minimum age required to calculate a mean von Bertalanffy growth curve that produces an ultimate length ($L\infty$) that fell within the 95% confidence limits (CL) of the $L\infty$ for female muskellunge from 10 Ontario sources. Approximate ultimate age at the ultimate length and minimum age relative to this ultimate age are also provided. Overall means and 95% CL are given. An *N* of at least six was required at each age

Source	*N*	Minimum age	Ultimate age—when $L\infty$ reached	Minimum age relative to ultimate age (%)
Tweed water bodies	22	4	19.7	20.3
Lake St. Clair	31	11	21.2	51.9
Georgian bay	42	7	22.8	30.7
St. Lawrence river	15	6	24.6	24.4
Lake of the woods	27	8	26.6	30.1
Indian lake	38	5	26.6	18.8
Maskinonge lake	34	10	29.0	34.5
Eagle–Wabigoon lakes	46	8	31.9	25.1
Lac Seul	11	10	39.5	25.3
Kawartha lakes	73	13	53.1	24.5
Mean	33.9	8.2	29.5	28.6
95% CL—lower	21.2	6.2	22.3	21.8
95% CL—upper	46.6	10.2	36.7	35.3

minimum age was only 28.6% of overall ultimate age, with 95% CL ranging from 21.8 to 35.3%.

Minimum number of samples required to produce an $L\infty$ that fell within the 95% CL for female muskellunge was examined for three Ontario sources used to examine the various confidence limits. An average of 32 samples was available; an adequate number to determine from six different sequential random removals the minimum number of samples required (Table 2). These samples and analyses indicated that the mean minimum number of samples required was 11.6, with 95% CL ranging from 7.6 to 15.6, rounded to 12 ± 4 (Table 2). Mean minimum age, which was also estimated in this analysis, was 10.1 years, ranging from 7.9 to 12.3, rounded to 10 ± 2 years (Table 2).

Mean and extreme ultimate sizes and ages

Ultimate length was calculated for female (Table 3) and male (Table 4) muskellunge from 14 Ontario sources for the mean, as well as extremes, using 99% CL. Appendices A and B present these data in inches and pounds. For a few sources, samples were inadequate to determine ultimate length directly, and estimates had to be made using concordance sum of squares (Casselman et al. 1999). For females, overall mean ultimate length was 123.0 ± 11.0 cm (Table 3), compared with 104.4 ± 7.6 cm for males (Table 4), 15.1% smaller. Also, mean maximum ultimate length for females was 134.1 ± 12.8 cm, compared with 116.0 ± 9.2 for males, 13.5% smaller. Mean minimum ultimate length for

Table 2 Minimum sample required to produce an ultimate length ($L\infty$) that fell within the 95% confidence limits (CL) for female muskellunge from three Ontario sources. Determined from six different sequential random removals. The ages of these subsamples are also provided. An *N* of at least six was required at each age

Source	*N*	Number			Age		
		Mean minimum	95% CL		Mean minimum	95% CL	
			Lower	Upper		Lower	Upper
Indian lake	38	10.5	4.5.	16.5	7.3	5.6	9.0
Lake of the woods	27	8.2	5.9	10.4	10.7	8.0	13.4
Lake St. Clair	31	16.0	12.3	19.8	12.3	10.2	14.5
Mean	32	11.6	7.6	15.6	10.1	7.9	12.3

Table 3 Growth and age data important for setting length limits (cm) based on reproductive size and growth potential for female muskellunge from 14 Ontario sources. Ultimate length (cm) provides a measure of growth potential, whereas reproductive potential is indicated by the upper 99% confidence limits (CL) at age seven (2 years after first maturity, age five). Minimum ultimate size limits with estimates of weight (kg) and predictions of age (youngest and oldest) are provided. The oldest observed age is given. Overall means and 95% confidence intervals (CI) are provided

Females		Ultimate length (cm)			Upper 99% CL at age seven (cm)	Minimum ultimate size limit				
		Mean	99% CL			Length (cm)	Estimated weight (kg)	Predicted age		Oldest observed
Source	*N*		Upper	Lower				Youngest	Oldest	
Lac Seul	11	140.0	148.6	135.6	88.9	135	17.5	16	31	23
Lake of the woods	27	136.4	148.8	130.3	97.5	130	15.7	14	30	24
Eagle-Wabigoon lakes	46	133.1	141.2	128.0	93.2	128	14.8	16	30	23
St. Lawrence River	15	139.2	156.7	126.7	101.3	126	13.9	10	19	20
Ottawa river[a]	4	133.9	140.7	128.7	96.3	128	13.9	14	26	17
Georgian bay	30	134.1	148.6	126.0	95.8	126	13.9	13	24	30
Lake St. Clair	31	124.2	130.8	119.1	97.3	119	11.4	12	20	19
Kawartha lakes	74	129.8	143.0[b]	116.6	89.9	116	10.7	11	20	22
French river	17	132.6	149.4[b]	115.8	101.3	115	10.7	9	20	22
Niagara river[a]	6	116.6	130.6	114.3	108.2	114	10.7	11	33	15
Tweed water bodies	20	121.9	134.6	111.3	100.3	111	9.3	8	19	19
Indian lake	38	116.6	135.1[b]	98.0	80.3	98	6.3	9	22	14
Nogies creek	21	81.8	81.8	83.6	73.9	83[c]	3.7	15	15	10
Maskinonge lake	39	81.3	88.1	75.7	68.1	75	2.5	8	14	14
Mean ± 95% CI		123.0 ± 11.0	134.1 ± 12.8	115.0 ± 10.3	85.9 ± 14.6	115	11.1 ± 2.6	11.9 ± 1.7	23.1 ± 3.5	19.4 ± 3.0

[a] Data were calculated from concordance sum of square comparisons (Casselman et al. 1999)

[b] Values were adjusted to an equal confidence interval because original values were unrealistically large

[c] Minimum length limit was suspect because mean ultimate length was less than the lower 99% CL

Table 4 Growth and age data important for setting length limits (cm) on a biological basis, considering reproductive size and growth potential for male muskellunge from 14 Ontario sources. Ultimate length (cm) provides a measure of growth potential, whereas reproductive potential is indicated by the upper 99% confidence limits (CL) at age six (2 years after first maturity, age four). Minimum ultimate size limits with estimates of weight (kg) and predictions of age (youngest and oldest) are provided. The oldest observed age is given. Overall means and 95% confidence intervals (CI) are provided

Males		Ultimate length (cm)			Upper 99% CL at age six (cm)	Minimum ultimate size limit				
		Mean	99% CL			Length (cm)	Estimated weight (kg)	Predicted age		Oldest observed
Source	*N*		Upper	Lower				Youngest	Oldest	
Lake of the woods[a]		114.3	122.9	110.0	82.6	110	9.2	8	22	
St. Lawrence river[b]	6	114.3	125.5	106.2	84.6	106	8.0	9	18	17
French river[b]	5	114.3	125.5	106.2	84.6	106	8.0	9	18	11
Lac Seul[b]	5	115.1	125.9	104.4	79.0	104	8.0	11	30	25
Eagle-Wabigoon lakes	17	115.1	125.9	104.4	79.0	104	8.0	11	30	18
Ottawa river[b]	3	106.7	111.8[c]	101.6	85.6	101	7.4	8	31	17
Georgian bay[b]	6	110.5	123.4	100.6	86.6	100	6.9	9	17	20
Lake St. Clair	25	107.2	118.6	96.8	82.8	96	6.3	8	26	17
Tweed water bodies[a]		103.9	112.8	96.3	85.3	96	5.8	7	13	
Indian lake	28	97.3	103.9	89.4	70.9	89	4.8	10	25	16
Kawartha lakes	21	104.9	130.3	85.1	85.6	86	4.4	6	12[d]	18
Niagara river[b]	3	104.9	130.3	85.1	85.6	86	4.4	6	12[d]	11
Nogies Creek	23	82.3	89.4	75.2	72.6	75	2.7	6	12	13
Maskinonge lake	14	70.6	77.2	66.3	63.0	66	1.9	6	15	14
Mean ± 95% CI		104.4 ± 7.6	116.0 ± 9.2	94.8 ± 7.5	80.6 ± 5.1	95	6.1 ± 1.3	8.1 ± 1.0	20.1 ± 2.5	16.4 ± 4.1

[a] Male data were not available for these water bodies, so male growth was estimated from regressions of the relationship between female and male growth (Casselman et al. 1999)

[b] Data were calculated from concordance sum of square comparisons (Casselman et al. 1999)

[c] Values were adjusted to an equal confidence interval because original values were unrealistically large

[d] Age at which the slowest-growing muskellunge reached the ultimate size, which was less than the minimum size limit because the minimum reproductive size limit was greater and was used rather than the minimum ultimate size limit

females was 115.0 ± 10.3 cm, compared with 94.8 ± 7.5 cm for males, 17.6% smaller. Female growth was much more variable than male growth.

Estimates of weight came from a length-weight relationship and are probably more variable than length but are provided for descriptive and comparative purposes and were not used for setting size limits that were based entirely on length. Estimates of mean ultimate weight were 11.1 ± 2.6 kg for females compared with 6.1 ± 1.3 kg for males, 44.6% smaller.

The predicted youngest and oldest mean ages at which individuals would reach minimum ultimate length ranged from 12 to 23 years for females, a difference of 11 (Table 3), and from 8 to 20 years for males, a difference of 12 (Table 4). Age of the oldest observed females in the samples ranged from 10 to 30 years (mean 19.4) and for males, 11 to 25 years (mean 16.4).

Setting minimum ultimate size limits

MUSL is set in relation to reproductive and growth potential. Fig. 1 illustrates these two important length-limit benchmarks for (A) relatively slow-growing fish that would reach a large ultimate length and (B) initially fast-growing fish that would reach a smaller ultimate length. MUS in slow-growing fish would invariably exceed length at maturity and minimum reproductive size by a considerable amount, whereas young, very fast-growing fish that result in a lower ultimate size might result in a minimum reproductive size that could exceed the MUS (Fig. 1B). For males from the Kawarthas and the Niagara River, this reproductive size exceeded the MUSL, so reproductive size became the recommended minimum upper size limit (Table 4).

MUSL, which is the lower 99% CL but rounded down to the nearest centimetre or inch, is used to ensure that fewer large fish are excluded. Average MUSL for females for these Ontario populations was 115 cm (Table 3) and for males, 95 cm (Table 4). These values vary substantially across water bodies, ranging from 75 to 135 cm for females and from 66 to 110 cm for males.

Trophy standards and comparable indices of relative size and status

MUS can be used to more precisely define a trophy muskellunge based on minimum standards for comparing within and among populations (Fig. 2), providing a more consistent definition of a trophy. This minimum, or trophy, standard varied considerably across water bodies and was strongly influenced by sex of the fish. Average length of a trophy female for 14 Ontario populations was 114.8 cm, with mean and maximum ultimate lengths 7% and 17% higher, respectively

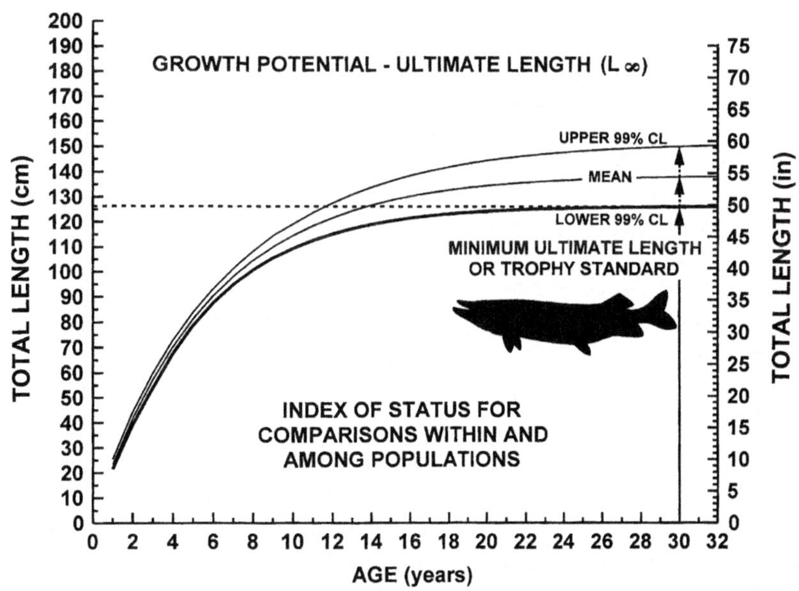

Fig. 2 Illustration of size-at-age growth trajectories for determining growth potential using minimum ultimate length (lower 99% confidence limits, dark curve) for developing trophy standards and indices for comparing relative size or growth status of individuals within and among populations

Table 5 Minimum ultimate length and estimated weight of female and male muskellunge, which provide a minimum ultimate, or trophy, standard for muskellunge from 14 sources in Ontario. Mean ultimate and maximum ultimate length (cm) and corresponding weights (kg) as a proportion of the minimum ultimate or trophy standard are provided. Overall means, as well as 95% and 99% confidence limits (CL), are also provided. Provides a basis for comparing size relative to growth potential and a trophy standard within and between muskellunge populations and individuals

Source	Females						Males					
	Length			Weight			Length			Weight		
	Minimum or trophy standard (cm)	Proportion of minimum		Minimum or trophy standard (kg)	Proportion of minimum		Minimum or trophy standard (cm)	Proportion of minimum		Minimum or trophy standard (kg)	Proportion of minimum	
		mean	maximum		mean	maximum		mean	maximum		mean	maximum
Lac Seul	135.6	1.03	1.10	18.0	1.10	1.33	104.4	1.10	1.21	7.9	1.37	1.82
Lake of the woods	130.3	1.05	1.14	15.9	1.15	1.51	110.0	1.04	1.12	9.3	1.14	1.43
Ottawa river	128.7	1.06	1.11	14.6	1.18	1.38	101.6	1.05	1.10	7.3	1.17	1.36
Eagle-Wabigoon lakes	128.0	1.04	1.10	15.1	1.13	1.36	104.4	1.10	1.21	7.9	1.37	1.82
St. Lawrence river	126.7	1.10	1.24	14.6	1.34	1.93	106.2	1.08	1.18	8.0	1.27	1.71
Georgian bay	126.0	1.07	1.19	14.3	1.22	1.69	100.6	1.10	1.23	6.8	1.35	1.93
Lake St. Clair	119.1	1.04	1.10	12.1	1.14	1.34	96.8	1.11	1.23	6.2	1.39	1.92
Kawartha lakes	116.6	1.11	1.23	11.3	1.39	1.88	85.1	1.23	1.53	4.1	1.95	3.91
French river	115.8	1.15	1.29	11.1	1.52	2.19	106.2	1.08	1.18	8.0	1.27	1.71
Niagara river	114.3	1.02	1.14	10.6	1.06	1.51	85.1	1.23	1.53	4.1	1.95	3.91
Tweed water bodies	111.3	1.10	1.21	9.8	1.33	1.81	96.3	1.08	1.17	6.1	1.27	1.66
Indian lake	98.0	1.19	1.38	6.6	1.71	2.69	89.4	1.09	1.16	4.8	1.31	1.62
Nogies Creek[a]	83.6	0.98	0.98	4.0	0.93	0.93	75.2	1.10	1.19	2.8	1.33	1.74
Maskinonge lake	75.7	1.07	1.16	3.0	1.26	1.62	66.3	1.07	1.17	1.9	1.22	1.61
Mean	114.8	1.07	1.17	11.5	1.25	1.66	94.7	1.10	1.23	6.1	1.38	2.01
95% CL—lower	104.6	1.04	1.11	8.9	1.13	1.40	87.1	1.07	1.15	4.8	1.24	1.54
95% CL–upper	125.0	1.10	1.23	14.1	1.36	1.91	102.6	1.14	1.31	7.4	1.53	2.48
99% CL—lower	100.6	1.03	1.09	7.9	1.09	1.31	84.3	1.06	1.12	4.5	1.18	1.35
99% CL–upper	129.0	1.12	1.25	15.1	1.41	2.00	105.2	1.15	1.34	7.9	1.59	2.67

[a] Because of the age and size of the samples, $L\infty$ for the von Bertalanffy growth relationships for the mean and maximum of females were less than the minimum

Table 6 Minimum ultimate length and estimated weight of muskellunge (sexes combined equally), which provide a minimum ultimate, or trophy, standard for muskellunge from 14 sources in Ontario. Mean ultimate and maximum ultimate length (cm) and corresponding weights (kg) as a proportion of the minimum ultimate or trophy standard are provided. Overall means and 95% and 99% confidence limits (CL) are also given

Source	Length			Weight		
	Minimum or trophy standard (cm)	Proportion of minimum		Minimum or trophy standard (kg)	Proportion of minimum	
		Mean	Maximum		Mean	Maximum
Lac Seul	120.1	1.06	1.14	13.0	1.18	1.48
Lake of the woods	120.1	1.04	1.13	12.7	1.15	1.48
St. Lawrence river	116.6	1.09	1.21	11.3	1.31	1.85
Eagle-Wabigoon lakes	116.3	1.07	1.15	11.5	1.21	1.52
Ottawa river	114.3	1.05	1.11	10.9	1.18	1.37
Georgian bay	113.3	1.08	1.20	10.6	1.26	1.77
French river	111.0	1.11	1.24	9.8	1.41	1.99
Lake St. Clair	108.0	1.07	1.16	9.2	1.22	1.53
Tweed water bodies	103.9	1.09	1.19	7.9	1.31	1.75
Kawartha lakes	100.8	1.16	1.36	7.7	1.54	2.42
Niagara river	99.8	1.11	1.31	7.4	1.31	2.18
Indian lake	93.7	1.14	1.28	5.7	1.54	2.24
Nogies creek	79.5	1.03	1.08	3.4	1.09	1.26
Maskinonge lake	71.1	1.07	1.17	2.4	1.25	1.61
Mean	104.9	1.08	1.20	8.4	1.28	1.75
95% CL–lower	96.3	1.06	1.15	6.9	1.20	1.54
95% CL–upper	113.5	1.10	1.24	10.7	1.36	1.95
99% CL–lower	93.0	1.05	1.13	6.2	1.17	1.46
99% CL–upper	106.8	1.11	1.26	11.4	1.39	2.03

(Table 5). In the case of males, mean minimum trophy standard for these populations was 94.7 cm, with mean and maximum ultimate lengths 10% and 23% higher, respectively (Table 5). It is possible to define a trophy standard in terms of either length or weight. However, weight can be quite variable and is probably less reliable than length. Sexual dimorphism in growth emphasized the importance of developing sex-specific trophy standards. But if sex is unknown or unidentified, an overall sex-independent trophy standard was developed that used equally weighted male and female data (Table 6). Appendix C presents these trophy standards for female and male muskellunge and sexes combined in inches and pounds.

Discussion

Growth potential, and particularly MUS or lower 99% CL, is an important benchmark not only for setting minimum size limits (MUSL) based on length but also for defining and evaluating size of trophy muskellunge. By refining growth analysis and developing the MUS, it is now possible to use a mathematically defined lower estimate of growth potential to develop trophy standards that specifically define trophy muskellunge and set sex-specific size limits. These can be used to evaluate and manage muskellunge populations and sport fisheries much more appropriately. Determining growth potential makes it possible to set size length limits that continue to use fish resources without unknowingly excluding large segments of the population.

Increasing length limits to MUS would reduce harvest mortality and protect fish to an older age. In Ontario populations, for females this would be at least from 8 to 16 years of age (mean 12). This would protect individuals for many more spawning seasons than would a reproductive size limit, hence reproductive capacity and natural recruitment of the population would be greatly enhanced. Although harvest would be reduced with this increased size limit, disproportionately more females than males would be harvested because of growth differential and ultimate size. Indeed, few if any males would be harvested at these higher length limits. This differential harvest would be eliminated only by applying sex-specific length regulations, which could be calculated or are now available (e.g., Tables 3, 4; Appendices A, B), and employing external sexing techniques. I believe that some day, length-limit regulations will be more precise and even sex-specific.

The lower 99% CL provides an MUS that most individuals with a reasonable life expectancy can reach. Calculations using 90%, 95%, and 99% CL at age indicated that $L\infty$ of the 99% limits produced length limits almost twice as broad as the 90% and 95% (Casselman unpublished data). This confirmed that the extreme limit produced the lowest value for minimum ultimate length. MUS provided by the 99% CL is broad, with the lower including a higher proportion of large fish than would any other confidence limit.

It is unfortunate that accurate age and growth determination of muskellunge requires killing the fish to obtain the cleithrum. Indeed, fewer muskellunge are killed now than ever before (Casselman et al. 1996). Luckily, relatively young muskellunge (age 8–10 ± 2 years) are old enough to provide a good approximation of ultimate size. However, more confidence could be placed in the results if older individuals were used. The fish used to calculate ultimate size should be as old as possible, since early growth can be variable and may change with age. It is especially fortunate that as few as 12 ± 4 fish are required to provide valid von Bertalanffy growth trajectories and ultimate length. Although for the present analysis, there were only enough samples in The Cleithrum Project to examine 14 Ontario populations, some examples exist for other water bodies, and it would be quite easy to acquire the few samples that would be needed to construct the growth analysis provided here. If an archive such as The Cleithrum Project is maintained, samples can be added and accumulated as they become available to acquire enough fish to describe a particular water body. Since size limits are currently set to protect faster-growing, reproductively important females, initially it would be most important to acquire samples from females. Indeed, if for some reason sex of an individual is unknown or undetermined, it is possible to identify sex by analyzing the growth trajectory, using von Bertalanffy growth parameters (Casselman et al. 1999). This, however, assumes that females are faster-growing and reach a larger ultimate size than males. Fortunately, sexually dimorphic growth is quite extreme in muskellunge (Casselman and Crossman 1986; Casselman et al. 1999).

Estimates of extreme ultimate size are useful indices not only for setting size limits on a biological basis but also for determining just how large a muskellunge might grow. For those interested in maximum size and growth potential of a population or water body, ultimate size associated with means and upper 99% CL provides mathematical estimates of potential capability. Even though these exceptional fish are rarely seen, these theoretical maximum estimates provide interesting insights.

In setting size limits, reproductive potential as well as growth potential should be taken into consideration (Fig. 1). Reproductive potential protects reproductive capacity of populations, whereas if size limits are increased to MUS, then this upper limit will greatly improve the quality of muskellunge populations and fisheries and decrease angling mortality through reduced harvest. However, it permits harvest of large individuals and sets a level low enough to allow eventual exploitation of virtually all females. Most importantly, this increase in upper size limit ensures increased reproductive capacity and many more years of egg production and spawning before harvest occurs. For the 14 Ontario populations, the predicted mean age when females reach this size was, for the youngest, 12 years; for the oldest, 23 years (Table 3); and for males, 8–20 years (Table 4).

Methods for calculating MUS, MUSL, and trophy standards developed here (Figs. 1, 2) could also be applied to other species. However, this growth analysis requires accurate age assessment and adequate samples. Age at first maturity and spawning experience would also need to be taken into consideration. Using a similar approach but for a slightly different purpose, these two benchmark size length limits were used to reduce harvest of large northern pike in the northeastern region of Ontario. As an alternative to a protected-size slot limit, daily harvest was reduced to two over the reproductive limit, one of which could be over MUS. Again, this used specific growth biology of the species to set length limits quite precisely, eliminating ambiguity and debate.

Increasing size limits has specific and well-proven advantages. In Largon Lake, Wisconsin, an increase from no length limit to a 32–inch minimum length limit resulted in a high-quality northern pike fishery (Benike 2004). Nordwall et al. (2000) used comparative modelling to examine size-limit strategies for exploited brook trout, *Salvelinus fontinalis* populations. They reviewed the effects of various types of size limits and confirmed the advantages of setting the largest sizes of fish in the population as a minimum size that can be legally retained. However, they were not specific as to how to set the limits and at what level. These observations reinforce the advantage of using MUS as one possible method for setting larger size limits. They documented, as expected, that with increased minimum size limits and harvest of progressively larger fish, there was a positive shift in size structure, with an increased proportion of larger fish in the population and, as had been demonstrated by Quinn et al. (1994), an increase in population size reduced exploitation.

In contrast, Power and Power (1996) suggested that protected slot limits were superior to minimum size limit in terms of population size, yield, and structure for brook trout. However, Nordwall et al. (2000) felt that the difference they found in their own study was due to lower density-dependent mortality, longer life spans, and larger size at first maturity, features considered to be associated with a "slower life history." These features are more typical of muskellunge populations.

Wright (1992), in a review of brook trout size limits, emphasized that minimum length limits should be set at a level that allows a full age class of females to spawn at least once to ensure maintenance of the population's reproductive potential but that this limit needed to be set at the upper end of the length-frequency distribution for the age at which the majority of the females were mature. It is encouraging that independently I have come to the same conclusions and recommendations for muskellunge.

Dunning et al. (1982) studied minimum size limits and their effects on northern pike, using a simulation modelling approach. Contrary to the expectation that increased size limits would increase yield of larger fish by reducing harvest of younger fish and increasing recruitment, the increased size limit actually decreased yield of older fish. In this case, intense fishing pressure shifted to older fish, reducing their relative abundance. Heavy selective exploitation of the vulnerable size group explained this result. This would probably not be a problem if size limits were set appropriately high, as is recommended with the minimum ultimate length limits proposed here for muskellunge.

I must emphasize that if minimum size limits are to be effective, hooking and handling mortality must be held at a very low level. Unfortunately, it is unknown just how low this level needs to be, nor it is measured in any quantifiable way for muskellunge populations. Furthermore, handling methods would be extremely variable among anglers and fish, making it difficult to generalize concerning catch-and-release mortality.

Using the growth analysis presented here, it is now possible to quite specifically define size of trophy muskellunge based on growth potential. If MUS is used as a trophy standard (Fig. 2), it is possible to determine the smallest, slowest-growing trophy-sized fish for each population. This trophy can be defined by either length or weight and can produce a standard that makes all fish comparable. It not only quantifies growth potential of a particular water body but allows individuals to be compared consistently across water bodies. It even makes males and females comparable in terms of ultimate, or trophy, size. The real advantage is that growth potential standardizes the definition, making it possible to evaluate different water bodies and body sizes relative to a universal trophy standard. This means that fish from small- and medium-bodied populations from water bodies with a lower growth potential can be related in a quantitative and comparable way to larger individuals from large-bodied populations. The advantage of this process is that growth data come directly from the population and growth potential is an empirical description of the fish and its water body.

Trophy standard provides a water-body-specific size description that quantitatively defines one trophy muskellunge unit. It can be evaluated on either length or weight by or independent of sex, and all fish can be compared to these units on a proportional and relative basis (e.g., males and females, Table 5, and sex unknown or not determined, Table 6; Appendix C). This trophy standard unit provides a specific quantitative value to assess catch that would be universally comparable among fish and water bodies. Most importantly, the standard makes it possible for anglers to evaluate catch on a relative basis, independent of absolute size.

Since anglers almost exclusively evaluate catch based on absolute size, it would be a progressive approach if muskellunge anglers accepted that smaller muskellunge from a population with a lower growth potential might also be defined as trophies. Indeed, muskellunge anglers are well informed and very much appreciate the limits of the resource they enjoy. As in the implementation of voluntary catch-and-release, they would probably be the first to realize that to restrict their impressions of success based on absolute size would greatly limit their appreciation of all muskellunge resources. With the inevitable increase in fishing pressure and the natural limits on fish resources, it is hoped that muskellunge anglers will maximize their use and enjoyment of all muskellunge populations by starting to compare and relate angling success based on the science and concepts of ultimate size and comparable trophy units.

It might be said that water-body-specific size limits would create significant enforcement problems. However, the experience in Ontario in recent decades, when several categories of size limits based on the MUS concept were applied across the province, indicates that almost without exception, anglers respect the intent, seeing the purpose when the science is adequately explained and transferred. In the case of Ontario, enforcement was far less a concern than had been anticipated. I believe that the rare problems, if they did exist, were probably well worth the benefits. If there is a need to emphasize compliance, violations can be easily tested and regulations enforced with genetic analyses to provide precedent-setting cases.

It is quite apparent from the number of mathematical estimates that were required to complete the comparison reported here for Ontario (e.g., use of concordance sum of squares and regression estimates by sex) that more samples are required, particularly for larger, older males from some water bodies. So biological data and cleithra should be obtained from every harvested muskellunge, not only to strengthen datasets but to monitor and, if necessary, check and from time to time re-evaluate trophy standards and minimum upper size limit regulations. Additional samples make it possible to evaluate whether growth is changing and limits need to be updated. It is important to monitor population status after implementing size-limit changes to make sure that with increased population size, density-dependent growth changes do not occur, a condition that is rarely seen in muskellunge populations unless fish are extremely abundant. Similarly, trophy size and status might change over time. There is strong evidence that with increasing water temperatures, muskellunge are now growing at a faster rate and reaching a larger size (Robinson 2005). Muskellunge populations should also be monitored for changes in maximum age and mortality rate because these provide important measures of population status (Casselman et al. 1996) and fish quality. Hence, every opportunity should be taken to collect biological data, length, weight, sex, and cleithra from each muskellunge that is harvested or found dead, regardless of origin (stocked or wild).

For example, a significant die-off of very large, mature muskellunge occurred in the upper St. Lawrence River, beginning in the fall of 2004, throughout 2005, and into the spring of 2006. Anglers, park wardens, biologists, and researchers worked together to collect carcasses, data, and cleithra from 61 muskellunge from Canadian waters (Casselman unpublished data) killed by an epizootic caused by viral hemorrhagic septicemia. This die-off also emphasizes an important concern that when muskellunge populations are being managed to produce disproportionately large numbers of large, old fish (present minimum size limit for the St. Lawrence River—122 cm, 48 in), they may be particularly vulnerable to stressors, disease, and catastrophic die-offs and quite ephemeral.

Samples used in this study were, for the most part, provided by muskellunge anglers in a cooperative and mutually beneficial project. Anglers provided cleithra and data from harvested, mounted fish, detailed growth data were extracted, and more appropriate size limits were developed that are being used to manage for quality muskellunge populations and fisheries, particularly in Ontario (Ontario Ministry of Natural Resources 1999). In addition, a new method that defines trophy muskellunge and indexes individuals in relation to trophy status has been detailed here for males and females (Table 5) and less precisely for fish of unknown sex (Table 6).

By working together, we have acquired detailed information, making it possible to set increased size limits. Quite importantly, minimum size and trophy standards give anglers the ability to appreciate each fish they have caught in relation to growth potential of the water body and a new trophy standard that involves either length or weight, as well as the ability to calculate an index to assess the status of individual fish in relation to lake-specific or any other trophy standard. Also, fish can be compared within and among populations, and sexes and comparable trophy units can even be accumulated. The real advantage of these new indices, however, is that they emphasize the quality of each fish and provide a relative way to evaluate and appreciate every fish for what it represents in relation to its specific population and origin, independent of its absolute size. All this has been done to provide a biological basis to better appreciate, evaluate, and maintain quality muskellunge populations and sports fisheries.

Dedication and acknowledgements The paper is dedicated to the late Dr. E.J. Crossman, a colleague and friend of all those interested in muskellunge. We worked jointly on The Cleithrum Project with enthusiasm to make sure that all possible information was collected when fish were harvested. Ed was particularly concerned that the science to manage "his fish" was appropriately assembled, conserved, and used to maintain and improve the quality of muskellunge populations and sport fisheries. It is obvious that in the province of Ontario over the past three decades, muskellunge populations have improved and fish have gotten bigger and more abundant, a turn of events that particularly pleased Ed.Thanks to the many anglers, guides, taxidermists, and members of organized muskellunge fishing groups, Muskies, Inc. and Muskies Canada, who assisted Ed and me in assembling this archive of samples and data. Chris Robinson assisted with some of the data analysis, and Jim Diana and Terry Margenau provided helpful reviews of the manuscript. Muskies, Inc. and Muskies Canada helped fund the initial extraction of data presented by Casselman et al. (1999), and these have been used to expand our understanding of muskellunge populations throughout Ontario by providing and using more precise data on growth potential and ultimate size.

Appendices

The three appendices that follow present, in inches and pounds, the ultimate length and growth data for setting length limits and trophy standards for female and male muskellunge from 14 Ontario populations. These correspond to the metric tables in text (Tables 3 and 4, as well as Tables 5 and 6 combined).

Appendix A Presentation of data in Table 3 in inches and pounds. Growth and age data important for setting length limits (in) based on reproductive size and growth potential for female muskellunge from 14 Ontario sources. Ultimate length (in) provides a measure of growth potential, whereas reproductive potential is indicated by the upper 99% confidence limits (CL) at age seven (2 years after first maturity, age 5). Minimum ultimate size limits with estimates of weight (lb) are provided. Overall means and 95% confidence intervals (CI) are provided

Females		Ultimate length (in)			Upper 99% CL at age seven (in)	Minimum ultimate size limit	
		Mean	99% CL			Length (in)	Estimated weight (lb)
Source	*N*		Upper	Lower			
Lac Seul	11	55.1	58.5	53.4	35.0	53	38.5
Lake of the woods	27	53.7	58.6	51.3	38.4	51	34.7
Eagle-Wabigoon lakes	46	52.4	55.6	50.4	36.7	50	32.6
St. Lawrence river	15	54.8	61.7	49.9	39.9	49	30.6
Ottawa river[a]	4	52.7	55.4	49.9	37.9	49	30.6
Georgian bay	30	52.8	58.5	49.6	37.7	49	30.6
Lake St. Clair	31	48.9	51.5	46.9	38.3	46	25.1
Kawartha lakes	74	51.1	56.3[b]	45.9	35.4	45	23.5
French river	17	52.2	58.8[b]	45.6	39.9	45	23.5
Niagara river[a]	6	45.9	51.4	45.0	42.6	45	23.5
Tweed water bodies	20	48.0	53.0	43.8	39.5	43	20.4
Indian lake	38	45.9	53.2[b]	38.6	31.6	38	13.8
Nogies creek	21	32.2	32.2	32.9	29.1	32[c]	8.1
Maskinonge lake	39	32.0	34.7	29.8	26.8	29	5.6
Mean ± 95% CI		48.4 ± 4.3	52.8 ± 5.1	45.2 ± 4.1	36.3 ± 2.6	45	24.4 ± 5.6

[a] Data were calculated from concordance sum of square comparisons (Casselman et al. 1999)

[b] Values were adjusted to an equal confidence interval because original values were unrealistically large

[c] Minimum length limit was suspect because mean ultimate length was less than the lower 99% CL

Appendix B Presentation of data in Table 4 in inches and pounds. Growth and age data important for setting length limits (in) on a biological basis, considering reproductive size and growth potential for male muskellunge from 14 Ontario sources. Ultimate length (in) provides a measure of growth potential, whereas reproductive potential is indicated by the upper 99% confidence limits (CL) at age six (2 years after first maturity, age four). Minimum ultimate size limits with estimates of weight (lb) are provided. Overall means and 95% confidence intervals (CI) are provided

Males		Ultimate length (in)			Upper 99% CL at age six (in)	Minimum ultimate size limit	
		Mean	99% CL			Length (in)	Estimated weight (lb)
Source	*N*		Upper	Lower			
Lake of the woods[a]		45.0	48.4	43.3	32.5	43	20.3
St. Lawrence river[b]	6	45.0	49.4	41.8	33.3	41	17.6
French river[b]	5	45.0	49.4	41.8	33.3	41	17.6
Lac Seul[b]	5	45.3	49.6	41.1	31.1	41	17.6
Eagle-Wabigoon lakes	17	45.3	49.6	41.1	31.1	41	17.6
Ottawa river[b]	3	42.0	44.0[c]	40.0	33.7	40	16.3
Georgian bay[b]	6	43.5	48.6	39.6	34.1	39	15.1
Lake St. Clair	25	42.2	46.7	38.1	32.6	38	13.9
Tweed water bodies[a]		40.9	44.4	37.9	33.6	37	12.7

Appendix B continued

Males		Ultimate length (in)			Upper 99% CL at age six (in)	Minimum ultimate size limit	
		Mean	99% CL			Length (in)	Estimated weight (lb)
Source	*N*		Upper	Lower			
Indian lake	28	38.3	40.9	35.2	27.9	35	10.6
Kawartha lakes	21	41.3	51.3	33.5	33.7	34	9.7
Niagara river[b]	3	41.3	51.3	33.5	33.7	34	9.7
Nogies creek	23	32.4	35.2	29.6	28.6	29	5.9
Maskinonge lake	14	27.8	30.4	26.1	24.8	26	4.2
Mean ± 95% CI		41.1 ± 3.0	45.7 ± 3.6	37.1 ± 2.9	31.7 ± 1.6	37	13.5 ± 2.8

[a] Male data were not available for these water bodies, so male growth was estimated from the regressions of the relationship between female and male growth (Casselman et al. 1999)

[b] Data were calculated from concordance sum of square comparisons (Casselman et al. 1999)

[c] Values were adjusted to an equal confidence interval because original values were unrealistically large

Appendix C Presentation of data in Tables 5 and 6 in inches and pounds. Minimum ultimate length (in) and estimated weight (lb) of female and male muskellunge, separately and combined, which provide a trophy standard, or minimum ultimate size, for muskellunge from 14 sources in Ontario. Overall means, as well as 95% and 99% confidence limits (CL), are also provided. Provides a basis for comparing size relative to growth potential and a trophy standard within and between muskellunge populations and individuals

Source	Females		Males		Sexes combined	
	Length (in)	Weight (lb)	Length (in)	Weight (lb)	Length (in)	Weight (lb)
Lac Seul	53.4	39.7	41.1	17.5	47.3	28.6
Lake of the woods	51.3	35.1	43.3	20.6	47.3	27.9
Eagle-Wabigoon lakes	50.4	33.2	41.1	17.5	45.9	25.0
St. Lawrence river	49.9	32.2	41.8	17.7	45.8	25.4
Ottawa river	49.9	32.2	40.0	16.0	45.0	24.1
Georgian bay	49.6	31.6	39.6	14.9	44.6	23.3
Lake St. Clair	46.9	26.6	38.1	13.7	43.7	21.1
Kawartha lakes	45.9	24.9	33.5	9.1	42.5	20.2
French river	45.6	24.4	41.8	17.7	40.9	17.5
Niagara river	45.0	23.4	33.5	9.1	39.7	17.0
Tweed water bodies	43.8	21.5	37.9	13.5	39.3	16.3
Indian lake	38.6	14.6	35.2	10.6	36.9	12.6
Nogies creek	32.9	8.9	29.6	6.1	31.3	7.5
Maskinonge lake	29.8	6.5	26.1	4.1	28.0	5.3
Mean	45.2	25.3	37.3	13.4	41.3	19.4
95% CL–lower	41.2	19.7	34.3	10.6	37.9	15.3
95% CL–upper	49.2	31.0	40.3	16.3	44.7	23.6
99% CL–lower	39.6	17.4	33.2	9.4	36.6	13.6
99% CL–upper	50.8	33.3	41.4	17.4	46.0	25.2

References

Benike HM (2004) Evaluation of a 32 inch minimum length limit for northern pike Largon Lake, Polk County, WI, MWBIC Code (2668100)

Casselman JM (1974) External sex determination of northern pike, *Esox lucius* Linnaeus. Trans Am Fish Soc 103:343–347

Casselman JM (1996) Age, growth and environmental requirements of pike, *Esox lucius*. In: Craig J (ed) Pike: biology and exploitation. Chapman and Hall, London, pp 69–101

Casselman JM (2002) Effects of temperature, global extremes, and climate change on year-class production of warmwater, coolwater, and coldwater fishes in the Great Lakes Basin, pp 39–59. In: McGinn NA (ed) Proceedings of American Fisheries Society Symposium 32, Fisheries in a changing climate. 295p

Casselman JM, Crossman EJ (1986) Size, age and growth of trophy muskellunge and muskellunge-northern pike hybrids—The Cleithrum Project, 1979–1983, pp 93–110. In: Hall GE (ed) Managing muskies—a treatise on the biology and propagation of muskellunge in North America. American Fisheries Society Special Publication 15

Casselman JM, Crossman EJ, Robinson CJ (1996) Assessing sustainability of trophy muskellunge fisheries, pp 29–39. In: Kerr SJ, Olver CH (eds) Managing muskies in the '90s. Proceedings Southern Region Science and Technoloty Transfer Unit Workshop, WP-007, Ontario Ministry of Natural Resources, Toronto, 169p

Casselman JM, Robinson CJ, Crossman EJ (1999) Growth and ultimate length of muskellunge from Ontario water bodies. North Am J Fish Manage 19:271–290

Casselman JM, Scott KA (2000) A general procedures manual for CSAGES—calcified structure age-growth data extraction software (Version 5.2). Spec. Publ. Glenora Fisheries Station, Aquatic Ecosystems Science Section, OMNR, Picton, Ont., 90p

Crossman EJ, Casselman JM (1996) The cleithrum project: an update to 1995. In: Kerr SJ, Olver CH (eds) Managing muskies in the '90s. Proc. South. Reg. Sci. Tech. Transfer Unit Workshop, WP-007, Ontario Ministry of Natural Resources, Toronto, pp 147–151

Dunning DJ, Ross Q, Gladden J (1982) Evaluation of minimum size limits for St. Lawrence river northern pike. North Am J Fish manage 2:171–175

Hunt RL (1970) A compendium of research on angling regulations for brook trout conducted at Lawrence Creek, Wisconsin. Wisconsin Department of Natural Resources, Research Report 54, Madison

Koppleman JB, Philipp DP (1986) Genetic application in muskellunge management. A Fish Soc Special Publ 15:111–121

Lebeau B, Pageau G (1989) Comparative urogenital morphology and external sex determination in muskellunge, *Esox masquinongy* Mitchill. Canadian J Zoo 67:1053–1060

Margenau TL, Hanson DA (1996) Survival and growth of stocked muskellunge: effects of genetics and environmental factors. Wisconsin Department of Natural Resources Research Report 172, May 1996

Nordwall F, Lundberg P, Eriksson T (2000) Comparing size-limit strategies for exploitation of a self-thinned stream fish population. Fish Manage Ecol 2000(7):413–424

Ontario Ministry of Natural Resources (1983) The identification of overexploitation. Report of SPOF working group No. 15. Strategic Planning for Ontario Fisheries Policy Development, Toronto, pp 84

Ontario Ministry of Natural Resources (1986) Report of the committee to review provincial muskellunge angling regulations. Ontario Ministry of Natural Resources Fisheries Policy Committee, Toronto

Ontario Ministry of Natural Resources (1999) Size limit regulations for Ontario muskellunge: a new approach. Ontario Ministry of Natural Resources Fisheries Policy Committee, Toronto

Power M, Power G (1996) Comparing minimum-size and slot limits for brook trout management. North Am J Fish Manage 16:49–62

Quinn NWS, Korver RM, Hicks FJ, Monroe BP, Hawkins RR (1994) An empirical model of lentic brook trout. North Am J Fish Manage 14:692–709

Robinson CJ (2005) Factors affecting estimates of year-class strength and growth of muskellunge (*Esox masquinongy*): a long-lived, low-density, trophy fish. Watershed Ecosystems Graduate Program, Trent University, Peterborough, pp 222

Wright S (1992) Guidelines for selecting regulations to manage open-access fisheries for natural populations of anadromous and resident trout in stream habitats. North Am J Fish Manage 12:517–527

Younk JR, Strand RF (1992) Performance evaluation of four muskellunge *Esox masquinongy* strains in two Minnesota lakes. Minnesota Department of Natural Resources Investigational Report 418

Environ Biol Fish (2007) 79:155–162
DOI 10.1007/s10641-006-9115-3

SPECIAL ISSUE-CROSSMAN

Effects of angling with a single-hook and live bait on muskellunge survival

Terry L. Margenau

Received: 23 September 2005 / Accepted: 6 July 2006 / Published online: 10 August 2006

Abstract Use of live bait for angling of muskellunge, *Esox masquinongy*, is popular in Wisconsin. A traditional method utilizes a large hook through the bait fish's snout, which requires the muskellunge to swallow the bait prior to hook set. Adult muskellunge (>76 cm; 30 in) were held in lined hatchery ponds and caught while fishing with live bait on 10/0 size single hooks. The leader was cut and the muskellunge was released when hooked in the stomach. Survival was monitored for up to 1 year. No immediate (<24 h) mortality occurred. However, 22% of hooked muskellunge died within 50 days and 83% died within 1 year. Necropsies revealed extensive trauma to the stomach and other organs from hooks, along with systemic bacterial infections. Highest mortality on both hooked and control fish occurred over winter through spring. This peak mortality may be associated with natural stressors that occur during the spring spawning period. Mortality rates observed in this study are considered unacceptable for trophy management of muskellunge. Although use of live bait for muskellunge is traditional in Wisconsin, terminal tackle such as quick-strike rigs that hook fish in the mouth or buccal cavity should enhance the chances a released muskellunge will survive.

T. L. Margenau (✉)
Wisconsin Department of Natural Resources, 810 West Maple Street, Spooner, WI, USA
e-mail: terry.margenau@dnr.state.wi.us

Keywords *Esox masquinongy* · Mortality · Catch and release

Introduction

Use of live bait, typically white sucker, *Catostomus commersoni*, is a popular method when fishing muskellunge, *Esox masquinongy*, in Wisconsin (Heiting 1992), particularly during fall months (Fig. 1). Nearly 40% of anglers who fish muskellunge in Wisconsin indicated they use live bait for muskellunge at certain times of the season (Margenau and Petchenik 2004). However, muskellunge anglers also have expressed concern with certain angling methods using live bait. Many (88%) muskellunge anglers felt that use of live bait with a single hook increased muskellunge mortality (Margenau and Petchenik 2004). This methodology involves placing a large single hook through a white sucker's snout. (White suckers used to fish large muskellunge are often 30–46 cm (12–18 in) in length). The white sucker is then either suspended from a bobber or allowed to swim freely. When a strike occurs, the angler allows the fish at least several minutes to swallow the bait before setting the hook. Muskellunge will typically grasp a prey item length-wise, then gradually turn the item before swallowing—usually with the head swallowed first (Fig. 2). Because of such feeding mechanics, a muskellunge is not likely to be hooked using this methodology until the bait is swallowed.

The management of muskellunge fisheries is usually geared toward the use of high minimum length limits in conjunction with voluntary live release to enhance the production of large (>40 in) trophy individuals (Simonson and Hewett 1999; Margenau and Petchenik 2004). For these management objectives it is particularly important to protect adult fish that have not reached the age/length to be considered a trophy by anglers. Even small losses to the population could jeopardize a lake's potential to produce trophy muskellunge. For instance, Casselman et al. (1996) suggested that with a 2% increase in mortality, recruitment would have to be doubled (natural or from stocking) to maintain the number of trophy muskellunge in a population. Hence, maximizing the survival of released muskellunge is an integral component of a successful management program.

Many hooking mortality studies have identified numerous factors that affect post-release survival of fish. Among these factors are species, size or age, environmental conditions (e.g. water temperature), handling time and air exposure, gear, and methods (Muoneke and Childress 1994). Previous hooking studies suggest that single hooks, especially those used with natural bait, result in higher mortalities compared to treble hooks because they are typically swallowed deeper and penetrate tissues farther (Muoneke and Childress 1994). However, studies on hooking mortality of muskellunge have been sparse and those that do exist have dealt with sub-adult fish

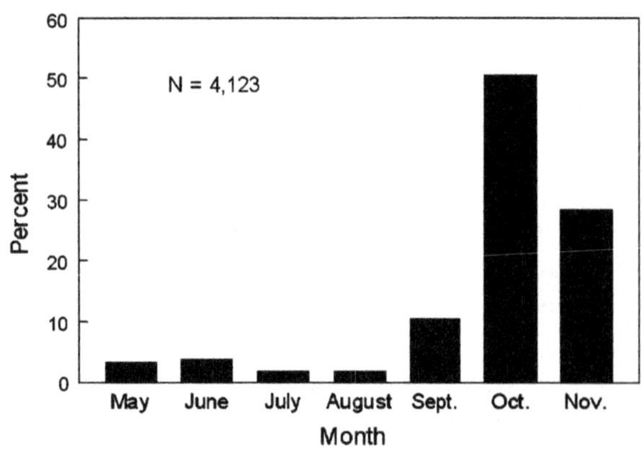

Fig. 1 Percent of muskellunge caught by month using live bait in Wisconsin, 1980–1999, (unpublished data; Muskies, Inc.)

Fig. 2 Illustration demonstrating muskellunge feeding mechanics and single hook methodology

and/or the use of artificial lures (Weithman and Anderson 1978; Newman and Storck 1986). Large muskellunge ≥76 cm (30 in) are difficult to handle and to subsequently remove a hook. In addition, large muskellunge may be exposed to additional stress as an angler attempts to determine if the fish exceeds a minimum length and handles the fish to photograph the catch (Beggs et al. 1980). The paucity of data on large muskellunge is due to low sample sizes and sparse populations. In addition, monitoring a test animal following a hooking event can be problematic. Use of telemetry provides an option for tracking a released muskellunge, but attaching transmitters may add additional stress to an already stressed fish (Bettoli and Osborne 1998). In addition, a failed/shed transmitter or lost signal makes interpretation of data difficult.

The objectives of this study were to determine the effects of an angling method using a single-hook and live bait on adult muskellunge. More specifically, I intended to gain insight on the use of a controlled system for holding muskellunge and to evaluate short and long-term survival following release of hooked muskellunge. I also monitored certain biological aspects important for long-term health, such as behavior, feeding, growth, and reproduction.

Methods

Adult muskellunge were collected by electroshocking during September 2001 and 2002 from

Deer Lake, Polk County, in northwestern Wisconsin. Deer Lake has a high density of adult muskellunge (0.94/surface acre ≥76 cm (30 in); Margenau and AveLallemant 2000) and is maintained by stocking (WDNR 1996). Muskellunge were transferred to Governor Thompson State Hatchery at Spooner, WI on hatchery trucks in water with a 1% salt solution. All muskellunge were measured, weighed and marked with individually numbered preopercle tags (Johnson 1971) before being released into a 0.4 ha (1-acre) lined pond. The pond had a constant water flow (approximately 568 l min^{-1}; 150 gal min^{-1}) and a maximum depth of 1.8 m (6 ft). Pond temperatures at transfer were 27–32°C (70–78°F).

Muskellunge were given 10 days to acclimate to the pond and to allow mortality associated with handling and transport to occur before beginning the experiment. Muskellunge were then angled with live bait. No forage was available in the pond during the angling period. Baits were white sucker, northern hog sucker, *Hypentilium nigricans*, and redhorse, *Moxostoma* spp., ranging in length from 27 to 32 cm (10.5–12.5 in). White suckers were used most frequently. Initially, bait was purchased from a commercial dealer, but low survival of these fish made it necessary to electroshock for fresh bait from the Yellow River (Washburn County). These fish had excellent survival and were much more vigorous. Terminal tackle consisted of a single hook (Mustad 10/0 carbon steel with bronze finish) attached to a 27 kg (60 pound) wire leader and 13.6 kg (30 pound) monofilament line. The baitfish was hooked through the snout and a bobber was used to suspend the bait below the surface of the water. Following strikes, muskellunge were given time to turn and swallow the bait. Researchers were often able to visually observe when the bait was swallowed and to set the hook accordingly. Hooked muskellunge were retrieved quickly to minimize stress from fatigue, and landed in a landing net. Fish were then transferred into a cradle net where tag number was recorded, hooking location noted, and fish condition determined. If the fish swallowed the hook or the hook was not visible, the leader was cut before releasing the fish back into the pond from which it was caught. In the second year of the study, released fish were placed back into a separate section of the pond. This precaution eliminated fish being hooked multiple times, as happened during the first year. Angling effort was completed by early October. Pond water temperatures during angling ranged from 17 to 9°C (62 to 48°F) when angling was completed. In both years, several unhooked or superficially hooked fish remained in the pond as control fish. Muskellunge were provided forage fishes after angling was completed and feeding occurred ad libitum.

In early November, the pond was drained and surviving muskellunge were removed and transferred into a similar size pond for overwintering. Transfer was necessary to avoid possible ice damage to a production pond. The lined pond for overwintering muskellunge is used by the hatchery as a settling pond for suspended solids, where water is passed from production ponds at times of high sediment load (typically harvest) prior to release into the Yellow River. Hence, the settling pond had sediment on the bottom (8–15 cm; 3–6 in) during the winter. The settling pond had no water flow from November until May. Because of these conditions, dissolved oxygen was monitored at regular intervals during both winters. Dissolved oxygen never dropped below 7.3 mg l^{-1} by late March in either year. In spring (May) muskellunge were removed from the settling pond using a seine and transferred back into a production pond (as described earlier) until their final removal at 1 year.

During the angling period SCUBA transects were swum every 48 h until 1 week after angling was concluded. Thereafter, transects were swum twice a week until November. Muskellunge that died during the study were removed from the pond as soon as possible and necropsies were performed on them. Fish were X-rayed to aid in determining location and position of hooks. Specific information recorded included the time from hooking until death, anatomical hook location, and qualitative observations on tissue damage caused by the hook. When a fish was recovered shortly after death (<24 h) tissue samples were collected from the hook puncture site to assess possible bacterial infection. Tissue samples were also collected from the liver, spleen, and kidney to identify systemic bacterial infections. Tissue samples were preserved in 10% buffered formalin or placed on ice and shipped to the Department of Natural Resources fish health lab in Madison for bacterial culture. Bacteria were isolated and identified at the Wisconsin State Lab of Hygiene in Madison. Because this study involved the holding of live animals, including death as a possible outcome, the study protocol was reviewed and approved by the WDNR Animal Care and Use Committee.

Muskellunge mortality was separated into four categories: immediate, short-term delayed, overwinter, and cumulative. Immediate mortality was defined as mortality that occurred within 24 h of hooking. Short-term delayed mortality occurred between 24 h and 50 days following hooking (approximate time of ice formation on the hatchery pond), whereas overwinter mortality covered the period from 51 days to early June of the following year. Cumulative mortality included all mortality up to 1 year.

Non-lethal effects of hooks on muskellunge were also investigated. Individual muskellunge were measured and weighed following each pond transfer and relative weight calculated (Neumann and Willis 1994). Relative weight (W_r) values were used to determine any change in fish condition (treatment and control) during the study using a paired *t*-test. A random sample of four hooked and released muskellunge were sacrificed about 1 month following hooking in 2001. Necropsies (including tissue samples for bacteria culture) were conducted on these fish, with comparisons made to mortally wounded fish during this period on quantifiable aspects from injuries, e.g. length of rips or tears in stomach from hooks.

Corrosion analysis of hooks was conducted by APL, Inc. (Milwaukee, WI) to measure dissolution rates of hooks in muskellunge stomachs. Hooks were sampled from several different muskellunge stomachs after approximately 1 month (31–33 days), again in the following spring about 7 months after initial hooking (208–239 days), and at 1 year. Metallographic cross-sections were used to estimate the extent of corrosion. Three cross-sections were made on each hook sampled and the maximum depth of the corrosion or pitting was determined by digital analysis. Cross-sections were not completed on hooks at 1 year as only three fish were sampled and hook corrosion was extensive or the hook was absent.

Results

A total of 40 muskellunge were transferred into the hatchery pond during the 2 year study period (2001, $N = 19$; 2002, $N = 21$). Muskellunge length averaged 91 cm (35.9 in) and ranged from 80 to 107 cm (31.5–42.3 in). While muskellunge were in the pond, their behavior appeared normal. This conclusion was based on visual observations, their resumption of feeding, and an increase in weight over a 1-year period while in the pond. Mean increase in weight for 11 muskellunge (treatment and control) retained for 1 year was 0.65 kg (SD = 0.27; 23 oz, SD = 9.4). Mean relative weight of these fish increased from 93 (SE = 2.4) to 98 (SE = 2.1; $t = 4.4$, $P = 0.001$, df = 10). At the conclusion of angling in 2001, three introductions at one-week intervals were made with forage (similar in length and species composition to that used during angling). Of a total of 121 prey items released, 41 were consumed by 15 muskellunge during the following month (mean 2.7 prey/muskellunge). In addition, during both years muskellunge successfully reproduced in the hatchery pond. Up to several hundred fingerlings were removed when the pond was drained in fall. No mortality was observed as a result of the field transfer.

Twenty-three muskellunge were hooked in the stomach using the single hook technique during the angling period (Table 1). Fourteen muskellunge, which were either never hooked or were hooked superficially (e.g., lip), were used as controls. Three muskellunge were excluded from the data. One additional control fish was excluded from the long-term component of the study when it was found dead on the pond dike during July after surviving the overwinter period. Muskellunge often used the pond liner to trap prey, and would frequently come out of the water onto the liner while in pursuit of a prey item. The plastic liner extended approximately 0.9 m (3 ft) from the water line to the dike.

Table 1 Summary of muskellunge used to determine effects of single hook and live bait on muskellunge survival

	Total fish	Mortalities					Surviving fish
		Immediate	Short-term	Sacrificed	Overwinter	Cumulative	
Hooked[a]	23	0	5	4	10	15	3
Control[a,b]	14	0	0	0	4	4	8
N/A[c]	3	–	–	–	–	–	–
Total	40	0	5	4	14	19	11

[a] Overwinter and cumulative estimates do not include muskellunge (one treatment, one control) returned to Deer Lake in November 2001

[b] Cumulative estimate does not include one muskellunge found dead on pond dike in July 2002

[c] Muskellunge that were held in hatchery ponds but not used in study results

Mean time from when the bait was taken until the hook set was 17 min (SD = 7.2, range 4–37 min, $N = 35$). Playing time, i.e. time from hook set until landing, was less than one minute (SD = 0.3, range 0.3–1.5 min, $N = 34$). Mean handling time (landing until release) was 5.7 min (SD = 1.9, range 2–10 min, $N = 35$). Muskellunge that were caught twice took a bait a second time shortly (mean 2.7 days, range 1–5 days) after the initial time they were caught. In all cases, the first encounter resulted in muskellunge not consuming the prey item, either because the sucker was shaken off after the hook was set, or the sucker came out of the stomach after the fish was landed. Hence, the initial experience of being hooked, landed, and handled did not seem to deter hungry muskellunge from attempting to feed again a short time later.

No immediate mortality occurred for any stomach-hooked muskellunge ($N = 23$; Table 1). All fish caught and handled appeared healthy and swam away upon release. No bleeding was observed for any of the hooked fish prior to release. Short-term delayed mortality was 22% (5 of 23 fish). The five fish that perished during this period succumbed at 5, 6, 12, 32, and 48 days following hooking. Mean length of hooked fish that perished (897 mm, SD = 53.5 (35.3 in, SD = 2.1) $N = 5$) was comparable to hooked fish surviving during that period (922 mm, SD = 65.0 (36.3 in, SD = 2.6) $N = 18$). During this period the largest hooked fish was 1074 mm (42.3 in) while the largest fish that died was 978 mm (38.5 in). Necropsies on these fish indicated considerable damage to the stomach from hooks. Tears or rips in the stomach averaged 63 mm (2.5 in), range 25–114 mm (1–4.5 in). The muskellunge with the largest tear (115 mm; 4.5 in) had consumed a 305 mm (12 in) white sucker following release from the hooking event. The white sucker exited the muskellunge's stomach through the tear and decomposed in the body cavity. In comparison, mean length of stomach tears from hooks in muskellunge that remained alive but were sacrificed (30–35 days) after hooking was 17 mm (0.7 in), range 3–51 mm (0.1–2 in). No mortality occurred on any control fish during this period.

Internal examination of fish that died from hooking, and sacrificed fish that remained alive indicated infections most likely were due to hook damage. Qualitative examinations indicated the hook barb penetrated the stomach wall in all cases. Bloody ascites along with pseudo-membranes and adhesions in the body cavity were commonly associated with the location of the hook. In several of the sacrificed muskellunge the hook barb was exposed outside the stomach without damaging other organs, in one fish the barb was imbedded into the liver, and in another fish the barb punctured the gas bladder. Bacterial cultures indicated that the dominant bacterium was *Aeromonas hydrophila*, a common waterborne organism that can become pathogenic in stressed fish. *A. hydrophila* was also cultured in samples of liver, spleen, and kidney indicating infections were systemic. Other bacteria identified in cultures included *Yersinia ruckeri*, *Halfnia alvei*, *A. sobria*, along with *Shewanella putrefaciens* and *Pseudomonas* spp. associated with the puncture site of the gas bladder where extensive inflammation was observed.

Hooks were generally lodged in the anterior portion of the stomach (mean 39% of distance from anterior to posterior length of stomach, range 14–49%), but rips, or tears were common in the stomach, indicating initial hook penetration also occurred in the posterior part of the stomach (Fig. 3).

Overwinter mortality (from ice formation to early June) of hooked fish was 77% (10 of 13 fish) but interpretation of these results is more difficult than immediate and short-term mortality (Table 1). During the same time period, 31% (4 of 13) of the control fish died. Most (10 of 14) fish that perished died shortly following ice-out in the pond, when water temperatures were rapidly increasing. This mortality was observed for both hooked fish and control fish in both years.

Cumulative mortality (hooking to 1 year) of hooked fish was 83% (15 of 18 fish; Table 1). Cumulative effects from hooking were dispersed over the 12-month period from 5 days after hooking until early the next summer, with a peak in mortality occurring during spring. Interestingly, no muskellunge died in either year after the first week in June. However, less than 20% of hooked fish survived the hooking experience a full year until the following fall. Cumulative mortality on control fish was 33% (4 of 12). All mortality for control fish occurred during the overwinter period (ice formation until early June). Assuming cumulative mortality of control fish represents natural mortality, annual natural mortality would be estimated at 33%. Annual hooking mortality (total, or cumulative mortality for hooked fish minus natural mortality) would be estimated at 50%.

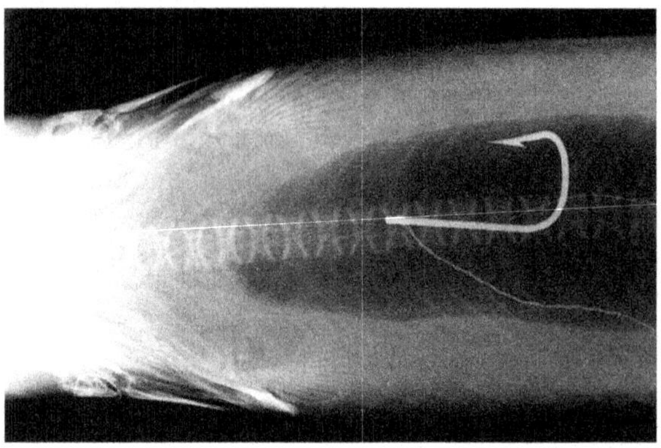

Fig. 3 Dorsal X-ray of 91 cm (35.8 in) muskellunge with Mustad 10/0 hook in stomach. Dark area is gas bladder

Hooks in muskellunge stomachs likely take a year or more to corrode totally. In the first month (31–33 days) after hooking, corrosion had penetrated up to an average of 9% (range = 7–12%) of the hook's radius (Fig. 4). By the following spring (208–239 days), corrosion reached up to an average of 29% (range = 15–45%) of the hook's radius (Fig. 5). The slowed rate of corrosion over the extended winter period is likely related to cold water and reduced fish metabolism. Hooks were nearly or completely dissolved by fall (1 year). In the three fish sampled, one no longer had a hook present. The hook was either completely dissolved or expelled from the body. Another fish had only a portion of the barb remaining. Interestingly, this part of the hook was found in the intestine. It is possible that with extensive corrosion this part of the hook began to pass through the fish's digestive system. It is unknown what complications the lodging of this part of the hook in the intestine may have caused. The third hook was intact in the stomach, but was corroded to an extent that it was easily broken. All three stomach hooked muskellunge that survived 1 year had good W_r values (range 95–99) and externally appeared healthy.

Discussion

Adult muskellunge up to 107 cm (42.3 in) in length were successfully held in captivity for up to

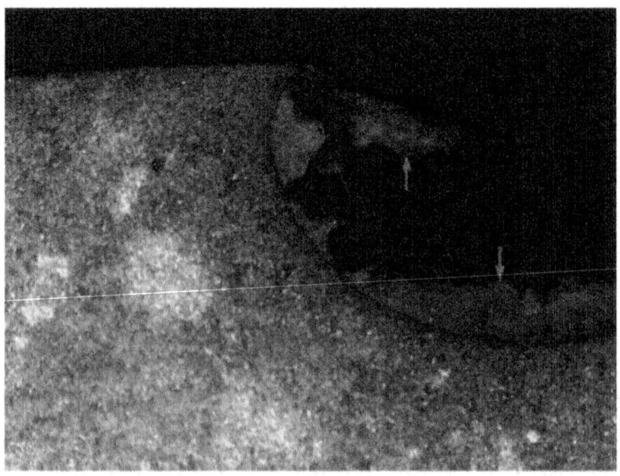

Fig. 4 Magnified (400×) cross section of Mustad 10/0 hook after 32 days in a muskellunge stomach showing corrosion pit along the outside diameter. Arrows indicate corrosion product

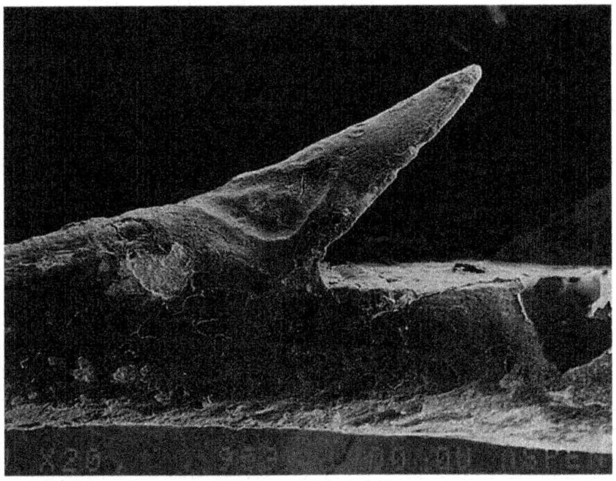

Fig. 5 Surface of Mustad 10/0 hook (magnified 20×) removed from a muskellunge stomach 211 days following hooking. Note the hook exhibits numerous areas of coating loss where pitting-type corrosion has initiated

1 year and normal behavior was observed. Muskellunge actively fed and not only maintained but improved their condition. In addition, muskellunge successfully reproduced in each of two springs after overwintering in ponds. While several control fish died during the overwinter period, it is unclear if this was a result of transfer to the settling pond during November, stress resulting from the spring spawning period, or a combination of both. Regardless, future studies on large predators such as muskellunge may include using lined hatchery ponds as a controlled environment in which to answer research questions.

Several important findings of this study can help direct fisheries biologists and anglers to make more informed decisions in management of trophy muskellunge fisheries. Because no immediate mortality occurred from the hooking experience, a well-intended angler could believe that survival is likely for deeply hooked muskellunge that are released. Conversely, results of this study indicated that a muskellunge with internal injuries might survive several days to weeks before dying.

In their summary of hooking mortality for recreational fisheries Mouneke and Childress (1994) state "Anatomical location of the hook wound, which is strongly influenced by fish behavior and terminal gear type, is apparently the most important factor in hooking mortality." The general nature of the single-hook live bait method used in this study makes hooking muskellunge in the esophagus or stomach unavoidable. The large hooks used puncture the stomach wall and often cause extensive trauma by tearing the stomach. Hooks also puncture and damage other organs in the body cavity such as the liver or gas bladder. When such extensive tissue damage occurred, muskellunge would not be expected to survive more that 50 days. When less extensive damage occurs, fish may still die from bacterial infections that became systemic during high stress periods such as spring warmup and spawning. Cold temperatures may suppress the effects of bacterial infections, which may become more damaging as spring arrives, metabolism increases, and additional stressors from spring warming and spawning are imposed on the fish. Both hooked and unhooked fish are exposed to increased bacterial action in spring. Some natural mortality is also expected for fish in a natural setting. Estimates of annual natural mortality rates from studies on adult muskellunge have ranged from 13 to 25% (Muir 1964; Spangler 1968; Harrison and Hadley 1978; Lyons and Margenau 1986). The mortality observed for control fish in this study (33%) is higher, yet comparable to that reported in the literature.

Acceptable hooking mortality varies among different sport fisheries. Muoneke and Childress (1994) suggested that mortality above 20% in a recreational fishery is generally considered high and deserving of management action. In this study, short-term mortality was 22%, and cumulatively (hooking plus natural mortality) less that 20% of muskellunge hooked survived 1 year. Regardless of the time period considered, hooking mortality observed on muskellunge from this single-hook rig is unacceptable for a species managed for trophy potential. These results suggest that while not all fish caught with single hook rigs are doomed to perish, the percent is great enough to warrant attention of those hoping to sustain or enhance muskellunge fisheries utilizing catch and release practices.

The use of live bait for muskellunge has been a traditional method of fishing for this species in Wisconsin (Margenau and Petchenik 2004). While regulation changes that prohibit such gear are possible, enforcement would be difficult. More likely, through education and angler awareness of the potential losses of muskellunge with these rigs, terminal tackle that utilizes less damaging rigging will replace the single-hook rig evaluated in this study. More sport anglers are utilizing gear such as quick-strike rigs that allow for the hook-set immediately when the bait is taken. These rigs use an arrangement of treble hooks that have a greater probability of hooking a fish in the mouth or buccal cavity. While the multiple prongs of a treble hook may require additional removal time, use of treble hooks in cool water temperatures and use of proper release practices should maximize the survival of a released muskellunge.

Acknowledgements Thanks are due the staff at Governor Tommy Thompson Fish Hatchery, and other Department of Natural Resources staff for their support and assistance during this study. Special thanks to Joel Howard for his assistance in all phases of the study. Muskellunge were X-rayed at the Spooner Veterinary Clinic in Spooner, WI. Sue Marcquenski (WDNR fish health lab) and Marty Collins (WSLH) conducted necropsies on muskellunge and collected samples for bacterial cultures. Ruth King prepared the drawing of the muskellunge. Funding for this study was provided in part by *Musky Hunter* magazine, the Federal Aid in Sport Fish Restoration Act (F-95-P), and the Wisconsin Department of Natural Resources.

References

Beggs GL, Holeton GF, Crossman EJ (1980) Some physiological consequences of angling stress in muskellunge, *Esox masquinongy* Mitchill. J Fish Biol 17:649–659

Bettoli PW, Osborne RS (1998) Hooking mortality and behavior of striped bass following catch and release angling. North Am J Fish Manag 18:609–615

Casselman JM, Crossman EJ, Robinson CJ (1996) Assessing sustainability of trophy muskellunge fisheries. In: Kerr SJ, Oliver CH (eds) Managing muskies in the '90s. Ontario Ministry of Natural Resources, Workshop Proceedings WP-007, Kemptville, pp 29–40

Harrison EJ, Hadley WF (1978) Ecological separation of sympatric muskellunge and northern pike. Am Fish Soc Special Publ 11:129–134

Heiting S (1992) Musky mastery – the techniques of top guides. Krause publications, Iola, Wisconsin

Johnson LD (1971) Growth of known-age muskellunge in Wisconsin and validation of age and growth determination methods. Wisconsin Department of Natural Resources Technical Bulletin 49. Madison, Wisconsin, USA

Lyons J, Margenau T (1986) Population dynamics of stocked adult muskellunge (*Esox masquinongy*) in Lac Court Oreilles, Wisconsin 1961–1977. Wisconsin Department of Natural Resource Technical Bulletin 160. Madison, Wisconsin, USA

Margenau TL, Petchenik JB (2004) Social aspects of muskellunge management in Wisconsin. North Am J Fish Manag 24:82–93

Margenau TL, AveLallemant SP (2000) Effects of a 40-inch minimum length limit on muskellunge in Wisconsin. North Am J Fish Manag 20:986–993

Muir BS (1964) Vital statistics of *Esox masquinongy* in Noiges Creek, Ontario. II. Population size, natural mortality and the effect of fishing. J Fish Res Board Can 21:727–746

Muoneke MI, Childress WM (1994) Hooking mortality: a review for recreational fisheries. Rev Fish Sci 2:123–156

Neumann RM, Willis DW (1994) Relative weight as a condition index for muskellunge. J Fresh Ecol 9:13–18

Newman DL, Storck TW (1986) Angler catch, growth, and hooking mortality of tiger muskellunge in small centrarchid-dominated impoundments. Am Fish Soc Special Publ 15:346–351

Simonson TD, Hewett SW (1999) Trends in Wisconsin's muskellunge fishery. North Am J Fish Manag 19:291–299

Spangler GR (1968) Angler harvest and mortality of *Esox masquinongy* in Pigeon and Sturgeon lakes, Ontario. J Fish Res Board Can 25:1145–1154

Weithman AS, Anderson RO (1978) Angling vulnerability of Esocidae. Proceedings of the 30th Annual Conference of Southeast Association Game and Fish Commissioners 30:99–102

WDNR (Wisconsin Department of Natural Resources). 1996. Wisconsin muskellunge waters. WDNR. Revision 1-3600 (82), Madison

Environ Biol Fish (2007) 79:163–169
DOI 10.1007/s10641-006-9092-6

ORIGINAL PAPER

A program for successful muskellunge management—A Minnesota success story

Paul J. Wingate · Jerry A. Younk

Received: 23 September 2005 / Accepted: 20 June 2006 / Published online: 29 July 2006

Abstract Minnesota has established a successful muskellunge, *Esox masquinongy*, stocking program as a result of a series of research studies that enabled area fisheries managers to make informed management decisions. The previous propagation and stocking program (pre-1982) reared muskellunge progeny from Shoepack Lake near the Minnesota–Ontario border, but these fish were not attaining trophy size. Protein electrophoresis showed there were two different muskellunge strains in the state. A controlled study was conducted on progeny from natural muskellunge populations from Leech and Shoepack lakes. Fish from these populations, together, were used to stock two other Minnesota lakes, and their growth rates, age of maturity, and maximum size attained were compared. Leech Lake muskellunge grew faster and attained a larger maximum size than the Shoepack Lake strain. As a result, a new stocking program based on the Leech Lake strain was developed. Because it was difficult to collect Leech Lake gametes, the Minnesota Department of Natural Resources created seven brood stock lakes using Leech Lake progeny. As a result of combining genetic considerations, new research on spawning sites, more restrictive harvest regulations, and catch-and-release fishing, anglers' catch of trophy-sized muskellunge in Minnesota has increased.

Keywords Stocking · Genetics · Research · Adaptive management

P. J. Wingate (✉) · J. A. Younk
Minnesota Department of Natural Resources, 500 Lafayette Road, St. Paul, MN 55155-4020, USA
e-mail: jack.wingate@dnr.state.mn.us

J. A. Younk
e-mail: jerry.youuk@dnr.state.mn.us

Introduction

Minnesota has 107 muskellunge *Esox masquinongy* lakes covering over 197,000 ha (486,000 acres) in all major drainage systems of the state (Mississippi River, Hudson Bay, and Lake Superior). Of these lakes, 44, that cover over 97,000 ha (240,000 acres), contain native populations (Younk and Pereira 2003). There are also six river systems with native populations. This extensive muskellunge resource places Minnesota among the top native muskellunge areas in North America, and is a valuable part of Minnesota's fishing heritage. The quality of Minnesota's muskellunge resource has been maintained and enhanced by the Minnesota Department of Natural Resources (MNDNR) by prudent use of management tools—restrictive regulations, voluntary catch and release, and the creation of new muskellunge waters by stocking. It was estimated that muskellunge anglers average 17.6 trips per year with an average trip length of 6.5 h (Younk and Pereira 2003). The number of muskellunge anglers was estimated to range from 31,100 to 78,900 (Younk and Pereira 2003).

The objective of this paper is to describe the approach used by the Minnesota Department of Natural Resources to successfully modify its muskellunge stocking and management program. A major part of this story is how a strong research program, and close collaboration with fisheries managers, contributed to management practices that led to quality muskellunge fishing for anglers.

Historical perspective

Historically, Minnesota has managed native and stocked populations of muskellunge. In the 1930s an expanded muskellunge stocking program was initiated in direct response to the large decline in native muskellunge stocks in northwestern Minnesota (Olson and Cunningham 1989). Catch of trophy-sized muskellunge continued to decline for the next three decades followed by a slight increase in the following 20 years (MNDNR unpublished data). The absence of larger fish was primarily in the stocked waters, but historic native waters (e.g. Leech Lake, Cass Lake, Lake Winnibigoshish) were also producing fewer large fish (MNDNR unpublished data). It was uncertain whether this was because of over-harvest, habitat loss, or a combination of both. Before the early 1950s, muskellunge stocking was sporadic, because of difficulties procuring mature muskellunge. In the early 1980s, average stocking was 20,000 advanced fingerlings (300–350 mm (12–14 in), total length) each year, with the culture program peaking in the mid-1990s at 40,000 fingerlings per year. Stocking by MNDNR currently averages 30,000 advanced muskellunge fingerlings each year at an estimated cost of $8.00 (2005 US Dollars) per fingerling (MNDNR unpublished data).

Additional management measures during this period (1930–1990s) included more restrictive regulations, increased stocking efforts, and the establishment of self-sustaining populations in new waters. In the 1930s, the possession limit was reduced to two (from unlimited) making it the same as the bag limit; in the 1950s the possession limit was reduced again, to one fish, and the opening of the season was delayed until mid-May with closing still in mid-February. A 762 mm (30 in) minimum length limit was imposed in the early 1960s; a 914 mm (36 in) minimum length limit and an early June opener was imposed in the early 1980s; and a 1016 mm (40 in) minimum length limit was enacted in 1993 (Younk and Pereira 2003). A 1219 mm (48 in) minimum length limit was also imposed on brood stock lakes for Leech Lake strain muskellunge in 1990.

Before 1953 spawn was taken from the upper Mississippi River drainage (Fig. 1); this was regarded as unreliable because of the difficulty of obtaining mature fish. There was, for example, no stocking with muskellunge for several years because of the inability to capture female fish. In 1953 muskellunge egg-taking operations were moved to Shoepack Lake, located near International Falls in northern Minnesota (Fig. 1). Shoepack Lake was chosen because it was easy to collect enough muskellunge for the egg-taking operation. This ability to collect the necessary gametes outweighed the need to fly to the site. Eggs were taken from Shoepack Lake, or lakes

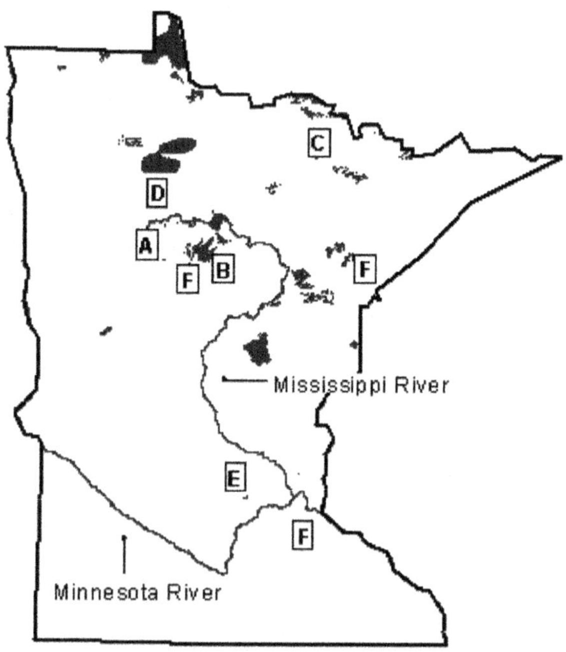

Fig. 1 Minnesota muskellunge waters: (*A*), Early egg collection sites; (*B*), Leech Lake; (*C*), Shoepack Lake; (*D*), Grace Lake; (*E*), Waconia Lake; and (*F*), Brood stock lakes

with their progeny, until well into the 1980s. Eddy and Surber (1943) and Eddy and Underhill (1974) described these fish as having characteristics typical of native Minnesota muskellunge. For over 30 years, muskellunge fingerlings from Shoepack Lake were used for stocking throughout Minnesota to create new fisheries, to supplement existing fisheries, and to reestablish lost fisheries (Younk and Strand 1992). Unfortunately, little formal evaluation of the stocking program was conducted during that period.

In the late 1970s evidence from accumulated MNDNR lake survey data revealed different growth patterns in lakes with introduced Shoepack Lake populations and lakes with naturally occurring populations (MNDNR unpublished data). Analysis of angler catch data supported these observations and indicated that very few trophy-sized fish were being caught (MNDNR unpublished data). The few trophy-sized fish caught were coming from waters containing native fish, for example Leech, Cass, and Winnibigoshish lakes. Questions were raised by biologists and anglers about habitat limitations, amount of harvest or over-harvest, and the possibility of genetically distinct populations which might limit the attainment of trophy size. In 1982, MNDNR reduced the length of the angling season for muskellunge and initiated work with Muskies, Inc. to promote catch and release. In addition, MNDNR decided to investigate the growth differences between two native muskellunge populations—Leech Lake and Shoepack Lake.

Leech Lake is a 45,134 ha (111,500 acres) mesotrophic lake in north-central Minnesota that was formed by a moraine dam in the Mississippi River drainage (Fig. 1). The fish community is dominated by percids and esocids (including northern pike *Esox lucius*); white sucker, *Catostomus commersoni*, cisco, *Coregonus artedi*, lake whitefish, *C. clupeaformis*, and rock bass, *Ambloplites rupestris* are also common. Muskellunge spawn in offshore locations in water depths of 1–2 m over a soft calcareous substrate where *Chara* spp. is the dominant vegetation. The substrate is flocculent marl of a low organic content that has oxygenated water at the mud/water interface (Strand 1986). Muskellunge spawned 7–14 days after the ice had melted when surface water temperatures reached 11.6°C or higher. Broodstock was typically >1000 mm (>39 in) total length. Shoepack Lake near International Falls in northern Minnesota is a 124 ha mesotrophic lake on the Kabetogama Peninsula within Voyageur National Park in the Hudson Bay drainage. The fish community is dominated by muskellunge (no northern pike), yellow perch, *Perca flavescens*, and white sucker. Muskellunge spawn near the shore or in backwater areas in water depths of < 1 m, areas typically regarded as northern pike spawning habitat. Shoepack Lake muskellunge usually moved into these areas to spawn when the ice had melted and temperatures were approximately 5°C. Broodstock was typically < 750 mm (29.5 in) total length (MNDNR unpublished data). It was originally thought the Shoepack Lake muskellunge were stunted and that this explained their small size.

Genetic considerations

In 1981, Minnesota participated with the Wisconsin Department of Natural Resources (WIDNR) and Muskies, Inc. to fund a study to analyze the genetic variation of the different muskellunge populations in Minnesota and Wisconsin. Two muskellunge populations from Minnesota, four from Wisconsin, and one from North Dakota (Federal Hatchery) were investigated using protein electrophoresis. The electrophoretic work seemed to indicate that the seven muskellunge populations could be separated into four distinct groups. Leech Lake and Shoepack Lake muskellunge seemed to be genetically (Hanson et al. 1983) and phenologically (spotted or barred) different. As a result, MNDNR considers the two populations as separate strains for management purposes. Strain, as used in this paper, means fish from, or descended from, their respective source populations, for example Leech Lake or Shoepack Lake. The Wisconsin populations from Lac Courte Oreilles Lake seemed to be in the same group as those from Leech Lake; those from Minocqua Lake seemed to be in the same group as those from Shoepack Lake. The other two groups were the other Wisconsin muskellunge populations and the North Dakota fish (Hanson et al. 1983). From historical sampling data, MNDNR knew that the Shoepack Lake muskellunge grew more slowly than the Leech Lake fish but had no definitive information on maximum size attained by either strain (MNDNR unpublished data).

We hypothesized that different growth rates, age of maturity, and maximum size were linked to genetic characteristics of the two populations. To test the differences between the growth of these strains, the next step was to pair stock muskellunge from Leech Lake and Shoepack Lake muskellunge populations in the same water body and monitor their growth under similar environmental conditions. Before such a study could be conducted, MNDNR had to find where the muskellunge spawned in Leech Lake so that mature fish could be collected. Strand (1986) used radio telemetry to establish where muskellunge were spawning in Leech Lake and discovered that one of the reasons MNDNR had previously had poor luck capturing adult muskellunge was that Leech Lake muskellunge spawned in offshore locations that had not previously been regarded as spawning habitat. This information enabled the DNR to collect eggs from Leech Lake and rear the progeny to advanced fingerlings for stocking.

Two lakes were selected on the basis of:

1 geographical location;
2 no previous record of muskellunge introduction;
3 an abundance of prey;
4 relatively low northern pike populations; and
5 at least one of the lakes was outside the native muskellunge range (Younk and Strand 1992).

Waconia Lake (on the western edge of the St. Paul—Minneapolis metropolitan area) (Fig. 1) was stocked with advanced muskellunge fingerlings derived from Leech Lake, Shoepack Lake, and Lac Courte Oreilles Lake (a Wisconsin strain provided by WIDNR) stock. Waconia Lake is a 1256 ha (31,000 acres) shallow eutrophic lake located in the transitional (forest–agriculture) zone of southeastern Minnesota (Fig. 1). Grace Lake in northwestern Minnesota was stocked with advanced muskellunge fingerlings derived from Leech Lake, Shoepack Lake, and Minocqua Lake (another Wisconsin strain provided by WIDNR) stock. Grace Lake is a moderately shallow 359 ha (890 acres) mesotrophic lake located in the forested zone of northwestern Minnesota. Both lakes are hard-water lakes with fish populations dominated by centrarchid–percid complexes (Younk and Strand 1992). Wisconsin muskellunge strains were included in the study because some Wisconsin strain muskellunge were being used for stocking by MNDNR. Grace Lake was stocked once with 500 muskellunge fingerlings of each strain; the average size was 279 mm (11 in) total length. Waconia Lake was stocked once with 1000 muskellunge fingerlings of each strain; the average size was 279 mm (11 in) total length.

Distinct differences in growth and age of maturity among the strains were observed for the paired stocks used, with the Shoepack Lake muskellunge clearly the smallest strain. Leech Lake muskellunge females matured at age 6 (1029 mm, 40.5 in) and attained a maximum calculated length of 1341 mm (52.8 in) whereas Shoepack Lake fish females matured at age 4 (768 mm, 30.2 in) and attained a maximum calculated length of 955 mm (37.6 in) (Table 1). These results confirmed the concerns we had about poor performance of the Shoepack Lake strain in many of our Mississippi basin waters. In response to these findings, MNDNR switched their entire muskellunge stocking program to Leech Lake strain fish.

Because of the extreme difficulty of consistently collecting muskellunge eggs from offshore spawning sites in Leech Lake, the MNDNR opted to create seven muskellunge brood stock lakes using progeny from Leech Lake (Fig. 1). To minimize concerns about introgression among introduced fish associated with a relatively small number of effective breeders, MNDNR worked with a geneticist from the University of Minnesota to stock in a way that slowed introgression in our brood stock lakes. This included taking muskellunge eggs from Leech Lake every 2–3 years and using the resulting advanced fingerlings for stocking the brood stock lakes, mixing gametes from different year classes in the brood stock lakes, and occasionally mixing gametes from different brood stock lakes.

Table 1 Mean and 95% confidence intervals of von Bertanlanffy growth data for four muskellunge strains in Grace and Waconia lakes

Strain/Sex	N	L_∞ (mm)	Confidence interval	
			Upper	Lower
Grace Lake				
Leach Lake				
Male	26	1092	1119	1066
Female	23	1201	1229	1173
Shoepack Lake				
Male	20	805	833	777
Female	20	889	914	865
Minocqua Lake				
Male	7	966	992	940
Female	5	1127	1204	1051
Waconia Lake				
Leach Lake				
Male	9	1105	1181	1028
Female	16	1341	1411	1271
Shoepack Lake				
Male	8	846	887	805
Female	7	955	1014	896
Lac Courte Oreilles				
Male	22	1057	1097	1016
Female	31	1194	1234	1153

Management strategies

With any new management scheme it is important to determine whether the management objectives are being met. The objectives of this modified muskellunge stocking program were acceptable survival rates of the stocked fish, increased angler success, and attainment of trophy-sized fish. From the strain comparison study it was discovered that the annual natural mortality of stocked Leech Lake fish was 31% whereas for Shoepack Lake fish it was 38% (Younk and Strand 1992). By analysis of muskellunge angler diaries Younk and Cook (1992) and Younk and Pereira (2003) showed that the proportion of successful anglers increased from 20.7% for all fish sizes during 1986–1989 to 27.8% during 1996–1998. The proportion of anglers successfully catching fish over 1015 mm (40 in) long also increased, from 3.8 to 9.3%. The median size of fish caught from 1986 to 1989 was 860 mm (33.8 in) compared with 920 mm (36.2 in) from 1996 to 1998.

All facets of this evaluation indicate the objectives of the modified muskellunge stocking program have been met. The combination of stocking with a more genetically appropriate strain in the Mississippi basin, the implementation of more restrictive harvest regulations, and the increased promotion of catch-and-release fishing have contributed to improved catch rates and population size structure for the muskellunge. It is unusual for a state management agency to maintain an adaptive approach for the length of time needed to evaluate and change its program. This adaptive approach included:

1 recognizing that the previous stocking program was not producing the desired fisheries;
2 evaluating the performance of the original strain compared with other potential strains;
3 recognizing the original strain was not the best;
4 modifying the stocking program to use the new strain;
5 using the new strain for statewide stocking; and
6 evaluating the performance of the new strain.

The necessary research took over 10 years and several more years passed before trophy-sized muskellunge began to appear in angler catches from lakes with introduced populations. During this time there were several changes of administration, and there were skeptics internally and externally who thought the changes were not needed.

Our objective in this paper was to describe a research/management process that was successful. By following defined research procedures we were able to provide fish managers with the information needed to provide anglers with more trophy fishing experiences while at the same time incurring no significant new expenses. Although initial investment was needed to create broodstock lakes, it is now easier and less expensive to obtain the necessary eggs. The new program is sounder genetically than the previous program, because most of the muskellunge water is in the Mississippi River watershed and the fish used for stocking are from this watershed.

When devising a new program, many researchers, managers, and anglers are inclined to use fish that reach the largest size and have the most trophy potential. There is, however, no guarantee fish will retain this advantage when they are used to stock new waters. Only by designing a series of sound research studies can the necessary information for evaluating new strain performance be understood. Fish strains have adapted to the environments in which they evolved, and often do not attain the same sizes or have the same growth rates in new waters (Koppelman and Philipp 1986). For example, a cooperative study conducted in the early 1980s in Minnesota by MNDNR and the Illinois Natural History Survey on Florida strain and northern strain largemouth bass, *Micropterus salmoides*, found that all Florida strain largemouth bass died during the first winter, because of the low water temperatures, whereas none of the northern strain bass succumbed to the cold (MNDNR unpublished data). Koppelman and Phillip (1986) state that managers and anglers have often damaged or altered native populations by introducing new populations thinking they will result in a better fishery. Genetic research in the late 1980s in Minnesota on native walleye populations that had undergone extensive stocking with fry and fingerlings indicated that any genetic changes in the native population that might have been expected were not detectable with protein electrophoresis (DNR unpublished data; Piller et al. 2005). Minnesota's muskellunge experience is an example of a sound procedure that produced the desired results and was more genetically sound. Some Leech Lake fish are being used to stock the Hudson Bay and Lake Superior watersheds; further work on stock structure among the major drainages would help clarify the appropriateness of this action.

Conclusions

As with any successful stocking and management program, there is a need to thoughtfully investigate which strain to use, to thoroughly study the different strains before adding to a stocking regime, and then to use the new information to design a management program to maximize the potential of the strain used for stocking. This new program must then be evaluated. Following a sound research protocol will ultimately lead to a sound management program. Minnesota's successful muskellunge program has generated increased interest in fishing for trophy muskellunge and created additional new opportunities in muskellunge fishing.

Acknowledgements The authors wish to thank the many Muskie, Inc. members who assisted us over the years in evaluating this new program. We also wish to thank R. Pierce, M. Drake, and D. Schupp, and four anonymous reviewers, for many helpful editorial comments on the draft manuscript.

References

Eddy S, Surber T (1943) Northern fishes. University of Minnesota Press, Minneapolis

Eddy S, Underhill JC (1974) Northern fishes, 3rd Edition. University of Minnesota Press, Minneapolis

Koppelman JB, Philipp DP (1986) Genetic applications in muskellunge management Am Fish Soc Special Public 15:111–121

Hanson D, Strand RF, Post DD, LeGrande WH, Fillbach S (1983) Muskellunge electrophoresis study Muskie 17:9–13

Olson DE, Cunningham PK (1989) Sport-fisheries trends shown by an annual Minnesota fishing contest over a 58-year period North Am J Fish Manage 9: 287–297

Piller KR, Wilson CC, Lee CE, Lyons J (2005) Conservation genetics of inland lake trout in the upper Mississippi River Basin: stocked or native ancestry? Trans Am Fish Soc 134:789–802

Strand RF (1986) Identification of principal spawning areas and seasonal distribution and movements of muskellunge in Leech Lake, Minnesota. Am Fish Soc Special Public 15:62–73

Younk JA, Cook M (1992) Application of an angler diary for muskellunge *Esox masquinongy*. Minnesota Department of Natural Resources, Division of Fish and Wildlife Investigational Report 420. St. Paul

Younk JA, Strand RF (1992) Performance evaluation of four muskellunge *Esox masquinongy* strains in two Minnesota lakes. Minnesota Department of Natural Resources, Division of Fish and Wildlife Investigational Report 418. St. Paul

Younk JA, Pereira DL (2003) An examination of Minnesota's muskellunge waters. Minnesota Department of Natural Resources, Division of Fish and Wildlife Investigational Report 498. St. Paul

Environ Biol Fish (2007) 79:171–177
DOI 10.1007/s10641-006-9169-2

ORIGINAL PAPER

Muskellunge in Missouri: a 10 year strategic plan

Mark F. Boone

Received: 9 September 2005 / Accepted: 11 October 2006 / Published online: 28 November 2006

Abstract In 1995, a 10-year strategic plan for muskellunge program management in Missouri was developed by the Missouri Department of Conservation. Our mission was to efficiently produce and manage muskellunge in order to provide high-quality angling opportunities in suitable, well-distributed lakes throughout the state. The first goal was to increase the number and improve the distribution of muskellunge fisheries in small lakes. In 1995, our three existing muskellunge fisheries were in west central, northwest, and north Missouri. Six additional lakes were stocked with muskellunge from 1995 to 2001 in southeast, northeast, southwest, and central Missouri, plus the St. Louis area. The second goal was to maintain density and size structure of muskellunge populations commensurate with muskellunge angler expectations. Our objective was for muskellunge anglers to catch a muskellunge at least 91 cm (36 in.) in 20–40 h of angling. The Show-Me Muskie Project, a volunteer angler diary program, was initiated in 1996 to help evaluate this objective. The third goal was to inform Missouri anglers about increased opportunities for high-quality muskellunge fishing. We developed a muskellunge web page, created a release poster, updated our muskellunge brochure, published magazine articles, and mailed biannual letters to our Show-Me Muskie Project cooperators.

Keywords Missouri · Muskellunge · Strategic plan · Accomplishments

Introduction

Muskellunge, *Esox masquinongy*, were first introduced into Missouri waters in 1966 when 19,323-ha (7,820 ac) Pomme de Terre Lake (US Army Corps of Engineers reservoir), in west central Missouri, was stocked with fry and fingerlings. The objective of the stocking was 'to provide Missourians with a trophy fishery utilizing a predator that could prey upon large non-game fishes such as gizzard shad, *Dorosoma cepedianum*, common carp, *Cyprinus carpio*, and various redhorses, *Moxostoma* spp.' (Dent 1986). Muskellunge fingerlings have been stocked nearly every year since 1966.

In 1973, the Missouri Department of Conservation's (MDC) 593-ha (240 ac) Pony Express Lake, located in northwest Missouri, was stocked with 87 muskellunge, along with 85 hybrid muskellunge (*E. masquinongy* × *Esox lucius*), and 62 northern pike, *E. lucius*. The objective was to establish a trophy esocid fishery that would yield

M. F. Boone (✉)
Missouri Department of Conservation, 2302 County Park Dr., Cape Girardeau, MO 63701, USA
e-mail: Mark.Boone@mdc.mo.gov

an annual estimated or documented catch of 10–20 fish exceeding 76 cm (30 in.) by 1990 (Eder 1987). Hybrid muskellunge were stocked again in 1974, 1980, and 1981. Periodic muskellunge stockings occurred through 2002 (Table 1).

In fall of 1983, the City of Kirksville's (north Missouri) new 1,310-ha (530 ac) water supply reservoir, Hazel Creek Lake, was stocked with 1,500 fingerling muskellunge. The objective was to establish and maintain a naturally reproducing population of large muskellunge for the primary purpose of helping to maintain adult gizzard shad at moderate densities and secondarily to establish a trophy muskellunge fishery to introduce elements of diversity, mystery, and surprise to the quality of fishing at the lake (Neuswanger 1984). Muskellunge were not stocked again for 7 years until it became clear that natural recruitment was non-existent (Table 1).

A Muskellunge Planning Committee of MDC staff was organized and charged with the responsibility of developing a 10-year strategic plan. They were to evaluate the status of Missouri's muskellunge fisheries; document what biologists had learned about muskellunge culture, ecology, and management; and clarify MDCs vision and develop an effective system for managing an efficient statewide program.

The Muskellunge Planning Committee conducted 'special interest workshops' in Kansas City and St. Louis during fall 1992 in order to gain the perspectives of an urban-biased sample of avid muskellunge anglers prior to drafting goals and objectives. The committee learned a number of important things in discussions with 28 anglers that attended the workshops. Most of them (85%) belonged to a muskellunge fishing club; 64% of the anglers were members of Muskies Inc., 21% belonged to the Mid America Muskie Association, and 14% were members of both organizations. Over two-thirds favored introducing muskellunge into several small public lakes over any other water-type if muskellunge fishing opportunities could become better distributed throughout Missouri. Muskellunge anglers strongly supported our lake-specific approach to muskellunge harvest regulations. When asked at what minimum length a Missouri muskellunge should be legally harvestable, the average response was 91 cm (36 in.).

In another opportunity to assess angler attitudes toward muskellunge management in Missouri, 17 non-MDC members of the Conservation Federation of Missouri's Fisheries and Water Resources Committee completed a questionnaire in February 1993. Most respondents (76%) had never fished for muskellunge. Their main reason for not fishing for muskellunge in Missouri was unfamiliarity with opportunities or lack of suitable opportunities (80%) rather than disinterest (20%). Two-thirds indicated that they would be 'excited about the diversity, mystery, or surprise that muskellunge might bring to fishing trips' at their favorite small lake. One-third indicated that they would be 'upset with the possibility that fishing for more preferred species might be adversely affected' (D. Neuswanger, unpublished data).

Table 1. Number of muskellunge fingerlings stocked in Missouri lakes, 1995–2004

Year	Pomme de Terre	Hazel Creek	Fellows	Henry sever	Lake 35	Pony express	Girardeau	Binder	Lower taum sauk
1995	2813	530					486		
1996	7000	530	820	474			243	384	
1997	5790					610			
1998	2210					594	486		200
1999	5419		2460	474		600		384	
2000	10611	684							
2001	2428	376			93	600	486		200
2002	6075		2990	474	112	616		384	
2003	4000								
2004	2040		730						
Size (ha)	19323	1310	2026	390	154	593	400	316	494

The MDC approved a 10-year strategic plan for managing muskellunge on April 6, 1995 (Neuswanger et al. 1994). Our mission was to efficiently produce and manage muskellunge in order to provide special, high-quality angling opportunities in suitable, well-distributed lakes where more traditional Missouri fisheries would not be adversely affected and might be enhanced.

The goals of the plan were to:

- Increase the number of opportunities and improve the distribution of opportunities for anglers to experience the thrill of fishing in a small lake, which contains muskellunge.
- Maintain density and size structure of muskellunge populations commensurate with muskellunge angler expectations without exceeding 1992 hatchery production.
- Ensure that Missouri anglers know about increased opportunities for high-quality muskellunge fishing in small lakes, and evaluate whether angler benefits generated by such opportunities justify agency cost.

This paper describes the primary accomplishments since the inception of the 1995 10-year strategic plan.

Goal 1: Increase the number of opportunities and improve the distribution of opportunities for anglers to experience the thrill of fishing in a small lake, which contains muskellunge

In 1995, the geographic distribution of Missouri's three muskellunge fisheries (Pomme de Terre, Pony Express, and Hazel Creek lakes) was inconvenient for St. Louis anglers and most others outside the greater Kansas City area (Fig. 1). It was unknown if Pony Express and Hazel Creek lakes were among the most suitable small lakes (<405 ha, 1,000 ac) in Missouri for muskellunge fisheries, nor had there been a systematic examination of other potential small public lakes.

Lakes selection

Minimum criteria were developed and used to select small lakes to be stocked with muskellunge. From 1995 to 2001, six additional lakes were nominated by fisheries biologists, approved by the Muskellunge Planning Committee, and stocked with muskellunge: Lake Girardeau, Fellows Lake, Henry Sever Lake, Binder Lake, Lower Taum Sauk Lake, and Busch Conservation Area (CA) Lake 35 (Fig. 1, Table 1). Lakes were evaluated and selected based upon the following criteria:

- Not subject to flooding by an adjacent stream.
- Low likelihood of dissolved oxygen or water temperature regimes that might threaten muskellunge survival or hinder their growth for prolonged periods of time. In our case, this meant that mid-summer dissolved oxygen must exceed 5 mg/l and water temperature must be less than 32°C (90°F) for muskellunge to survive and less than 29°C (84°F) for them to grow.
- Water transparency (Secchi disk) almost always at least 61 cm (24 in.) to facilitate muskellunge growth.
- Gizzard shad present to serve as primary prey for muskellunge.
- Substantial areas of macrophytes or dense flooded timber to provide protective cover for stocked fingerlings and facilitate angler contact with muskellunge. Neither minimum nor maximum aerial coverage of macrophytes were specified in the plan.
- Low likelihood of emigration, accomplished through either a physical barrier (e.g., outlet or emergency spillway structure) or low watershed : lake area ratio (15:1 or less) that would limit the frequency and magnitude of flushing flows.
- No ongoing program involving the stocking of walleye *Sander vitreus*, striped bass *Morone saxatilis*, hybrid striped bass *M. saxatilis* × *Morone chrysops*, or flathead catfish *Pylodictis olivaris*; and no recent program of stocking these large piscivores that would confound evaluation of the effects of stocking muskellunge.
- Low likelihood of changes to the watershed that could cause rapid eutrophication or other water quality problems.
- Open to fishing all day throughout the year.

Fig. 1 Map of Missouri muskie lakes

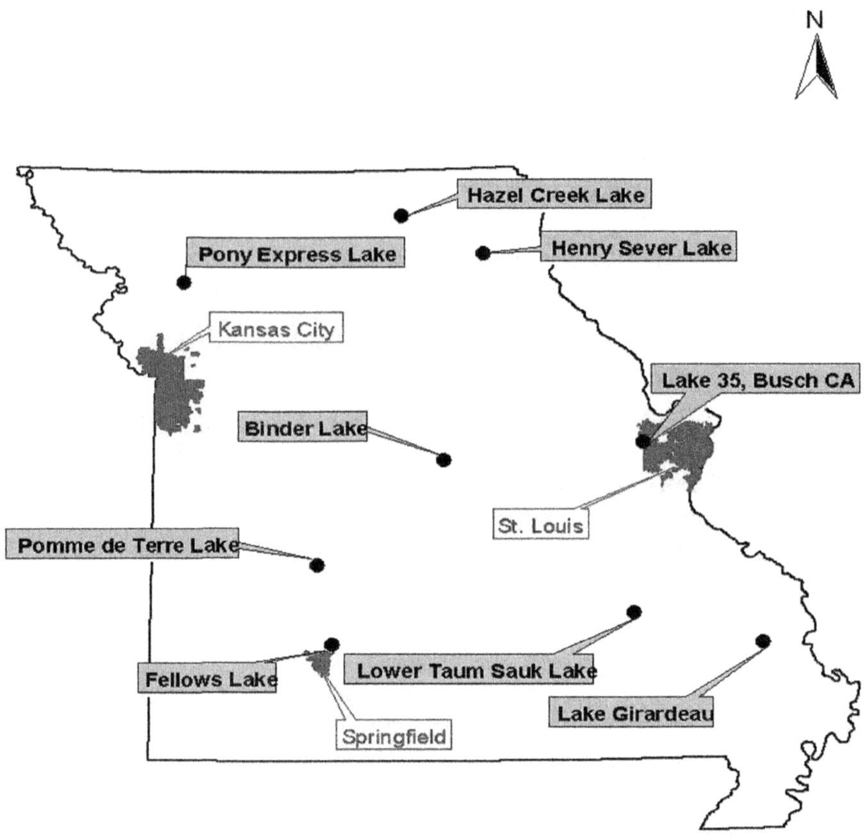

Goal 2: Maintain density and size structure of muskellunge populations commensurate with muskellunge angler expectations without exceeding 1992 hatchery production

One of the dilemmas we had was determining the best method to evaluate our muskellunge fisheries. We decided to evaluate our fisheries based on angling success and that the most efficient method of gathering information would be through an angler diary program similar to that developed in Minnesota (Younk and Cook 1992). The Show-Me Muskie Project, a statewide volunteer angler diary program, was initiated in 1996 to help evaluate our muskellunge program. Muskellunge anglers (cooperators), of all skill levels, volunteered to participate in the project.

Our objective was for anglers to catch at least one muskellunge ≥ 91 cm in 20–40 h of angling. During two workshops held in 1992, the participants agreed that this objective was desirable. We also wanted to provide higher angler catch rates than documented in other states. In comparison, mean specific catch rates averaged one fish per 70 h in Chautauqua Lake, New York (Mooradian and Shepherd 1973), one fish per 71 h in eight northern Wisconsin lakes (Hanson 1986), and one fish per 91 h in Minnesota waters (Younk and Cook 1992). Axon (1978) reported that anglers harvested one muskellunge per 48–82 h in Cave Run Lake, Kentucky.

To help meet our angler catch objective, we wanted to protect muskellunge less than 91 cm from harvest. In 1996, the statewide minimum length limit (MLL) for muskellunge was increased from 76 to 91 cm. Hazel Creek Lake is regulated with a 107-cm MLL.

We used spring fyke netting to gather catch-per-unit-of-effort (CPUE) and size structure data on muskellunge populations. Sampling was conducted during the spring spawning period (late March through early April) when the water temperatures were 10–13°C (50–55°F). Fyke nets had 2.2 cm square nylon mesh and were constructed with two $1.2 \times 1.8\text{-m}^2$ frames with center braces and four 1.2-m diameter hoops. Each net had a single lead that was 22.9 m long, 1.2 m deep, and 2.2 cm square mesh. We documented trends in CPUE as the number of muskellunge captured per fyke net per day (FND) and size structure as determined by relative stock density 91 cm (RSD91) (Anderson and Gutreuter 1983).

Show-Me Muskie project results

At Pomme de Terre Lake we met our angler catch objective (20–40 h per muskellunge ≥ 91 cm) in 3 years during the period 1996–2005 (Table 2). From 1996 to 2003, the hours to catch a muskellunge at least 91 cm ranged from 29 to 98 h, meeting the objective once in 1997 (Table 2). In 2004 and 2005, it took our cooperators 23 and 32 h, respectively, to catch a muskellunge at least 91 cm (Table 2).

From 1996 to 2004, we met or exceeded our angler catch objective at Hazel Creek Lake, with a range of 8–38 h to catch a muskellunge at least 91 cm (Table 2). At Fellows Lake, our angler catch objective has been met since 2002 (Table 2). The hours to catch a muskellunge longer than 91 cm has risen from 10 h in 2002 to 33 h in 2005. Few muskellunge anglers fished Pony Express, Girardeau, Binder, Hazel Creek, Binder, Lake 35, and Lower Taum Sauk lakes, so no results are reported in Table 2.

Fyke net results

The spring fyke net CPUE at Pomme de Terre Lake ranged from 2.0 to 6.6 muskellunge FND-1 from 1996 to 2003 (Table 3). The CPUE increased to 11.0 fish FND-1 in 2004, 14.9 fish FND-1 in 2005, and 10.6 fish FND-1 in 2006. The CPUE increased as the large 2000-year class became older and more vulnerable to fyke nets. From 1996 to 2002, the RSD91 ranged from 34 to 49%. It decreased in 2003 due to recruitment of the 2000-year class. At Hazel Creek Lake, the CPUE ranged from 2.8 fish FND-1 in 2001 to 11.2 fish FND-1 in 1998 (Table 3). CPUE increased to 6.4 FND-1 in 2006. RSD91 ranged from 25 to 64%. During 2000–2006, Fellows Lake CPUE ranged from 1.0 to 3.4 fish FND-1 (Table 3). Since 2000, RSD91 has exceeded 21%. At Henry Sever Lake, CPUE was 1.8 fish FND-1 in 2001, and then decreased in the subsequent 3 years to 0.8, 0.6, and 0.3 fish FND-1 (Table 3). CPUE increased to 2.4 fish FND-1 in 2005 and 5.7 FND-1 in 2006. Numerous muskellunge (numbers unknown) emigrated over the spillway during a heavy spring rain in 2001. RSD91 has increased to 24%. Lake 35 fyke net CPUE in 2004 (2.0 fish FND-1) and 2005 (2.1 fish FND-1) was consistent (Table 3). RSD91 increased to 7% in 2005 and 19% in 2006. CPUE at Pony Express Lake ranged from 2.4 to 4.0 fish FND-1 during 2000–2003 (Table 3). RSD91 ranged from 0 to 30%.

Fyke net results indicated that a viable population did not develop at Girardeau, Binder, and Lower Taum Sauk lakes. In fact, CPUE were less than one fish FND-1. Results from the Show-Me Muskie Project (Table 2) and spring fyke netting (Table 3) were used to decide the future (more stocking) of these fisheries. Pomme de Terre

Table 2. Number of fishing trips and hours to catch a muskellunge ≥ 91 cm at for Pomme de Terre, Hazel Creek, and Fellows lakes, 1996–2005

Year	Pomme de Terre		Hazel Creek		Fellows	
	Number of trips	Hours	Number of trips	Hours	Number of trips	Hours
1996	162	52	99	14		
1997	113	29	130	24		
1998	50	98	58	18		
1999	53	53	72	16		
2000	205	50	107	16	0	0
2001	138	48	26	8	6	0
2002	221	79	23	22	18	10
2003	294	81	17	22	47	17
2004	407	23	31	38	45	26
2005	348	32	17	102	85	33

Table 3. Fyke net CPUE (no. muskellunge FND-1) and RSD91at eight Missouri lakes, 1996–2006

Year	Pomme de Terre	Hazel Creek	Fellows	Henry sever	Lake 35	Pony express	Girardeau	Binder	Lower taum sauk
CPUE (no. FND-1)									
1996	2.4	5.2							
1997	4.5	6.8							
1998	6.6	11.2							
1999	2.7	9.5							
2000	2.0	NS	3.4	0.5		2.4	0.1	0.4	
2001	2.7	2.8	1.9	1.8		2.9	0.2	0.3	
2002	2.5	4.3	3.3	0.8			0.2		0.3
2003	4.0	3.7	2.1	0.6		4.0	0.2	0.8	
2004	11.0	4.0	2.2	0.3	2.0				
2005	14.9	6.8	1.0	2.4	2.1		0.1		
2006	10.6	6.4	3.0	5.7	1.7				
RSD91 (%)									
1996	37	33							
1997	41	56							
1998	42	64							
1999	39	55							
2000	39	NS	36	0		0	0	15	
2001	49	36	51	0		30	27	40	
2002	34	47	36	0			25		0
2003	12	40	27	0		2	60	7	
2004	10	25	44	0	0				
2005	15	48	21	5	7		0		
2006	25	57	31	24	19				

Lake, Hazel Creek Lake, Fellows Lake, Henry Sever Lake, and Lake 35 will continue to be stocked with muskellunge. Pomme de Terre Lake, Hazel Creek Lake, and Fellows Lake are meeting objectives and providing quality muskellunge fisheries. We will continue stocking muskellunge in Henry Sever Lake, install a fish barrier, and then evaluate the future. Lake 35 will continue to be stocked because its proximity to St. Louis will provide unique angling opportunities. Pony Express, Girardeau, Binder, and Lower Taum Sauk lakes were removed from the muskellunge program in 2003 because quality muskellunge populations did not develop and because of low angler-use for muskellunge.

Goal 3: Ensure that missouri anglers know about increased opportunities for high-quality muskellunge fishing in small lakes, and evaluate whether angler benefits generated by such opportunities justify agency cost

A designated 'Muskellunge Program Advisor' (now called Muskellunge Program Coordinator) was appointed in 1996. This individual established the biological basis of various muskellunge management strategies and communicated this information to the public. According to Behnke (1987), '...an agency should have an authoritative spokesperson, thoroughly knowledgeable about the factors determining the successes and failures of special regulations, who is admired and respected by the anglers, and who makes frequent contact with angler groups to get the message across.' It is essential that coordinators for state agencies are assigned to administer such programs.

Improving public awareness of Missouri's muskellunge fishing opportunities was achieved in several ways. We developed a 'Muskellunge Release Poster' in 2002 that helps educate anglers on how to handle and release caught muskellunge. Also in 2002, a muskellunge web page was developed and was updated annually. We provided a variety of information to the Pomme de Terre Chapter of Muskies Inc. for posting on their web page. In 2003 and 2006, our muskellunge fishing brochure was updated, which discussed muskellunge life history, lake information, and conservation. Magazine articles discussing Missouri's muskellunge management program were published in 2004 in *Esox Angler, Musky Hunter*, and *Outdoor Life* magazines. Twice annually our Muskellunge Program Coordinator sent letters to our Show-Me Project cooperators with a variety of information concerning muskellunge angling success and management in Missouri. Oehmcke (1986) and Younk and Cook (1992) both stressed the importance of frequent communication in making such a program work.

References

Anderson RO, Gutreuter SJ (1983) Length, weight, and associated structural indices. In: Nielson LA, Johnson DA, Lampton SS (eds) Fisheries techniques. American Fisheries Society, Bethesda, MD, pp. 283–300

Axon JR (1978) An evaluation of the muskellunge fishery in cave run lake, Kentucky. Am Fish Soc Spec Publ 11:328–333

Behnke RJ (1987) Catch and release: the last word. In: Barnhart RA, Roelofs TD (eds) Catch-and-release fishing: A decade of experience. Proceedings of a national sport fishing symposium, Humboldt State University, Arcata, CA, 30 September –1 October 1987, pp. 291–299

Dent RJ (1986) A case history of voluntary catch and release of muskellunge with management implications for fishery programs. Am Fish Soc Spec Publ 15:316–322

Eder LS (1987) Pony express lake management plan. Missouri Department of Conservation, St. Joseph, MO

Hanson DA (1986) Population characteristics and angler use of muskellunge in eight northern Wisconsin lakes. Am Fish Soc Spec Publ 15:238–248

Mooradian SR, Shepherd WF (1973) Management of muskellunge in Chautauqua Lake. NY Fish Game J 20:152–157

Neuswanger DJ (1984) Hazel creek lake fishery management plan. Missouri Department of Conservation, Kirksville, MO

Neuswanger DJ, Weithman AS, Kruse MS, Meade R, Suppes VC (1994) Muskellunge in Missouri: a ten-year strategic plan for program management. Missouri Department of Conservation, Cape Girardeau, MO, pp. 47

Oehmcke AA (1986) Panel 2—private sector participation: the role of anglers and private organizations in muskellunge management. Am Fish Soc Spec Publ 15:323–334

Younk JA, Cook MF (1992) Applications of an angler diary for muskellunge *Esox masquinongy*. Minnesota Department of Natural Resources, Investigational Report 420

Environ Biol Fish (2007) 79:179–185
DOI 10.1007/s10641-006-9137-x

ORIGINAL PAPER

History of muskellunge management in Nebraska

Daryl L. Bauer

Received: 7 September 2005 / Accepted: 11 August 2006 / Published online: 20 September 2006

Abstract Muskellunge, *Esox masquinongy*, were first introduced to Nebraska waters in 1958. By 1967 anglers had caught the first muskellunge from Nebraska waters. Hybrid muskellunge, *Esox masquinongy* × *E. lucius*, culture and stocking started in 1976. Harvest regulations for muskellunge and hybrid muskellunge were first established in 1977 with a daily bag limit of one muskellunge or hybrid muskellunge of at least 762 mm (30 inches) in length. In 2000 the minimum length limit for muskellunge and hybrid muskellunge was increased to 1016 mm (40 inches). Hybrid muskellunge have been phased out, none have been stocked in Nebraska waters since 2003. Since 1998 muskellunge fingerlings have been held in Nebraska hatcheries until age-1 before stocking. Yearling muskellunge are raised on minnows and other small fish in Nebraska hatcheries and stocking success apparently is very good. Management and research of Nebraska's muskellunge and hybrid muskellunge fisheries has been limited to routine sampling that has documented the presence of those fish in waters where they were stocked. Fishable populations of muskellunge can now be found in nine Nebraska waters across the state, but most Nebraska anglers are relatively uneducated in fishing techniques for muskellunge, identification, and proper handling.

Keywords Management · Stocking · Regulations

Introduction

Since Nebraska is geographically centered in the United States, it has been described as a place where "east" meets "west," and "north" meets "south". Because of its location in this transition zone, the fauna of the state is diverse. Over one hundred species of fish ranging from cold-water species to warm-water species are found in Nebraska waters. Grass pickerel, *Esox americanus*, and northern pike, *E. lucius*, are believed to have been native to Nebraska (Jones 1963). With the diversity of aquatic habitats in Nebraska, fisheries managers have tried to provide a diversity of fishing opportunities. Many attempts to establish these fisheries have consisted of stocking fish into habitats that fisheries managers believed were acceptable.

Culture and stocking

Muskellunge were first stocked in Nebraska in 1958 at Lake Minatare, an irrigation reservoir

D. L. Bauer (✉)
Nebraska Game and Parks Commission,
2200 North 33rd Street,
Lincoln, NE 68503, USA
e-mail: daryl.bauer@ngpc.ne.gov

Table 1 Surface area at full pool and years of stocking of muskellunge and hybrid muskellunge in Nebraska waters

Water body	Area (ha)	Years stocked	
		Hybrid muskellunge	Muskellunge
Alexandria East	8	1986–1992	
Alexandria West	9	1986–1992	
Branched Oak	729	1976, 79, 83 (a)	
Calamus Reservoir	2,076		2000, 02–05 (s, a, F)
Chalkrock	16	1988–1992, 96–97 (s, a)	
Champion	5	1999	
Cheyenne	7	1988–1994 (s)	
Cochran	18	1988–1991, 93 (a)	
Conestoga	93	1985–1987	
Cottontail	12	1987	
Cottonwood	14	1991 (s)	1986 (s)
Cottonwood-Steverson	275		1998–2000, 02, 04 (s, F)
Crystal Lake	4	1996	
Crystal Springs Park	4	1985–1987	
Cub Creek Lake	11	1985–1986, 88–89, 91–92 (s, a)	
Dad's Lake	414		1969
Duck Lake	27		1972
East Twin	85	1990–1993, 95, 99 (s, a)	1994 (s)
Elwood	538	1983, 91–92	1986, 95–97, 99, 02, 04 (s, a, F)
Fremont 7–8	4		1998–2000, 03, 05 (s, F)
Frye	98	1995, 97–00 (s, a)	1999, 02
Grand Island Eagle Scout	17	1995–1997 (s)	
Grand Island L.E. Ray	12	1991–1997 (s, a)	1999, 02, 04
Grand Island Wastewater	3	1991–1997	
Grove	20	1991–1992 (a)	
Hayes Center	20	1993	
Hull	2	1985–1986	
Killdeer	8	1989	1976
Kilpatrick	65	1992, 96, 98 (a)	
Lake Winters Creek	95		1988, 00 (s)
Long	63		1974
Louisville 1	2	1985–1987	1992
Louisville 1a	1	1991 (a)	1992
Louisville 2	7	1985–1991	
Mayberry	10	1993–1995 (s)	
McConaughy	12,146	1983–1984, 88–93 (a)	
Meadowlark	22	1986	
Melham Park	1	1985–1986	
Merritt	1,176		1964, 77, 79–84, 87, 89, 91–99, 03, 05 (s, a, F)
Minatare	874		1958
Missouri River	N.A.		1983–1985 (s, a, F)
Mormon Island East	9	1989–1991, 93 (s, a, F)	1992, 94–96 (F)
Mormon Island Middle	8	1989–1991, 93 (s, a)	1992, 94–97, 99, 02, 04 (s, a)
Offutt	65	1992	
Oliver	101	1986–1994, 96–00, 02–03, 05 (s, a, F)	2001, 04
Pawnee	300		1978
Pibel	10	1988–1990	1992 (s)
Ravenna	6	1991	
Red Willow	659	1993 (s, a)	
Rock Creek Lake	20	1991 (a)	
Schuyler City Lake	5	1986	

Table 1 continued

Water body	Area (ha)	Years stocked	
		Hybrid muskellunge	Muskellunge
Skyview	20	1991–1992	
Smith	117	1990–1991 (a)	
Snus	50	1991	
Swanson	2,014	1983 (a)	
Timber Point	11	1986–1995, 99–00 (a)	2002–2003
Two Rivers 1	2	1986–1987, 93	
Two Rivers 2	1	1993	
Two Rivers 3	2	1986–1987, 93	
Two Rivers 4	2	1986–1987, 93 (a)	
Verdon	12	1985–1988	
Victoria Springs Lake	2	1984–1987	
Wagon Train	128	1976, 79	2002–2005 (s, F)
Walgren	46	1983, 87–88 (a)	
Watts	93		1978–1981, 83, 85–88, 96–97 (a)
Wellfleet	20		1962
West Wood River Pit	6	1988–1994 (s)	
West Twin	24	1986	
Wild Plum	6	1986–1987	
Willow Creek Reservoir	283	1983–1988, 91–93, 99 (a)	
Zorinsky	103	1990–1991, 93, 95 (s, a)	

Letters in parentheses designate that records exist of fish later being collected in fish community samples on those waters (s) or records exist of fish being caught by anglers from those waters (a). Fishable populations that are present now are designated by (F)

(Table 1). Since that time, muskellunge or hybrid muskellunge have been stocked in 70 standing waters and one major river in Nebraska. A variety of habitats from small impoundments and borrow pits to the largest reservoir in Nebraska have received muskellunge or hybrid muskellunge stockings; these waters have ranged from 1 ha (2.5 acres) to over 12,000 ha (29,652 acres). Many stockings were made infrequently and no muskellunge or hybrid muskellunge fisheries ever developed in many waters (Table 1).

Muskellunge stockings in Nebraska were sporadic from 1958 through 1974 (Fig. 1). During this period some stockings were done in hopes of establishing a source of brood fish for hatchery production of muskellunge for future stockings (Bouc 1975). The original muskellunge stockings were made in six different bodies of water. In the years to follow, a total of 27 bodies of water received muskellunge stockings. The number of water bodies stocked in a given year with muskellunge has generally been limited; in 2002 a total of 8 different bodies of water were stocked (Fig. 2).

Fourteen of 27 water bodies stocked with muskellunge have also received hybrid muskellunge stockings at some time. The first stocking of hybrid muskellunge in Nebraska waters occurred in 1976 (Fig. 1), and then the next stocking of hybrid muskellunge occurred in 1979. At least some hybrid muskellunge were stocked in Nebraska waters each year 1983–2000, with a peak of 27 different water bodies being stocked in 1991 (Fig. 2). Hybrid muskellunge have been

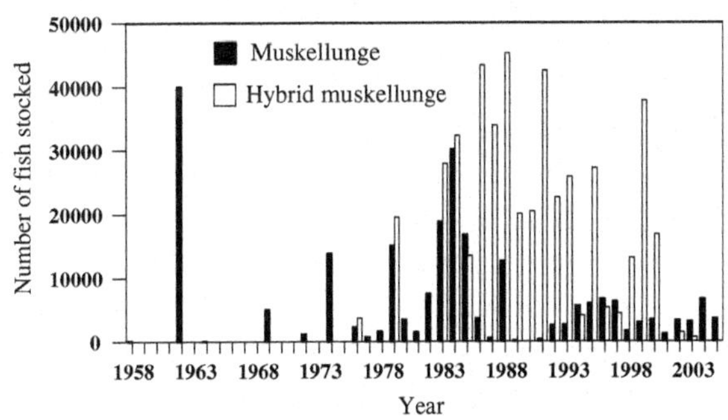

Fig. 1 Number of muskellunge and hybrid muskellunge, fingerlings and larger, stocked in Nebraska annually, 1958–2005

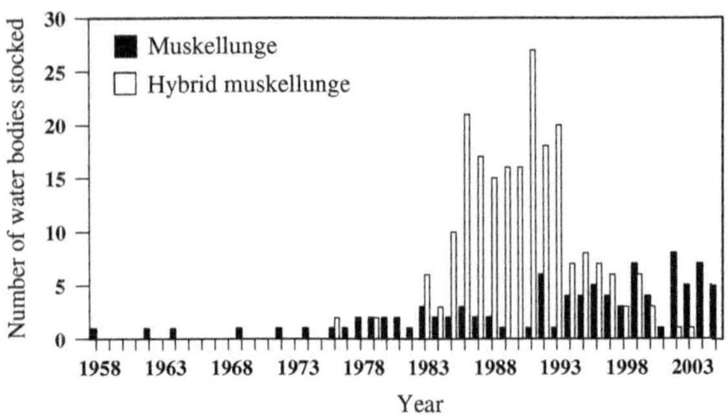

Fig. 2 Number of Nebraska water bodies stocked with muskellunge and hybrid muskellunge annually, 1958–2005

stocked in a total of 58 different water bodies in the state. The last hybrid muskellunge stockings in Nebraska waters occurred in 2003. Nebraska fishery managers discontinued hybrid muskellunge stockings because it was believed that muskellunge are more tolerant of warm water (Bevelhimer et al. 1985), and therefore were more appropriate for Nebraska waters. In addition, development of a trophy fishery is one of the management objectives when stocking muskellunge in Nebraska, so another reason muskellunge have been favored over hybrid muskellunge is they tend to be less susceptible to angler harvest and therefore survive longer and reach a larger size (Weithman and Anderson 1976; Brege 1986).

Muskellunge fry were stocked on one occasion in 1987 (6500 fry), and hybrid muskellunge fry were stocked in 1985 (10,000 fry), 1989 (100,000 fry), and 1990 (562,142 fry). All other muskellunge and hybrid muskellunge stocked in Nebraska waters have been fingerlings ranging from 25–330 mm (1–13 inches). Many of the early stockings in the state consisted of fingerlings less than 254 mm (10 inches). The number of muskellunge stocked in Nebraska peaked in 1962 when 40,000 fingerlings were stocked (Fig. 1). Hybrid muskellunge fingerlings stocked in Nebraska peaked in 1988 (Fig. 1). Since 1998 muskellunge raised in Nebraska hatcheries have been grown to age-1 and 280–330 mm (11–13 inches) before stocking, and those fish have been reared in hatchery ponds on minnows or other small fish. Stocking success of age-1 advanced fingerling muskellunge has apparently been much higher than other stocking strategies tried in Nebraska. Stocking of large muskellunge fingerlings has generally been found to be most successful (Johnson 1978; Carline et al. 1986; Hanson 1986; Larscheid et al. 1999), and that prompted the switch to stocking age-1 muskellunge in Nebraska. It is expected that hatchery production of age-1, advanced fingerling muskellunge will continue and will be sufficient to maintain the muskellunge fisheries now found in Nebraska. Limited natural reproduction of muskellunge has been documented in Nebraska and it is expected that most muskellunge fisheries in the state need to be maintained by stocking.

On occasion Nebraska has produced its own muskellunge and hybrid muskellunge for stocking from collections of brood fish at Merritt Reservoir and Watts Lake. The last attempt to produce muskellunge or hybrid muskellunge from Nebraska brood fish occurred in 1989. Most of the muskellunge and hybrid muskellunge stocked in Nebraska have been fingerlings obtained from hatcheries outside the state. Once obtained, fingerlings from other states have been reared in Nebraska hatcheries to stocking size. Muskellunge stocked in Nebraska waters have come from North Dakota, Ohio, Iowa, and Illinois. Hybrid muskellunge have come from North Dakota, Iowa, Illinois, Colorado, Michigan, and Pennsylvania.

Management and research

Justifications for introducing muskellunge, *Esox masquinongy*, or hybrid muskellunge, *Esox masquinongy* × *E. lucius*, to Nebraska waters have included providing a trophy fishery, and providing an additional predator to utilize rough fish (e.g. common carp, *Cyprinus carpio*), panfish (e.g. bluegill, *Lepomis macrochirus*) or prey fish (e.g. gizzard shad, *Dorosoma cepedianum*). It is known that esocids prefer soft-rayed fish (Beyerle and Williams 1968; Weithman and Anderson 1977), in particular gizzard shad (Carline et al. 1986), and they will consume centrarchids as well (Gillen et al. 1981). However, little or no evaluation has been done on the success of muskellunge or hybrid muskellunge as predators in any Nebraska waters.

Few concrete management goals for the introduction of muskellunge or hybrid muskellunge into Nebraska waters have been established. Evaluations of the success of those introductions primarily have consisted of documenting the occurrence of muskellunge or hybrid muskellunge in waters where they were introduced. By those standards, 13 of 27 (48%) muskellunge stockings in Nebraska have been successful and 28 of 58 (48%) hybrid muskellunge stockings have been successful (Table 1).

Muskellunge and hybrid muskellunge have been collected during routine fish population monitoring conducted by the Nebraska Game & Parks Commission, but no sampling efforts specifically targeted at collecting muskellunge or hybrid muskellunge have been done on Nebraska waters. A variety of gear types have been used in the sampling including boat electrofishing, frame nets and experimental gill nets (Zuerlein and Taylor 1985). Since 1995 muskellunge have been sampled from 13 different water bodies in Nebraska; muskellunge had been stocked in 12 of those water bodies. A total of 154 muskellunge were sampled from those waters ranging in size from 140 to 979 mm (5.5–38.5 inches). The highest catch rates of muskellunge sampled from Nebraska waters have been 68.0 muskellunge/hour of boat electrofishing observed at Fremont Pits 7–8 in 1999, 1.0 muskellunge per frame net-night observed at Fremont Pits 7–8 in 2000, and 2.8 muskellunge per gill net-night observed at Lake Winters Creek in 2000. Since 1993 hybrid muskellunge have been sampled from 16 different water bodies in Nebraska, and they had been stocked in 15 of those water bodies. A total of 212 hybrid muskellunge were sampled from those waters ranging in size from 190 to 1009 mm (7.5–39.7 inches). The highest catch rates of hybrid muskellunge sampled from Nebraska waters have been 7.0 hybrid muskellunge/hour of boat electrofishing observed at Chalkrock Reservoir in 1997, 2.2 hybrid muskellunge per frame net-night observed at Frye Lake in 1999, and 4.5 hybrid muskellunge per gill net-night also observed at Frye Lake in 1999.

All aging of muskellunge and hybrid muskellunge collected from Nebraska waters has been done by examining scales (DeVries and Frie 1996). Of 50 muskellunge and 138 hybrid muskellunge that have been aged, the oldest fish have been 6 years old. The mean length at age-5 of aged Nebraska muskellunge has been 669 mm (26.3 inches) and the mean length at age-5 of hybrid muskellunge has been 674 mm (26.5 inches). These apparent growth rates are slower than the average reported by Casselman and Crossman (1986).

Regulations have been set and changed over the years according to what was believed to be appropriate for managing Nebraska's muskellunge and hybrid muskellunge as trophy fish. The first regulation of sportfishing for muskellunge in Nebraska occurred in 1977 when a daily bag limit of one muskellunge was established. A possession limit of two muskellunge and minimum length limit of 762 mm (30 inches) total length (TL) were also enacted at that time. In 1982 the possession limit was reduced to one muskellunge. The minimum length limit remained at 762 mm (30 inches) TL for muskellunge and hybrid muskellunge through 1985. In 1986 the minimum length limit for hybrid muskellunge was reduced to 610 mm (24 inches) TL and at that time corresponded to the same minimum length limit in effect for northern pike in eastern Nebraska. The minimum length limit for hybrid muskellunge was again set at 762 mm (30 inches) TL in 1991 and corresponded to the minimum length limit that had been in effect for muskellunge since 1977. In 1992 the possession limit for muskellunge and hybrid muskellunge was increased to two. Finally, in 2000 the minimum length limit for muskellunge and hybrid muskellunge was increased to 1016 mm (40 inches) TL.

Angling

The first hook and line state record muskellunge caught from Nebraska waters was recognized in 1967. That original record fish was a 6.1 kg (13.4 pounds) muskellunge caught from Merritt Reservoir. Subsequently, the muskellunge state record has been broken six times and all of those fish were caught from Merritt Reservoir. The current record muskellunge is an 18.8 kg (41.5 pounds) fish caught in 1992. The first hook and line state record hybrid muskellunge was recognized in 1987—a 7.0 kg (15.4 pounds) fish caught from Lake McConaughy. The hybrid muskellunge record has also been broken six times since the original record was established. Subsequent hybrid muskellunge records were caught either from Lake McConaughy or Oliver Reservoir until the current state record of 15.1 kg (33.3 pounds) was caught from Johnson Reservoir. Although no hybrid muskellunge ever were stocked in Johnson Reservoir, it is on a canal system downstream from Lake McConaughy.

Nebraska has an angler recognition program for anglers to register notable catches, and that program has been very popular (over 2400 total entries in 2004). The first muskellunge in the Master Angler Program were entered in 1969 (Fig. 3). Early muskellunge entries in the Master Angler Program were any muskellunge caught, until 1977 when muskellunge or hybrid muskellunge had to weigh a minimum of 4.5 kg (10 pounds) to qualify for a Master Angler Award. The 4.5 kg (10 pounds) weight was the same as the minimum weight required for a northern pike to qualify for a Master Angler Award. Starting in 1986 a muskellunge or hybrid muskellunge could qualify for a Master Angler Award by being at least 889 mm (35 inches) TL

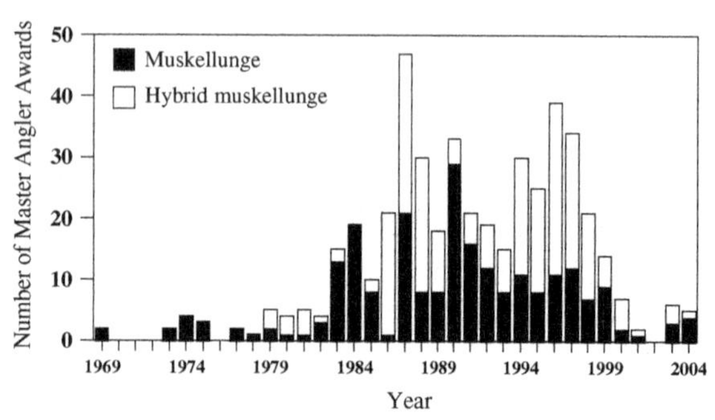

Fig. 3 Number of muskellunge and hybrid muskellunge entered in Nebraska's angler recognition (i.e. Master Angler) program, 1969–2004. Any muskellunge caught prior to 1977 qualified for entry in the Master Angler program. From 1977 until 2000 a muskellunge or hybrid muskellunge could qualify for an award by weighing a minimum of 4540 gm (10 pounds). Beginning in 1986 fish also could be measured and then immediately released and qualify for an award by being a minimum of 889 mm (35 inches) long. In 2000 the minimum length to qualify for a Master Angler Award for a released fish was increased to 1016 mm (40 inches) and the minimum weight to qualify a harvested fish was increased to 6804 gm (15 pounds)

or 4.5 kg (10 pounds). The minimum length standard allowed anglers to qualify for a Master Angler Award by measuring a fish and then releasing it. In 2000 the minimum standard for a muskellunge or hybrid muskellunge to qualify for a Master Angler Award was increased to 1016 mm (40 inches) TL and the minimum weight was increased to 6.8 kg (15 pounds). Those standards matched the statewide minimum length limit that was implemented in 2000. Another alteration to the Master Angler Awards program was made in 1992 when catch and release of trophy fish was further encouraged; anglers that measured the length of their fish and immediately released them received a catch and release hat pin in addition to the Master Angler certificate. In 2004 two-thirds of the trophy muskellunge and hybrid muskellunge entered in the program were measured and released.

Fish entered in the Master Angler program have been the primary documentation of angler-caught muskellunge in Nebraska. Using Master Angler entries, anglers have caught muskellunge from six water bodies (Table 1). Five of those water bodies were originally stocked with muskellunge while one, Lake Ericson, never had any muskellunge stocked. It is possible that the muskellunge recorded as being caught from Lake Ericson was misidentified. Master Angler entries exist for angler-caught hybrid muskellunge from twenty-six water bodies (Table 1). Three water bodies where hybrid muskellunge have been captured by anglers have never had any hybrid muskellunge or muskellunge stocked; however all three of those water bodies lie downstream from waters where hybrid muskellunge were stocked.

Future

Early attempts to establish muskellunge fisheries in Nebraska were made in many types and sizes of water bodies. Recently, muskellunge stockings have been more targeted at a few select waters. Waters currently being stocked range from sandpits as small as 4 ha (9.9 acres) to irrigation reservoirs as large as 2076 ha (5130 acres). There are additional waters in Nebraska that may support muskellunge fisheries in the future, but the stocking of over 70 bodies of water is a thing of the past. Muskellunge fisheries are most likely to be established in larger bodies of water, but low water levels in irrigation reservoirs are a threat to the development of muskellunge fisheries in those waters. For example, Merritt Reservoir, the oldest muskellunge fishery in Nebraska, experiences average annual drawdowns where water levels decline over 5 m (16 feet) annually; a developing muskellunge fishery at Elwood Reservoir is currently threatened by water levels that are over 12 m (40 feet) below full pool. Muskellunge fisheries now exist at Calamus Reservoir, Cottonwood-Steverson lakes, Elwood Reservoir, Fremont pits 7–8, Merritt Reservoir, Missouri River, Mormon Island pits, Oliver Reservoir, and Wagon Train Reservoir (Table 1).

Because of man-power constraints, additional muskellunge research or specific evaluations of Nebraska's muskellunge fisheries are not likely to be done in the future. Muskellunge will continue to be sampled as part of regular monitoring of fish communities in the waters where they have been stocked. During that monitoring, collection of cleithra or otoliths from any muskellunge that perish would allow for more accurate age and growth determinations of Nebraska muskellunge. Likewise, additional sampling of the muskellunge sport fishery in Nebraska is not anticipated in the future. However, as more anglers become interested in pursuing muskellunge in Nebraska, it may become possible to monitor their effort and catch through an angler diary program.

Nebraska anglers are still learning about muskellunge and muskellunge fishing. Misidentification of muskellunge, usually mistaking them for northern pike, is common, and Nebraska anglers are not well-versed in the handling of muskellunge. The establishment of a muskellunge angler organization would be very beneficial in education of anglers, promotion of Nebraska's muskellunge fisheries, and raising awareness of the challenges facing muskellunge management in Nebraska.

Acknowledgements Special thanks to all of the district fisheries management staff of the Nebraska Game and Parks Commission who collected and analyzed samples over the years. Also thanks are extended to Jay Francis who developed a computer program that aided in compiling these data, and Donna Waller for the compilation of Master Angler data over the years. Partial funding provided by Federal Aid in Sport Fish Restoration Grant F-87-R to the Nebraska Game and Parks Commission.

References

Bevelhimer MS, Stein RA, Carline RF (1985) Assessing significance of physiological differences among three esocids with a bioenergetics model. Can J Fish Aquat Sci 42:57–69

Beyerle GB, Williams JE (1968) Some observations of food selectivity by northern pike in aquaria. Trans Am Fish Soc 97:28–31

Bouc K (1975) Muskies in the making. Nebraskaland 53(2):35

Brege DA (1986) A comparison of muskellunge and hybrid muskellunge in a southern Wisconsin lake. In: Hall GE (ed) Managing Muskies, Am Fish Soc Spec Pub 15:203–207

Carline RF, Stein RA, Riley LM (1986) Effects of size at stocking, season, largemouth bass predation and forage abundance on survival of tiger muskellunge. In: Hall GE (ed) Managing Muskies, Am Fish Soc Spec Pub 15:151–167

Casselman JM, Crossman EJ (1986) Size, age and growth of trophy muskellunge and muskellunge-northern pike hybrids – the cleithrum project, 1979–1983. In: Hall GE (ed) Managing Muskies, Am Fish Soc Spec Pub 15:93–110

DeVries DR, Frie RV (1996) Determination of age and growth. In: Murphy BR, Willis DW (eds) Fisheries techniques, 2nd edn. Am Fish Soc, Bethesda, Maryland

Gillen AL, Stein RA, Carline RF (1981) Predation by pellet-reared tiger muskellunge on minnows and bluegills in experimental systems. Trans Am Fish Soc 110:197–209

Hanson DA (1986) Population characteristics and angler use of muskellunge in eight northern Wisconsin lakes. In: Hall GE (ed) Managing Muskies, Am Fish Soc Spec Pub 15:238–248

Johnson LD (1978) Evaluation of esocid stocking program in Wisconsin. In: Kendall RL (ed) Selected coolwater fishes of North America, Am Fish Soc Spec Pub 11:298–301

Jones DJ (1963) A history of Nebraska's fishery resources Nebraska Game and Parks Commission, Lincoln

Larscheid J, Christianson J, Gengerke T, Jorgensen W (1999) Survival, growth, and abundance of pellet-reared and minnow-reared muskellunge stocked in northwestern Iowa. N Am J Fish Man 19:230–237

Weithman AS, Anderson RO (1976) Angling vulnerability of Esocidae. Proc Con SE Assoc Fish Wild Agen 30:99–102

Weithman AS, Anderson RO (1977) Survival, growth, and prey of esocidae in experimental systems. Trans Am Fish Soc 106:424–430

Zuerlein GJ, Taylor MW (1985) Standard survey guidelines for sampling lake fishery resources. Nebraska Game and Parks Commission, Lincoln

CPSIA information can be obtained at www.ICGtesting.com
Printed in the USA
BVOW10*0234161215

430421BV00004B/6/P

9 781402 060489